DEFINING THE DELTA

DEFINING THE DELTA

MULTIDISCIPLINARY PERSPECTIVES ON THE LOWER MISSISSIPPI RIVER DELTA

Edited by Janelle Collins

The University of Arkansas Press
Fayetteville • 2015

Manufactured in the United States of America
ISBN: 978-1-55728-687-1 (paper)
ISBN: 978-1-55728-688-8 (cloth)
e-ISBN: 978-1-61075-574-0

19 18 17 16 15 5 4 3 2 1

Text design by Ellen Beeler

♾ The paper used in this publication meets the minimum requirements of the American National Standard for Permanence of Paper for Printed Library Materials Z39.48-1984.

Library of Congress Control Number: 2015948063

Contents

Acknowledgments

This book was four years in the making, and I am tremendously grateful to the contributors in this collection for their enthusiasm for and commitment to the project. Many signed on when it was in its embryonic stage, and their contributions influenced the shape of the book. Nearly everyone I contacted during the solicitation stage found the project intriguing, relevant, and worthwhile. Those who could not contribute gave me contact information for others in their field. I was inspired and gratified by the generosity of many Delta scholars.

I owe special thanks to William Clements for his mentorship when I became general editor of *Arkansas Review: A Journal of Delta Studies*; in addition to his general guidance, Bill suggested that I revive the "What Is the Delta?" series in the journal with the eventual goal of editing a collection of definitional essays.

I also appreciate those who assisted at key moments. Guy Lancaster, editor of the *Encyclopedia of Arkansas History and Culture*, generous with his friendship and contact list, suggested potential scholars for the project. Amber Strother copyedited an early draft of the manuscript. Marcus Tribbett checked and corrected citations in the final draft. Aaron Smithers, of the Southern Folklife Collection, Wilson Library, Chapel Hill, retrieved and scanned William Ferris's photographs when we needed them on short notice.

Many people and divisions at Arkansas State University have provided resources to *Arkansas Review* and by extension to *Defining the Delta*. The Faculty Research Awards Committee helped directly by awarding me release time to begin the project. I appreciate both the individual and institutional support.

The folks at the University of Arkansas Press have been wonderful to work with from the beginning. Larry Malley expressed immediate enthusiasm for the project; Mike Bieker moved the manuscript forward with excitement and efficiency; and David Scott Cunningham and Deena Owens answered countless questions and guided the manuscript into its final version. Thanks also to the peer reviewers who offered valuable feedback that prompted significant chapter additions.

My personal thanks extend beyond any particular project. My parents, Jack and Jackie Collins, shaped and continued to share my life even when I moved from the West Coast to the Delta, nearly two thousand miles away from home. My husband, Marcus Tribbett, and daughter, Sierra Tribbett-Collins, support, delight, and amuse me every day of my life.

CHAPTER 1

Defining the Delta

A RIVER, A FLOODPLAIN, AND GEOPOLITICAL BOUNDARIES

JANELLE COLLINS

> The Mississippi is well worth reading about. It is not a commonplace river, but on the contrary is in all ways remarkable. . . . [I]t is the longest river in the world—four thousand three hundred miles. It seems safe to say that it is also the crookedest river in the world, since in one part of its journey it uses up one thousand three hundred miles to cover the same ground that the crow would fly over in six hundred and seventy-five. . . . No other river has so vast a drainage basin . . . and almost all this wide region is fertile; the Mississippi valley, proper, is exceptionally so.
>
> —*Mark Twain,* Life on the Mississippi, *Part 1*

> [The Mississippi River] remains what it always was—a kind of huge rope, no matter with what knots and frayings, tying the United States together. It is the Nile of the Western Hemisphere.
>
> —*John Gunther,* Inside USA: 50th Anniversary Edition

Regional definitions are debatable, disputable, and often controversial. They begin with boundaries, which themselves are variable and porous. The brightly colored jigsaw-puzzle pattern of a United States map indicates state borders, but other maps of the same land mass might erase the political markers and illustrate boundaries based on physical features, geology, climate zones, ecoregions, transportation routes, or natural resources. Regardless of the parameters used, defining a region will engender debate. Rarely, however, are regional definitions arbitrary. They stem from purpose and perspective.

The Mississippi River, as Mark Twain, John Gunther, and countless others have observed, is a remarkable waterway that bisects the continental United States and has played a significant role in the nation's subsistence, exploration, settlement, transportation, and economy.[1] It is the longest river in North America, over four thousand miles from its headwaters in Minnesota to its outflow in the Gulf of Mexico, traversing through ten states. The river system has numerous tributaries and confluences and three subdivisions: the river known as the upper Mississippi begins at Lake Itasca in Minnesota and ends at St. Louis, Missouri; from its confluence with the Missouri River to its confluence with the Ohio River in Cairo, Illinois, the river is called the middle Mississippi; and the final segment, the lower Mississippi, flows from Cairo to the Gulf of Mexico. Although it is more accurate to call it a river system rather than simply one long river, there is little controversy in naming its sections.

Defining the Mississippi River Delta, however, is a much more difficult task. If etymology is considered, the literal definition is the most narrow: the Mississippi River Delta is the land built up by alluvium where the Mississippi River enters the Gulf of Mexico in Louisiana. This physiographic feature, however, is not the customary meaning of the Delta. The region more typically associated with the term is not actually a river delta but is the land that lies along the Mississippi River north of the river delta, which includes the Mississippi embayment, the sedimentary basin that is part of the larger alluvial plain created by the Mississippi River. Perhaps what most commonly springs to mind when the Delta is mentioned is the region specifically known as the Yazoo-Mississippi Delta, the subject of James Cobb's *The Most Southern Place on Earth.*[2] In geographic terms, the Yazoo-Mississippi Delta is the floodplain of the Mississippi and Yazoo Rivers; in cultural terms, it follows the oft-repeated adage of David L. Cohn as the region that "begins in the lobby of the Peabody Hotel in Memphis and ends on Catfish Row in Vicksburg."[3]

Yet the Lower Mississippi Delta is not limited to the Mississippi side of the river. There is also a distinct Arkansas Delta with the river functioning as the state line between Mississippi and Arkansas. Of course, within the states are several regions other than the Delta: the Mississippi Delta is one of five regions in Mississippi and the Arkansas Delta is one of eight regions in Arkansas. Both Deltas extend along the Mississippi River to their northern and southern state borders and are the primary agricultural regions of their states. The Mississippi alluvial plain, however, stretches north past these two states into the western edge of both Tennessee and Kentucky as well as into southeastern Missouri and southern Illinois. Thus a broad definition of the region is no longer limited to "the South's south," as Richard Ford has called the Mississippi Delta, but also includes portions of two border states and a northern one.[4]

Any discussion of the Delta, then, has to first articulate which Delta (or Deltas) is the subject of inquiry. Since the terminology is not precise in physiographic terms—the river delta is one small portion of the region under discussion and the Mississippi-Yazoo Delta, the most commonly understood meaning of the Delta, is actually an alluvial floodplain of two rivers—defining the Delta is a complicated and recurring requirement in Delta scholarship.

Delta Studies

Arkansas Review became an interdisciplinary journal of Delta studies in 1998 because a committed group of scholars believed in the significance and value of both regional studies and interdisciplinary scholarship. The journal, however, was not started from scratch. It had previous incarnations as an interdisciplinary journal of Midwest studies and an internationally acclaimed literary magazine, both under the name of *Kansas Quarterly* in its original home at Kansas State University. It moved to the Department of English and Philosophy at Arkansas State University in 1995 (after retirements of several key editorial members at Kansas State), underwent the name change to *Arkansas Review/Kansas Quarterly*, and gained a new editor, creative writing professor Norman Lavers. Even with a much smaller staff than the journal had at Kansas State, Lavers was able to continue publication as a literary magazine for another two years. Since it received some funding from the Delta Studies Program at ASU, the journal saved space in each issue for creative materials related to the Mississippi Delta. When finally overwhelmed with reading hundreds of manuscripts each month, Lavers gave up the editorship in 1997. With the journal's future in jeopardy, folklorist William Clements was approached by the dean of the College of Arts and Sciences, Richard McGhee, and chair of the English Department, Charles Carr, to serve as editor and return the journal to its interdisciplinary roots, this time with a regional focus reflective of its relatively new home. Arkansas State University, with its growing commitment to Delta studies, promised long-term support for the journal.

As the new general editor, William Clements assembled an advisory board of national scholars and an editorial board of Arkansas State University faculty members. Both boards were filled with representatives from multiple and diverse disciplines. The next step was to identify the journal's regional scope and call for submissions from disciplines in the humanities and social sciences such as history, geography, literature, music, political science, and sociology, among others. The editors mapped out the journal's focus as the swath of land that the river cuts through from St. Louis, Missouri, to the Gulf of Mexico, encompassing 214 counties and parishes in seven states. In practice, *Arkansas Review* takes as its subject of inquiry the region that begins at the confluence of the Mississippi and Missouri Rivers and includes portions of Missouri, Illinois, Kentucky, and Tennessee, and all of Arkansas, Mississippi, and Louisiana. For *Arkansas Review*'s purposes, the Delta is thus defined broadly, spreading out further than the physical or geographic reach of the lower alluvial valley of the Mississippi River.[5]

Arkansas Review's genesis as a journal of Delta studies was part of the burgeoning interest in the Delta that followed the creation of the 1988 Lower Mississippi Delta Development Commission (LMDDC). Congress established the commission in October 1988 to develop recommendations that would address the economic, educational, and social needs of people living in the seven-state region, which included some of the poorest and most economically depressed counties in the United States. The commission's 1990 report included recommendations for addressing health, education, housing, infrastructure, and economic development. Because the Delta—despite its poverty, limited economic opportunities, and racial inequity—has a rich cultural heritage, tourism and preservation of cultural resources were identified as central to the region's economic development.[6] As Luther Brown, associate dean for Delta Regional Development and director of the Delta Center for Culture and Learning at Delta State University, asserts, "The River bore the alluvial plain that is the Mississippi Delta, and the Delta bore fruit."[7]

Arkansas State University's Board of Trustees responded to the LMDDC's report with a resolution dated December 17, 1994, to establish the Delta Studies Center on campus to foster study on the Lower Mississippi River Delta. The center's charge was to encourage "interdisciplinary studies, research, teaching, and extension activities directed toward exploring the people, the institutions, the economy, the culture, the arts, and the biological and physical environments and characteristics of the Delta."[8]

The Department of English and Philosophy at Arkansas State University was a forerunner in the university's growing commitment to Delta-related scholarship. With support from the chair of the department and dean of the college, a committee of ASU faculty from several disciplines organized and hosted the Delta Studies Symposium in April 1995. The committee chose as its conference topic the Delta's most prominent cultural product, the blues, a topic that particularly encouraged interdisciplinary perspectives. The symposium brought together speakers, panelists, musicians, and scholars. Because of its success and the significance of its subject matter, the department, college, and university committed to making the symposium an annual event.[9]

Arkansas Review's history parallels that of the Delta Studies Symposium. The annual gathering of scholars—which extended well beyond its original musical focus—suggested that interdisciplinary and multidisciplinary scholarship focused on the region was alive and well and could be and should be disseminated through a journal with a Delta focus. When the journal formerly known as *Kansas Quarterly* lost its editor as a literary magazine, its shift to an interdisciplinary and Delta-focused journal was both instinctive and inspired. The journal's first issue appeared in 1998 when the Delta Studies Symposium was in its fourth year, and both outlets for Delta studies have remained in a mutually beneficial relationship with continued support by Arkansas State University faculty and administration.

Arkansas State University was not the only institution to respond to the LMDDC's report. State and federal agencies responded in a variety of ways, particularly with initiatives related to the recommendation to preserve and develop a tourism industry around the Delta's cultural resources. Under the aegis of the National Park Service, representatives from government, business, and communities conducted the Lower Mississippi Delta Region Heritage Study, which included a symposium in Memphis, Tennessee, on June 4, 5, and 6, 1996, that brought together experts in history, culture, economics, and the natural environment to identify the stories and sites that would bring national attention to the Delta.[10] The Heritage study greatly influenced perspectives on economic development in the Delta, and heritage or cultural tourism remains both a source of revenue for Delta communities as well as a source of scholarship on the Delta, where historically significant sites are preserved for research.

The issues identified in the Lower Mississippi Delta Development Commission's 1990 report continue to inform and animate Delta scholarship. Twenty-five years of Delta studies have demonstrated the richness of the topic and provided insights into the political, economic, and cultural issues that continue to impact the Delta's inhabitants. *Defining the Delta* reflects and contributes to the critical conversation about the region and argues that Delta studies, like the land itself, remain fertile fields.

"What Is the Delta?": Disciplinary Perspectives

When William Clements took on the role of general editor, he and his editorial board reorganized and refocused *Arkansas Review* in order to explore the Mississippi River Delta region through scholarship, creative works, regular features such as "Delta Sources and Resources," and book reviews. Because the cultural and political influence of the Delta extends beyond the Mississippi alluvial plain, the topics covered by the journal represent tremendous diversity in Delta environments, people, and cultural products.

As a consequence of this diversity, one of the features introduced by Clements in his first year as general editor, was a series that asked scholars to answer the question "What Is the Mississippi River Delta?" from their specific disciplinary perspective. The first to answer Clements's call was geologist Randel Tom Cox. Other scholars, such as historian Jeannie M. Whayne and archaeologist David H. Dye, provided articles that explored and defined the Delta from the perspectives unique to their disciplines. In addition to the articles that helped to define the Delta, the journal also published articles that examined economics, environmental science, and agriculture in the Delta.

Some of the original essays have been updated and expanded for this book. The definitional essays appear in the first part of the book as each author views the Delta from his or her disciplinary lens. The first five essays are based in the materiality of the Delta—its physical properties, boundaries, and climate—and include disciplines that define the Delta not only in terms of region or boundaries but also in temporal terms. Geologists, archaeologists, and environmental historians, in particular, take a chronologically elongated view of the Delta through their study of prehistory as well as recorded history.

Thus the collection begins with perspectives that focus significantly on the physical Delta, which are followed by perspectives from the social sciences and humanities that more directly explore the connection between the land, its people, and their cultural products.

As editor, I asked each scholar to define the Delta through his or her disciplinary lens as well as his or her personal specialization within the discipline.

The selected essays are definitional but do not presume to be definitive: they represent disciplinary viewpoints considered significant by the individual contributors. Some of the essays take a bird's eye view of the Delta to pull back and thus expose a long and broad perspective. Others focus tightly on an essential element of Delta life. Each of the essays presents a unique snapshot of the Delta; together they form a photomosaic that represents a collective portrait of the historical and present Lower Mississippi River Delta.

Delta Land and People

As Randel Tom Cox points out, in geologic terms, the region we refer to as the Lower Mississippi River Delta was once a vast bay that was gradually filled with sediment from the ancestral Mississippi. The story of the Delta, as Cox tells it, begins nearly six hundred million years ago and includes continental drift, a volcanic hotspot, an asteroid collision, Ice Ages, and earthquakes. The physical formation of what we now know as the Delta has a long and eventful history that only geologists can illuminate.

Like many geologists, archaeologists find their source material in prehistory. As we move forward on the geologic timeline to the moment when human settlement began in the Delta, archaeologists provide a perspective on a Delta that no longer exists. David H. Dye's chapter describes how the archaeological record of the Lower Mississippi Valley provides insights about the cultures that existed in the region before European contact. While archaeological excavations have yielded a great deal of information, Dye points out that there is more to learn. Preserving the Delta's heritage includes protecting its archaeological sites because technological advances in the field will engender further discovery as long as the sites can be utilized.

The chapter by environmental historian and Finnish scholar Mikko Saikku indicates that scholarly interest in the Delta is global, ranging beyond regional institutions and national borders. Interdisciplinary in its methods, environmental history is the study of the interaction between human beings and nature. Saikku, the author of *This Delta, This Land: An Environmental History of the Yazoo-Mississippi Floodplain*, extends the discussion in the preceding chapters beginning with the Delta's prehistory and progressing through time to explore the shaping of the Delta's environment by successive human populations.

Fiona Davidson and Tom Paradise describe and illustrate the multiple parameters and theoretical constructs through which geographers define regions. As a vernacular region, the Delta is perceptually—not geographically—unique. Davidson and Paradise expand the geographic definition of the Delta by including it in the broader formal and functional regions that extend well beyond the vernacular region known as the Yazoo-Mississippi Delta. The maps in the chapter provide a visual representation of the multiple factors by which the Delta extends beyond the borders of the Yazoo and Mississippi Rivers.

As a subfield of geography, climatology is the study of climate and weather patterns. Mary Sue Passe-Smith provides an overview of weather patterns in the Delta, explaining why the region is particularly susceptible to a variety of destructive weather events. By examining weather patterns over time, Passe-Smith reveals the vulnerability of Delta inhabitants as the various subregions in the Delta deal with flooding along the river, hurricanes along the Gulf Coast, ice storms in the north, and tornadoes across the region.

John J. Green, Tracy Greever-Rice, and Gary D. Glass Jr. offer a sociodemographic perspective on the Delta through a "livelihoods framework." Green, Greever-Rice, and Glass begin with a historical overview of the Delta's agricultural development, including the predominant plantation system that influenced much subsequent sociodemographic, economic, and cultural development. After establishing the historical context, the authors provide contemporary sociodemographic snapshots focusing on the multiple Deltas—upper, lower, rural, urban, and core—within the larger region.

Historian Jeannie M. Whayne, one of the original contributors to the journal's "What Is the Delta?" series, focuses on the Delta as an agricultural region, tracing the changes in farming through four distinct periods: antebellum, postbellum, neo-plantation, and portfolio plantation. Agricultural production not only shaped the land, but also determined the relationship between whites and blacks, which originated in slavery with white landowners and enslaved black labor. Iterations of that race and class dynamic continued in successive eras of agricultural production, sustaining the economic disparity in the region.

Political scientist Seth C. McKee examines the racially polarized voting patterns in the sixteen-county Delta region of northwestern Mississippi. Because of its plantation history, the Mississippi Delta is marked by an imbalance between the majority African American population and minority white political and economic power. McKee traces Delta politics from Reconstruction through the systematic suppression of the black vote to the re-enfranchisement of black Mississippians and their success in gaining local control through county and municipal elections. Although the democratization of the Delta is a remarkable story, economic injustice remains an intractable issue that cannot be resolved by elected black officials.

Delta Expressive Culture

Folklore, arguably the ur-interdisciplinary discipline, is also the discipline most closely aligned with regional studies. Thus it is not surprising that folklorist William Clements became the editor who reshaped *Arkansas Review* into a journal of Delta studies. Clements contributes an essay to this collection that explains how folklorists define regions and analyze the forms of expressive culture that shape and reflect regional identity.

It will come as no surprise to readers that this volume dedicates a fair number of pages to musicians and their music, particularly though not exclusively to the blues. Musicologist William L. Ellis identifies two primary musical traditions, New Orleans jazz and Mississippi blues. He notes the irony that while New Orleans and the surrounding regions near the mouth of the Mississippi are considered the river delta proper, it is a distinctly different musical tradition from what we commonly consider Delta music. Ellis's essay focuses on Delta blues, including its history and legacy as a music that has spread well beyond Delta borders, however they may be defined.

Folklorist William Ferris has spent a lifetime gathering and preserving interviews, music, and photographs of Delta residents, particularly African Americans. For this collection, Ferris contributes iconic images of blues musicians taken in the course of his extensive fieldwork.

Communication scholar Stephen A. King analyzes the tourism industry that has evolved from the late-twentieth-century blues revival, demonstrating how the music that originated in response to oppression and exploitation has become a valued commodity for Delta cities and towns. The names and images of bluesmen like Muddy Waters, Robert Johnson, and B. B. King are well known to those with even a passing familiarity with the Delta. King explains how preserving the history of the blues includes developing tourist sites that (problematically) promise an authentic blues experience.

Despite the presence of such blues women as Memphis Minnie, Bessie Smith, and Sister Rosetta Tharpe, who were born or played music in the Delta, men are overly represented in the scholarship about Delta musicians, a trend not limited to the blues. L. Dyann Arthur and her husband, Rick, set out in 2010 to crisscross the country interviewing, recording, and videotaping women roots musicians. Their trip through the Delta states is documented in Arthur's chapter, "Contemporary Women Musicians in the Delta." From Lafayette Parish, the Arthurs traveled through Louisiana to Mississippi and into Tennessee and Arkansas, finding musicians who played all genres of Delta music including Creole, Cajun, blues, and other traditional roots music. Arthur's essay is revelatory rather than definitional: Arthur and her husband recorded words, music, and images of the musicians while on their tour through the Delta, and the essay significantly expands our understanding of contemporary Delta music.

Tourism became a primary focus of economic development in the Delta after the Lower Mississippi Delta Development Commission's 1990 report. Ruth A. Hawkins, director of Arkansas State University's Heritage Sites, describes the variety and scope of heritage tourism in the Delta. Hawkins identifies the wide range of tourist destinations that reflect and showcase the history and heritage of the region, from its archaeological and architectural sites to its Civil War and civil rights battles, from its river and land to its literary and musical figures. In the two decades since the LMDDC report, communities have worked singly and together to chisel away at the disparity between the rich land and its poor people by preserving its histories—even those reflective of the region's economic inequity—and thus bring visitors to the Delta and money into local economies.

One of the most concrete methods of illuminating regional and ethnic diversity in the Delta is through an examination of its foodways. Jennifer Jensen Wallach's essay begins with pre-contact native societies and continues through European colonization, African enslavement, and waves of ethnically specific immigration to the region. Wallach tempers the celebratory story of southern foodways with reminders that the region's racial divide appears in its food customs, revealing both abundance and want and reflecting historical inequities.

Literary scholar Lisa Hinrichsen offers a comprehensive overview of writing produced in and about the region, illuminating Delta literature's remarkable diversity in language, ethnicity, race, culture, and genre. Hinrichsen situates the works in their historical context from the earliest travel writings through the southern renaissance to contemporary postmodern texts.

Defining the Delta ultimately suggests that just as there is not one Delta, neither is there one Delta story. The lands and lives in this region are myriad. Shifting from one perspective to the next is a little like turning the cylinder of a kaleidoscope: while the elements that are refracted in the mirrors are the same, each turn reveals a new pattern. Each of the following chapters takes the elements of the Delta—the river, the land, the people, their cultures—and offers a unique perspective. Like the Mississippi River, the Delta is well worth reading about, to borrow Twain's sentiment. Every turn of the page reveals a distinctive image, another way of seeing the seven-state region that is bisected by and dependent on the Mississippi River, the Nile of the Western Hemisphere.

Notes

1. Mark Twain, *Life on the Mississippi, Part 1* (New York: Harper and Brother Publishers, 1901), 1. John Gunther, *Inside USA*: *50th Anniversary Edition* (New York: New Press, 1997), 274. In the 1940s popular journalist John Gunther set out to explore the Unites States. Chicago-born Gunther had been a foreign correspondent who put together three previous books, each of which was a readable and entertaining combination of reportage, travelogue, and history, presented through a first-person narrator. Using a country-by-country survey format, Gunther published *Inside Europe* in 1936, *Inside Asia* in 1939, and *Inside Latin America* in 1942. With the successful format in place, Gunther turned to his home country and prepared extensively for a state-by-state trip that began in the summer of 1944. *Inside USA* was published in 1944 as a comprehensive portrait of the country barely into its postwar years. The thousand-page book offered a broad overview of the states and regions that made up the country. It was immensely popular in its day, and a fiftieth anniversary edition was published in 1997. It was Gunther who coined the term "Nile of the Western Hemisphere" to describe the Mississippi River, a phrase that appears often in discussions of the river and its Delta region. The National Park Service uses a variation for its Heritage Study website, *Nile of the New World*.

2. James C. Cobb, *The Most Southern Place on Earth: The Mississippi Delta and the Roots of Regional Identity (*New York: Oxford University Press, 1994).
3. David L. Cohn, *Where I Was Born and Raised* (New York: Houghton Mifflin, 1948), 12.
4. Richard Ford is quoted in Cobb's *The Most Southern Place on Earth*, 325.
5. *Arkansas Review* loosely follows the Delta initiatives parameters described in the National Park Service's Heritage Study: "As defined by the Delta Initiatives the study area encompasses all or part of seven states and 308 counties and parishes. This area includes all of Louisiana, Arkansas, and Mississippi; 29 counties in southeast Missouri; 16 counties in southern Illinois; 21 counties in western Kentucky; and 21 counties in western Tennessee. . . . For the most part it is bound together by its ties to the Mississippi River drainage system; however, there are portions of the study area in Louisiana, Mississippi, and Arkansas that are outside the direct influence of the Mississippi River. The 1994 legislation, under which this heritage study has been prepared, added to the LMDDC's original study area by requiring that states with more than 50 percent of their geographic area encompassed by the Delta Region be included. The primary focus of this heritage study has been on the areas of the original LMDDC's work while acknowledging the importance of the geographic, social, and natural influences of the additional counties of the legislatively defined region." For the full text as well as a map of the study area, see "Nile of the New World, Draft Heritage Study and Environmental Assessment," *National Park Service*, last modified March 14, 2001, http://www.cr.nps.gov/delta/volume1/intro.htm#study.
6. The Lower Mississippi Delta Development Commission sponsored a conference in Greenville, Mississippi, on November 30 and December 1, 1989, focused on the relationship between heritage and tourism in the Delta. The conference was cohosted by the Center for the Study of Southern Culture (University of Mississippi) and the Center for Southern Folklore (Memphis) and drew representatives from 214 Delta counties in the seven states. Preserving and Promoting Our Heritage: A Conference on Cultural Tourism in the Lower Mississippi Delta Region, November 30 and December 1, 1989, The Lower Mississippi Delta Development Commission Papers (collection no. M04E, box 2, nos. 58 and 59), Delta State University Library Archives. The report is online as a PDF from *BluesHighway.org* here: http://www.blueshighway.org/mdnhagreenville.pdf, accessed March 23, 2015.
7. Luther Brown, "The Delta Center for Culture and Learning," *Delta State University*, last modified July 22, 2013, http://www.blueshighway.org/.
8. "ASU Board of Trustees Resolution 94-103," Delta Studies Center, *Arkansas State University*, December 17, 1994, http://www.astate.edu/a/deltastudies/resolution-and-articles/index.dot.
9. Arkansas State University hosted its twentieth anniversary celebration of the Delta Symposium in April 2014.
10. The National Park Service's *Nile of the New World* website includes the Heritage Study and Environmental Assessment with a full explanation of the legislative requirements, purpose, study area, and methodology.

CHAPTER 2

A Geologist's Perspective on the Mississippi Delta

RANDEL TOM COX

A corridor of flat, fertile farmland up to one hundred miles across, from the Ozark Mountains on the west to the Chickasaw Bluffs of Tennessee and Mississippi on the east, is commonly referred to as the Mississippi Delta. The Delta runs from Cairo, Illinois, four hundred miles south to the Gulf of Mexico, where it merges imperceptibly with the coastal plain of Louisiana. This physiographic region is a distinct cultural province of the central and southern United States and through the decades has been the focus of literary, musical, sociological, economic, and ecological discussions in academic circles. From a geologic perspective, the Delta is remarkable for its flatness in the interior of a continent extending far from coastal lowlands. Typically, continents are visited by repeated episodes of uplift that steepen the slope of the land, quickening the flow of rivers and streams that in their turn carve a rolling landscape of valleys and ridges such as the Tennessee hills and the Ozarks. Bisecting the interior hills of North America, the flatlands of the Mississippi Delta offer a stark contrast to what is expected. These flatlands are even flatter than the often referred to flatlands of Kansas or Nebraska.

In a strict geological sense, a "delta" is a pile of gravel, sand, and mud that is being dumped at a continent's edge at the mouth of a major river—as the land south of New Orleans is now being built by the Mississippi River. From this usage it may seem a misnomer to call the Mississippi River lowlands in their entirety a "delta" rather than just a "flood plain" or "bottomlands." However, geologic investigations of the Mississippi Valley have revealed that this region was once a great bay of the Gulf of Mexico, and for the last 90 million years this bay was gradually filled from north to south with delta sediments of the ancestral Mississippi River. The western shore of this bay extended from what is now Texarkana to Little Rock, Arkansas, to Cape Girardeau, Missouri. Its eastern shore ran from Montgomery to Tuscaloosa, Alabama, along the present course of the Tennessee River in Tennessee and north to Paducah, Kentucky. In others words, the delta of the Mississippi River began building southward from the north end of this great bay at what is now Cairo, Illinois.

How and why this enormous body of water, termed the "Mississippi Embayment," formed and later filled with sediment to become the Delta region is an interesting story involving the breakup of ancestral North America, continental drift, a volcanic hotspot, an asteroid collision with Earth, frigid Ice Ages, and earthquakes *(fig. 2.1)*. This story begins a little less than 600 million years ago, when the opening of an ancient ocean basin (termed the Iapetus Ocean by geologists) split apart a supercontinent (termed Rodinia) that contained the landmass that is the present-day North American continent. Tremendous fissures developed in the first stage of breakup of this supercontinent. Many fissures linked together and widened, and after eruption of vast quantities of lava in the bottom of the fissure system, it became a narrow, early ocean basin, resembling the Red Sea opening between Africa and the Arabian Peninsula today. (We can "see" North America and other continents moving today as the Atlantic Ocean basin widens by about two inches per year because we can track slight changes in the locations of continents using satellites of the Global Positioning System.) Other fractures branched from this new ocean and propagated hundreds of miles into each of the parting continental blocks before stopping, including into the southern edge of North America. One of these "dead-end" fracture zones now lies beneath the Mississippi Delta and is called the Reelfoot Rift after a lake in northwestern Tennessee created by the great earthquakes of 1811 and 1812. This fracture zone, which contains

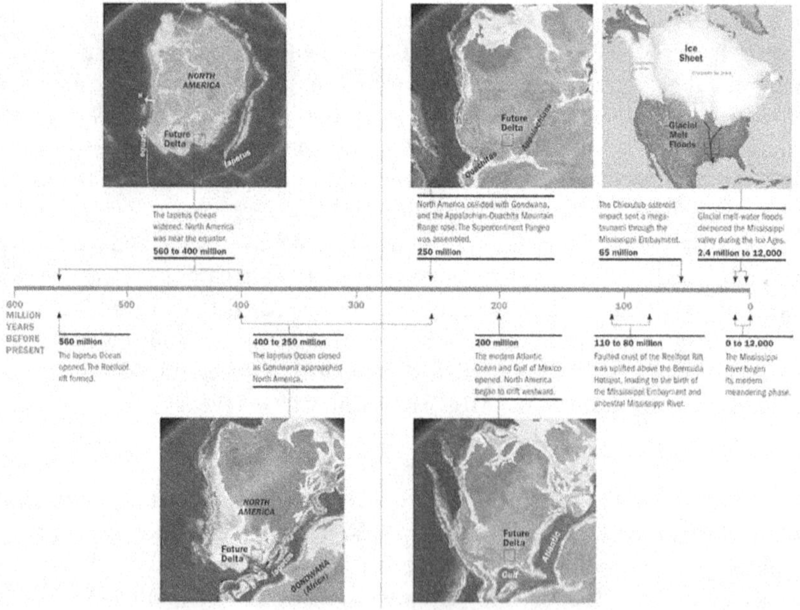

FIGURE 2.1. Mississippi Delta geologic timeline.

the New Madrid fault that is responsible for those great earthquakes, in large measure has governed the geologic history of the region.

Around 300 million years ago, before the age of dinosaurs, ocean basin growth had pushed North American thousands of miles, and it had drifted into warm, tropical climates at the equator, and another landmass (termed Gondwana) containing Africa and South America approached from the southeast. The Iapetus Ocean lay between the two continents, and as they closed on one another, the now old ocean floor suffered the ultimate fate of all old, cold ocean floor rock, sinking back into the interior of the earth (a process called "subduction"). The terrible collision of North America and Gondwana that ensued 250 million years ago raised the Appalachian and Ouachita mountain range along the eastern and southern margins of North America. This mountain range marks the line of suture that assembled Pangea (meaning "all land"), a supercontinent comprised of all continents. Pangea didn't last long geologically speaking. It broke apart when the modern Atlantic Ocean and Gulf of Mexico began to open about 200 million years ago. Remnants of Gondwana that were left attached to North America during the breakup of Pangea are now large parts of Florida and Louisiana as well as portions of other coastal states.

The Appalachian and Ouachita Mountains began as a continuous range of high peaks stretching from Nova Scotia to Mexico, similar to more recently uplifted ranges such as the Himalayas or Alps. This mountain range continued unbroken across the region where the Delta lowlands later developed. Erosion has slowly lowered these mountains to the gentle ridges we see today (although there is growing evidence that high elevations of the Blue Ridge in the Appalachians are due to a minor pulse-renewed uplift within the last few million years). A mountainous connection between the Appalachians and Ouachitas persisted across what is now Delta lowland in southeastern Arkansas and western Mississippi until about 100 million years ago, and until that time the Mississippi River Valley did not exist. Because no large river dumped mud, silt, and sand into the Gulf, a vast coral reef flourished in clear, warm, tropical waters along the coast, offshore from the mountains. This biological reef rivaled any on Earth today, including the Great Barrier Reef of northeast Australia.

About 180 million years ago, as dinosaurs roamed the site of the future Delta, North America started drifting slowly westward with the widening of the modern Atlantic Ocean. During this journey, the continent passed over several volcanic hotspots, including what we now refer to as the Bermuda Hotspot (more recently it made the western Atlantic volcanic island of Bermuda). The crust of the Earth is thickest at continents, and melted rock from a hotspot typically cannot work its way to the surface and erupt as a line of volcanoes, as routinely occurs in regions of thin oceanic crust (for example, the Hawaiian Island volcanic chain has built progressively toward the southeast as the

Pacific Ocean floor moves northwest across a hotspot). However, where continents are fractured by deep fault zones, rising hotspot melt can dynamically lift the weakened crust and melted rock can move up along fractures to erupt and spill lava and ash onto the surrounding landscape. As the faulted region of the Reelfoot Rift passed westward over the Bermuda Hotspot about 100 million years ago, volcanoes began to erupt in central Arkansas. Diamond-bearing volcanics at Murfreesboro and related igneous rocks near Hot Springs, Benton, and Little Rock were formed at the western margin of the rift zone by the hotspot at this time. In the course of drilling for petroleum exploration, samples have been taken from rock underneath sediments filling the Mississippi Valley. From these samples a line of eruptions can be recognized from central Arkansas to Jackson, Mississippi, where the volcanic activity ceased about 75 million years ago. For reasons not well understood, volcanoes worldwide began an episode of greatly reduced activity at about this time (known as the end of the "Cretaceous superplume event").

The Bermuda Hotspot lifted the crust within the northeast-trending Reelfoot Rift and created a highland standing several thousand feet above sea level. Rivers and streams draining this highland would have quickened their flow and cut deep canyons, creating a landscape much different from the present Delta, more like the Cumberland or Ozark Plateaus today. However, due to the heat from the volcanic hotspot, this region was peppered with volcanoes and probably geysers, like the modern continental hotspot region of Yellowstone.

Sediment from the erosion of canyons was carried away to the coast by the mountain rivers. There the sediment poured into the Gulf of Mexico, turning the clear waters muddy and choking the immense coral reef into oblivion, never to return to the northern Gulf Coast. The coral skeleton of this ancient reef is now buried by thousands of feet of sediment beneath southern Arkansas, Louisiana, Mississippi, and other coastal states. This buried reef, like a typical tropical reef, was filled with much open space within the hard skeletal framework, and over the ages these spaces collected petroleum as it seeped upward through the sediments. Tremendous quantities of oil and gas that have been pumped out of this reef greatly boosted the economies of the Delta, for example, the prolific Black Lake oil and gas field of Louisiana and the Waveland oil and gas field of Mississippi.

Elevation of the highland dropped rapidly as mountain rivers eroded it away, and by 75 million years ago the region was a low, rolling plain. Traveling west with North America as the Atlantic Ocean widened, the Reelfoot Rift had moved off of the Bermuda Hotspot, and the crustal block that had been lifted by the heat began to sink back to its original level. But because the highland had been eroded, the crust was thinner, and as it sank back down a great trough was created that flooded with seawater where the highland had been eroded away. Thus, the ancient bay of the ocean called the Mississippi

Embayment was born (fig. 2.2). The ancient river that flowed into the north end of this great bay, the ancestral Mississippi, soon began to fill the bay with sediment, and the true Mississippi Delta started its growth southward. It was a landscape of palms and ferns, and dinosaurs still reigned. Many herds of the behemoths likely roamed the Delta marshes and shores of this tropical seaway.

Then one day 65 million years ago, the sky fell: a six-mile-wide asteroid hit the Earth offshore of the Yucatan Peninsula in the southern Gulf of Mexico. The asteroid hit the ocean at a low angle from the south, sending a towering tsunami northward across the Gulf and into the Mississippi Embayment. This wave washed many hundreds of miles inland, drowning all creatures in its wake. This wave was many times taller than any in recorded history, over three hundred feet high, surpassing even the most fantastic Hollywood special effects. As it washed back to the sea, it carried huge boulders from distant hills down toward the shore of the Mississippi Embayment, where some still rest today on a hillside near Arkadelphia, Arkansas, and on a mountaintop near Rosie, Arkansas (fig. 2.3). The most immediate devastation from this extraterrestrial collision was in what has been termed the "kill zone" of the Gulf of Mexico and southern North America, but effects were felt worldwide, leading to climate deterioration and a global demise of dinosaurs.

Life on the Delta slowly recovered, with mammals replacing dinosaurs as the dominant land animals and flowering plants proliferating. By about 40 million years ago river sediment had built the Gulf shore southward to central Louisiana. (These river sediments now provide drinking water for Delta communities from artesian aquifers including the Sparta Aquifer and the Memphis

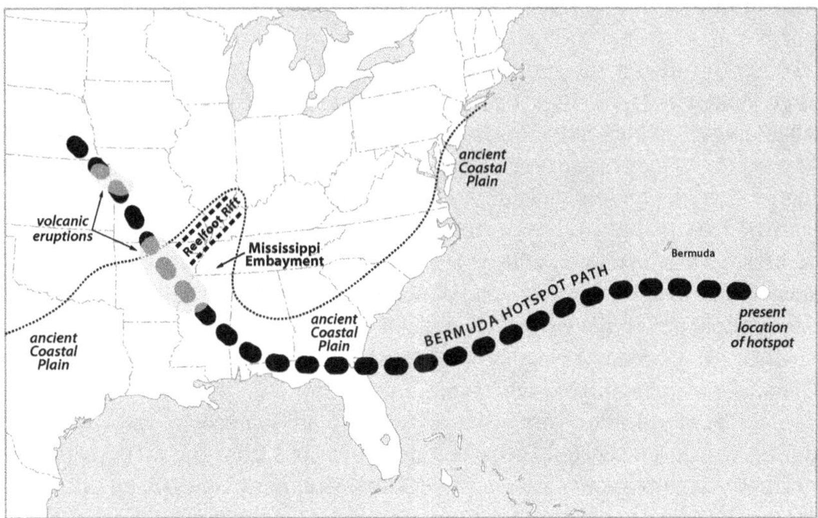

FIGURE 2.2. Bermuda Hotspot path on North America.

FIGURE 2.3. Tsunami boulder at Rosie, Arkansas.

Sand.) For a variety of reasons, the entire Earth's climate began to cool at that time, and by 4 million years ago large glacial ice caps were building at the north and south poles. The tropical climates and forests that had characterized North America for hundreds of millions of years disappeared, and a very large, ancestral Mississippi River flowed from at least as far north as central Alberta and Saskatchewan southward through the Delta region to the Gulf of Mexico (fig. 2.4). This ancestral river flowed in a valley over two hundred feet higher than the present elevation of the Delta (deepened by later erosion), and it carried six to eight times as much water as the modern Mississippi. Some of its large channels can still be seen on the landscape in the north of the state of Mississippi. Relatively small streams now follow these old channels, for example, Arkabutla Creek and parts of the Coldwater River. The gravel laid down by the ancestral Mississippi, mined widely in the uplands above the Delta, is a major source of aggregate for the region.

The last 2 million years of Earth's history are marked by great Ice Ages, during which ice sheets two miles thick advanced from the arctic region of northern Canada southward over North America many times, deep into what is now the Midwestern United States. These great ice sheets came as far south as the Missouri and Ohio Rivers, all the way to Cairo, Illinois, at the head of the Delta region. Their bulldozing action reshaped the Mississippi River and other drainage basins and carved out the Great Lakes. The southward advances of ice ushered in a landscape of tundra and spruce/jackpine forest that extended as far south as the Tennessee-Mississippi border, replacing the oak/hickory/southern pine forests that flourish in the Delta and vicinity in warmer times.

During the Ice Ages, the Delta lowlands were occupied by two large rivers, the Ice-Age Mississippi River on the west and the Ice-Age Ohio River on the east. At one time, their confluence was as far south as the present site of Natchez, Mississippi, where they combined into one great river. Large volumes of glacial melt water flowing down these two rivers increased erosion and caused deepening of their valleys (fig. 2.5).

The Ice-Age Mississippi and Ohio Rivers followed separate courses because ancient faults of the Reelfoot Rift were slowly shifting through time, causing blocks of land to rise within the Delta, tilting the land and deflecting the Mississippi River course to the west and the Ohio to the east. We know these elevated fault blocks as Crowley's Ridge. Over the ages the Mississippi River eroded through Crowley's Ridge at various points and joined with the Ohio progressively farther north until their present confluence was established at Cairo, Illinois, about eleven thousand years ago. Some geologists have speculated that earthquake fault ruptures created gaps in the ridge through which the Mississippi River diverted.

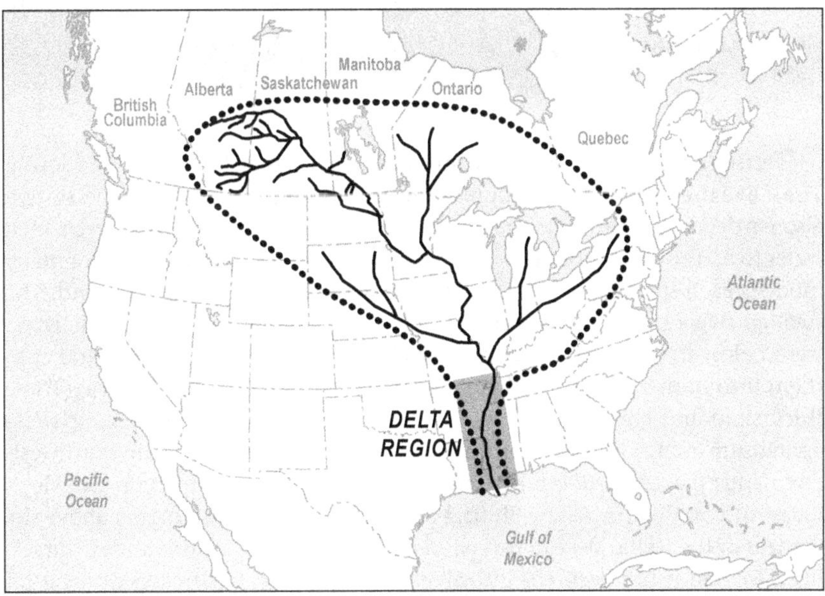

FIGURE 2.4. Mississippi River system before the Ice Age. *Cartography by Tom Paradise.*

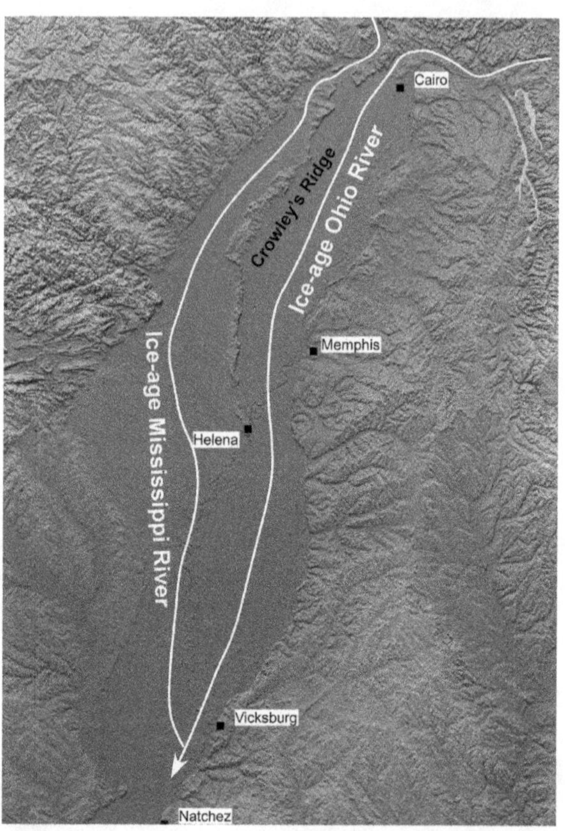

FIGURE 2.5. Ice-Age Delta rivers.

During climatic warming, vast amounts of glacial melt water flowed south. Thus, like the ancestral Mississippi earlier, the Ice-Age Mississippi/Ohio system also carried many times its present volume of water. There was enough melt water to fill the valley from east to west (bluff to bluff) with flood water for many months each spring, leaving Crowley's Ridge as an isolated high ground. The swollen rivers carried large volumes of gravel, sand, and silt. When the rivers were below flood stage, sand and gravel bars choked the channels and split them into numerous chutes that gave the rivers a braided appearance. (These thick sand and gravel bars provide a shallow groundwater aquifer for Delta agriculture today.) In drier months after waters had receded, strong southwesterly winds picked up flood-deposited silt and blew it into drifts on Crowley's Ridge and on the Chickasaw Bluffs, building these landforms higher above the flatness of the Delta. We call this wind-blown silt "loess" (pronounced "lurse"; fig. 2.6), and it has been cut into steep, deep hollows by streams flowing off the bluffs. These Ice-Age river- and wind-blown sediments provide rich soils for much of today's Delta agriculture.

FIGURE 2.6. Wind-blown silt capping the Chickasaw Bluffs.

During warmer episodes of the Ice Ages, tundra and forests of spruce and jackpine followed the shrinking ice sheets and retreated northward, and hardwood forests of oak and hickory expanded to replace them in the hills surrounding the Mississippi Valley. However, frigid glacial melt water still flowing down the Mississippi and Ohio Rivers chilled the valley air and permitted spruce forests to remain dominant in the Delta as far south as present-day New Orleans well after the climate warmed there. The earliest Native Americans lived in this strange country, wandering between cool, spruce forests along the rivers and temperate, oak-hickory forests in the uplands. The most recent Mississippi Valley spruce forest died out about ten thousand years ago, well after the coldest glacial conditions eighteen thousand years ago. Today, North American spruce forests are only found in cool climates of Canada and higher elevations of the Appalachians and Rockies.

After the last Ice Age, a smaller, slower, meandering Mississippi replaced the great, braided, melt water channels. Meandering river channels moving back and forth across the Delta landscape cut low, broad floodplains into the higher, older Ice-Age river terraces (sometimes called the second bottoms). Soils of these low floodplains provide the richest agricultural resource of the Delta. Today the Delta is a spectacularly flat region composed of a mosaic of ancient terraces of the braided channels of the Mississippi and Ohio Rivers formed during the Ice Ages and of modern flood plains of the meandering Mississippi and its tributaries (fig. 2.7).

Occasionally, about every five hundred years or so, the faults of the Reelfoot Rift shift and cause great earthquakes. This was witnessed by a sparse population of settlers in the winter of 1811–1812 when three large earthquakes (one in December 1811, one in January 1812, and one in February) struck the upper end of the Delta in the Missouri Bootheel and northwest Tennessee. These quakes destroyed the river community of New Madrid, Missouri, and were big enough to be felt on the East Coast. Fault movement during these quakes created a ridge that temporarily dammed the powerful Mississippi River and caused its waters to back up for a few hours (but not for three days as stated in folklore accounts). Memphis, Tennessee, Paducah,

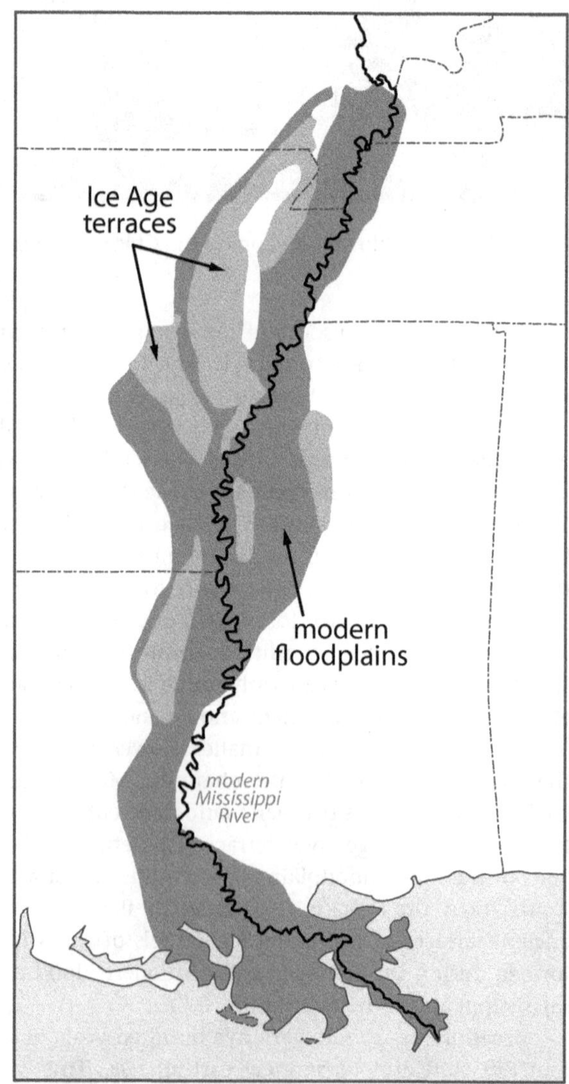

FIGURE 2.7. Delta mosaic of Ice-Age and post Ice-Age river sediments. *Cartography by Tom Paradise.*

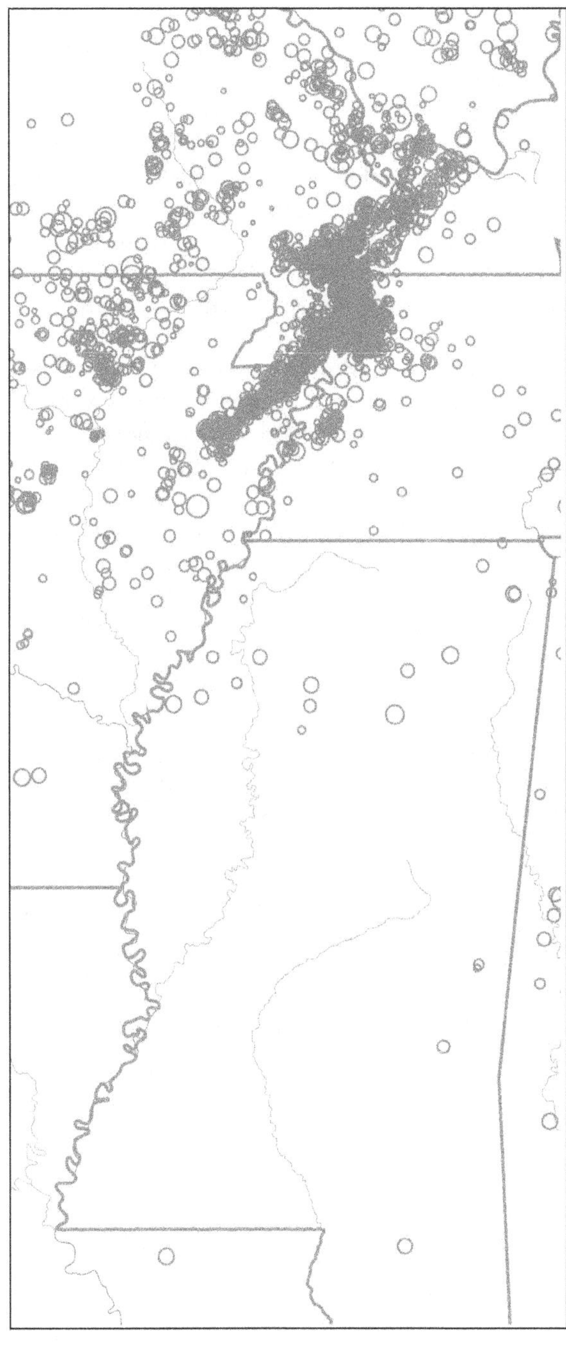

FIGURE 2.8. Earthquakes in the Delta region.

Kentucky, and other communities in the northern Delta region have invested significant resources in emergency response preparations in case of another powerful earthquake. Many small to moderate natural earthquakes are caused by the New Madrid fault system in the northern end of the Delta region (fig. 2.8), but there are numerous slowly sliding faults that rarely cause earthquakes in the southern Delta along the coast. The slow and steady ocean-ward sliding on these faults is driven by the weight of the sediment the Mississippi River dumps at the coast year after year. Creeping movement on these southern faults contributes to sinking of the ground and loss of land near the coast and is an important control on long-term environmental trends and stability of the Delta coastal marsh.

The idea of "geology" typically conjures up images of rocky peaks, volcanoes, waterfalls, and deep mountain gorges; a stark contrast to croplands and bayou swamps of the Delta. Although these flat lands may not make the casual observer think of Earth-shaping geologic forces, to a geologist they bear witness to a history as spectacular as that of the Alps or of the Grand Canyon. Beneath the rich Delta soil, deep below the mud of the great Mississippi River, within the rocks hidden there are the remains of volcanoes, lava fields, and geysers. Deeper still are the eroded roots of a lofty mountain range, and under these the buried landscape of a once great rift valley. So, next time you travel through the Delta, don't forget that its history is as geologically rich as that of any national park in the land.

References

Blum, Michael, and Mark Pecha. "M-d-Cretaceous to Paleocene North American Drainage Reorganization from Detrital Zircons." *Geology* 42, no. 7 (2014): 607–10.

Cox, Randel Tom, and Roy B. Van Arsdale. "Hotspot origin of the Mississippi Embayment and Its Possible Impact on Contemporary Seismicity." *Engineering Geology* 46, no. 3 (1997): 5–12.

Cox, Randel Tom, and Roy B. Van Arsdale. "The Mississippi Embayment, North America: A First Order Continental Structure Generated by the Cretaceous Superplume Mantle Event." *Journal of Geodynamics* 34, no. 2 (2002): 163–76.

Cox, Randel Tom, David N. Lumsden, and Roy B. Van Arsdale. "Possible Relict Meanders of the Pliocene Mississippi River and Their Implications." *Journal of Geology* 122, no. 5 (2014): 609–22.

Cupples, William, and Roy B. Van Arsdale. "The Preglacial 'Pliocene' Mississippi River." *Journal of Geology* 122, no. 1 (2014): 1–15.

Gallen, Sean F., Karl W. Wegmann, and DelWayne R. Bohnenstiehl. "Miocene Rejuvenation of Topographic Relief in the Southern Appalachians." *GSA Today* 23, no. 2 (2013): 4–10.

Galloway, William E. "Depositional Evolution of the Gulf of Mexico Sedimentary Basin." In *Sedimentary Basins of the World*, vol. 5, edited by K. J. Hsu, 505–49. Amsterdam, The Netherlands: Elsevier, 2008.

Johnston, Arch C. "The Rift, the River and the Earthquake." *Earth*, January 1992, 34–43.

Kidder, Tristram R., Katherine A. Adelsberger, Lee J. Arco, and Timothy M. Schilling. "Basin-Scale Reconstruction of the Geological Context of Human Settlement: An

Example from the Lower Mississippi Valley, USA." *Quaternary Science Reviews* 27, nos. 11–12 (2008): 1255–70.

Larson, Roger L. "The Mid-Cretaceous Superplume Episode." *Scientific American* 272, no. 2 (1995): 82–86.

McCulloh, Richard P., and Paul V. Heinrich. "Surface Faults of the South Louisiana Growth-Fault Province." Geological Society of America Special Paper 493. In *Recent Advances in North American Paleoseismology and Neotectonics East of the Rockies*, edited by R. T. Cox, M. P. Tuttle, O. S. Boyd, and J. Locat, 37–50. Boulder, CO: Geological Society of America, 2013.

Rittenour, Tammy M., Michael D. Blum, and Ronald J. Goble. "Fluvial Evolution of the Lower Mississippi River Valley during the Last 100 K.Y. Glacial Cycle: Response to Glaciation and Sea-Level Change." *Geological Society of America Bulletin* 119, nos. 5–6 (2007): 586–608.

Saucier, Roger T. *Geomorphology and Quaternary Geologic History of the Lower Mississippi Valley.* US Army Corps of Engineers, Waterways Experiment Station, Vicksburg, MS, 1994.

Schulte, Peter, Laia Alegret, Ignacio Arenillas, et al. "The Chicxulub Asteroid Impact and Mass Extinction at the Cretaceous-Paleogene Boundary." *Science*, March 2010, 1214–18.

Van Arsdale, Roy B. *Adventures through Deep Time: The Central Mississippi River Valley and Its Earthquakes*. Geological Society of America Special Paper 455. Boulder, CO: Geological Society of America, 2009.

Van Arsdale, Roy B., and Randel Tom Cox. "The Mississippi River's Curious Origins." *Scientific American*, January 2007, 76–82B.

Van Arsdale, Roy B., Robert A. Williams, et al. "The Origin of Crowley's Ridge, Northeastern Arkansas: Erosional Remnant or Tectonic Uplift?" *Bulletin of the Seismological Society of America* 85, no. 4 (1995): 963–85.

Yuill, Brendan, Dawn Lavoie, and Denise J. Reed. "Understanding Subsidence Processes in Coastal Louisiana." *Journal of Coastal Research* 54 (2009): 23–36.

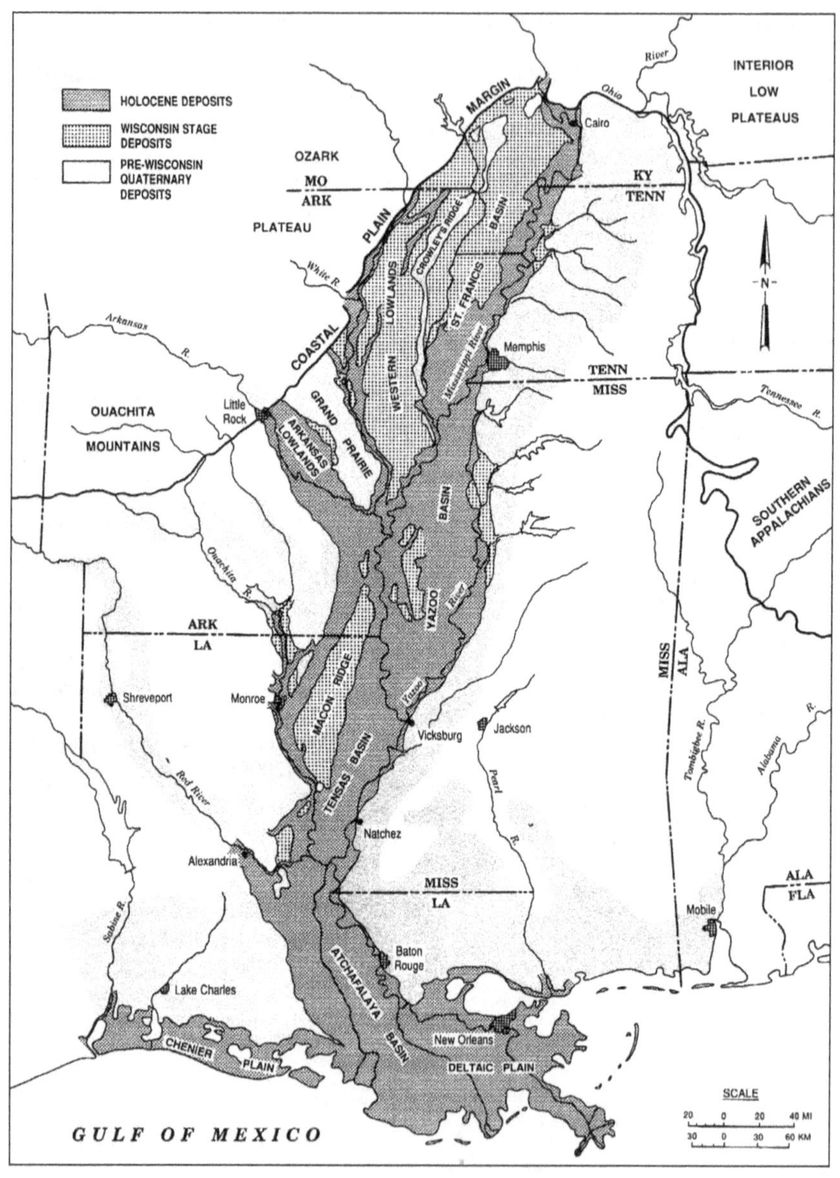

FIGURE 3.1. The Lower Mississippi Valley. *Courtesy of the US Army Corps of Engineers, Waterways Experiment Station, Vicksburg, MS.*

CHAPTER 3

An Archaeologist's Perspective of the Lower Mississippi Valley

David H. Dye

When Hernando de Soto, his colleagues, and their train of Native American interpreters and slaves traversed the Lower Mississippi Valley (LMV) in 1541, they were astounded at the grand and immense Mississippi River and the patchwork of bottomland cornfields and numerous, palisaded towns. Although de Soto was searching for gold and other riches, he inadvertently came upon an agrarian society of conservative and deeply religious farmers. These agricultural communities originated from Paleolithic hunter-gatherers who crossed the Bering land bridge into the Western Hemisphere some thirteen thousand years ago. Our appreciation and awareness of the LMV cultures and their remarkable history is the result of archaeological research that has garnered solid scientific evidence of the people who once called the LMV their home.[1]

The story of LMV Native Americans, from the earliest Ice-Age hunter-gatherers to the farmers encountered by de Soto, is an account of human adaptation and survival in one of the most fertile regions of North America. Their world was dominated by towering bottomland hardwood forests, immense oxbow lakes, and extensive backswamps teeming with abundant animals and plants. The LMV is rich, not in the gold de Soto sought, but in its fertile soils and plentiful resources. In this chapter I present an outline of Native American cultural development in the LMV that has been pieced together by archaeologists over the past century through painstaking excavation and detailed analyses.

To understand these indigenous societies and how they developed through the ages, archaeologists study cultures with similar developmental trajectories. Early documents written by European adventurers, colonists, and priests—Luys Hernández de Biedma, Pierre Le Moyne d'Iberville, Louis Jolliet, Robert de La Salle, Jacques Marquette, and Rodrigo Ranjel—add important details to the story of prehistoric life garnered through archaeological investigations. These avenues of research allow archaeologists to propose hypotheses to be tested through future fieldwork and laboratory analyses. New emphases on

TABLE 3.1: Chronological chart of archaeological periods and climate change

	PERIOD	NORTHERN LMV	SOUTHERN LMV	COASTAL LMV
1700	Late Protohistoric	Tunican	Nathchezan	Chitimacha
	Middle Protohistoric	Mississippian	Plaquemine	Plaquemine
1541	Early Protohistoric	Mississippian	Plaquemine	Plaquemine
AD 1400	Late Mississippi	Mississippian	Plaquemine	Plaquemine
AD 1200	Middle Mississippi	Mississippian	Plaquemine	Plaquemine
AD 1000	Early Mississippi	Plum Bayou/	Plaquemine	Plaquemine Mississippian
AD 800	Terminal Woodland	Plum Bayou/k	Coles Creek Coles Creek	Coastal Coles Creek
AD 600	Late Woodland	Baytown	Troyville	Coastal Troyville
200 BC	Middle Woodland	Marksville	Issaquena/ Marksville	Issaquena/ Marksville
800 BC	Early Woodland	Lake Cormorant	Tchefuncte	Tchefuncte
1700 BC	Late Archaic	Poverty Point	Poverty Point	Poverty Point
4000 BC	Middle Archaic			
8500 BC	Early Archaic	Dalton	San Patrice	
11,000 BC	Late Paleoindian	Folsom	Folsom	Folsom?
11,500 BC	Early Paleoindian	Clovis	Clovis	Clovis

iconography, politics, religion, ritual, and social organization also play an important role in interpretations that are continuously evaluated and revised.

The past two decades have witnessed significant improvements in chronological methods and a more nuanced understanding of the relationship among climate, culture change, and landscape. The connections between archaeological units and linguistic groups have come into sharper focus as researchers continue to analyze early European documents critically and to evaluate their correspondence with artifacts, community organization, and settlement patterns. The LMV has been relatively well studied by archaeologists, although gaps remain in key areas. Coupled with the establishment of a well-defined Holocene chronology, basic regional archaeological and geological processes are reasonably well understood, although intermediate-scale analyses to integrate archaeological and geologic data and to connect regional interpretations with specific site and locality settings are still lacking.[2]

For an archaeologist, the LMV is not simply an array of artifacts and habitation sites, but rather it is the location of ancient cultures that adapted, changed, and evolved over time. The surviving material culture and settlements provide important scientific documentation for understanding the LMV's prehistory. Archaeology differs from other scientific disciplines in that its practitioners provide a long-term view into ancient cultures by tracking cultural patterns and trends over thousands of years. We know that past hunting and gathering societies were not static cultures but changed over time from small family groups to gardeners and farmers who continued to hunt and gather while building massive earthworks and exchanging highly crafted, exotic goods over great distances.

Archaeology also differs from other disciplines in that its subject of study is non-renewable and has been undergoing destruction since the early days of European colonization. Many archaeological sites in the LMV, and the information they contain, have been destroyed through extensive land modification and rampant looting. Ten-thousand-year-old archaeological sites cannot be recreated. Once they are gone, the record of that culture's existence is lost forever. Archaeologists continue to impress upon the public the urgent message that we must be vigilant as stewards of the past to preserve and protect our non-renewable heritage, and to preserve cultural resources for future generations.

The archaeological record of the LMV extends from initial colonization by Paleolithic Ice-Age foragers to the village horticulturists encountered by Spanish explorers in the sixteenth century. Based on extensive research, archaeologists now recognize six major time periods for the LMV, each characterized by distinctive material culture and settlement patterns. The Paleoindian period (11,500–8500 BC) references foragers who settled in the Western Hemisphere as the last Pleistocene glaciers were melting. The Archaic period (8500–800 BC) represents the transition from the last Ice Age (Pleistocene) to the Recent (Holocene) age and the development of new cultural innovations as foraging societies became more affluent and began to domesticate indigenous plants. Between 800 BC and AD 1000 Woodland period foragers gradually added indigenous domesticated plants to their diet, bringing about changes in political, religious, and social practices, especially the Middle Woodland mortuary programs and their long-distance exchange of exotic, well-crafted materials. The Mississippi period (AD 1000–1541) witnessed the spread of corn agriculture and in the northern section the development of palisaded towns and endemic warfare. The de Soto expedition crossed the Mississippi Valley in the mid-sixteenth century, ushering in the Protohistoric period (AD 1541–1700). Spanish observations of indigenous cultural practices provide important documentation for Mississippian society. English and French colonists in the early Colonial or Historic period (AD 1700–1800) introduced epidemic diseases and slaving expeditions, causing local groups to collapse or to create new political and social organizations encountered by the rush of early nineteenth-century settlers from the east and south.

Location and Description

Geologists divide the Mississippi Valley into three broad regions: Lower, Middle, and Upper.[3] The upper segment begins at the headwaters in Minnesota and continues downstream to the Missouri River; the middle section is the area from the Missouri River to the Ohio River, and the lower reach extends for almost one thousand kilometers (about 620 miles) to the Gulf Coast, including the "bird-foot" delta below New Orleans. The broad, lower valley alluvial plain is known as the "Delta" in common parlance. The LMV is often split into three parts by archaeologists: a northern portion extending from the Ohio River south to the Arkansas River, a southern region from the Arkansas River to the Yazoo River, and a Gulf Coastal section south of New Orleans. Archaeologists sometimes refer to the northern half of the LMV as the Central Mississippi Valley.[4] It should be noted however, that the Central Mississippi Valley is also the term used by archaeologists to refer to the American Bottom, a wide floodplain along the Mississippi River in west-central Illinois opposite St. Louis, Missouri.

The LMV's biotic communities are dominated by rich floodplain forests and isolated prairies. Prior to the first settlers, the LMV was almost entirely covered by dense hardwood forests, cypress-tupelo swamps, and cane or palmetto thickets. From the nineteenth century to today, there have been massive landscape alterations to the original bottomland habitat, including extensive timbering and the creation of cotton, rice, soybean, and sugarcane fields. By the late 1930s about half of the original forest had been removed as a result of logging, and by the late 1970s only 20 percent of the floodplain forest remained. Not only have the LMV's biotic communities been largely eradicated, but thousands of archaeological sites have been destroyed in the process of deforestation and cultivation.

Paleoindian Period (11,500–8500 BC)

The first people to live in the LMV were Pleistocene foragers whose ancestors migrated over the frozen steppes of central Asia and across an Arctic land bridge known as Beringia. These hunter-gatherers entered North America by negotiating the Pacific Coast or by traveling along rivers into the continent's interior. By the end of the Paleoindian period early hunter-gatherer groups were thinly scattered across the entire Western Hemisphere.

At the height of the last glaciation, the massive Laurentide ice sheet in the northern hemisphere, which was approximately the size of present Antarctica, trapped large amounts of water in the form of ice. Eighteen thousand years ago, sea level had drawn ocean levels 120 meters (394 feet) lower than present, creating a "land bridge" 1,600 kilometers (994 miles) wide from north to south between Siberia and Alaska.[5] Many early hunter-gatherers followed interior continental game trails and southeastward-flowing rivers into the heart of the continent, especially the Missouri River with its headwaters in the northern Rockies, or passed through eastern corridors such as the Ohio River. Upon their arrival in the Mississippi Valley, they would have followed these alluvial corridors that provided primary routes for further entry into the eastern woodlands.[6] Archaeologists refer to this initial human settlement as the Paleoindian period and subdivide the time of early foragers into two or three subunits based on changes in toolkits and in spear-point morphology. Clovis and Folsom are two prominent and well-documented Paleoindian "cultures."

Fluted spear points are diagnostic of the early Paleoindian period. The earliest variety of fluted point is the Clovis type found throughout North America between about 11,500 and 11,000 BC. Earlier groups who left few diagnostic artifacts may have entered North America prior to Clovis: a controversy that continues to plague archaeologists. The Clovis people were highly mobile foragers who left few traces, other than distinctive spear points, in the LMV. Surface finds are rare and widely scattered, indicating either small or transient populations on the landscape. Clovis points are usually made of non-local cherts from northern and western source areas, suggesting populations were migrating from the north or west, moving along the axis of the Mississippi Valley.[7] The LMV may have been relatively inhospitable for early human occupation because of frequent flooding caused by glacial meltwater, which may have reduced fish and mammal populations.[8]

During a cold snap between 10,850 and 9550 BC—a time known as the Younger Dryas, or "the Big Freeze"—some thirty-five genera of Pleistocene megafauna became extinct, perhaps due to a combination of insufficient adaptability to climate, human predation, and vegetative changes.[9] The bones of large, extinct mammals (such as the American lion, giant bison, ground sloth, horse, mammoth, mastodon, musk-ox, paleo-llama, saber-toothed tiger, short-faced bear, and tapir) have been found on Gulf Coast salt domes and gravel bars adjacent to the modern Mississippi River course, where they eroded from the loess bluffs east of the river. By 8500 BC these large Pleistocene mammals no longer roamed the LMV's forests and grasslands. The Younger Dryas corresponds to the end of Clovis and the beginning of the Folsom culture. By the end of the Younger Dryas, vegetation in the Eastern Lowlands had changed from spruce-willow woodland to cypress–tupelo gum forest in permanent backswamps, sweetgum-elm forests on seasonal floodplains, and willow thickets and canebrakes along levee crests.[10]

Paleoindian groups were loosely integrated as independent, mobile, and small family units, connected by extensive exchange and kin networks. Although there is no direct archaeological evidence from the LMV, the ethnography of recent foragers suggests that each group probably numbered about 25 or so individuals, making up three to four extended families. The overall population of Paleoindian hunter-gatherers in the LMV was small, and family groups would have been isolated from their neighbors for periods of a few months to several years. In northeastern Arkansas, for example, there may have been no more than 75 to 150 people who formed a few bands foraging within a 30,000-square-kilometer (11,580-square-mile) territory.[11] Social contacts among neighboring groups for information, mates, and raw materials would have been critical to continued existence and survival. Alternating base camp locations seasonally, they exploited local floodplains and the lower terraces of braided streams, fishing, gathering, and hunting for a wide range of animals and plants.

Archaic Period (8500–800 BC)

Dalton (8500–7900 BC)

Sometime around 8500 BC, hunter-gatherer groups became increasingly focused on riverine resources, especially those associated with braided streams west of Crowley's Ridge and east of Maçon Ridge. A distinctive dart point, the Dalton point, is especially widespread in the northern portion of the LMV, while the related San Patrice point is found along Maçon Ridge in northeastern Louisiana and southeastern Arkansas. The Dalton period marks the transition from Pleistocene, Paleoindian foragers to Archaic, Holocene hunter-gatherers. Dalton people enjoyed the bounty of temperate forests that began to dominate much of the LMV after the last glaciation. In contrast to earlier habitations, Dalton sites abound in the Western Lowlands of eastern Arkansas and southeastern Missouri, where base settlements and temporary camps are found along abandoned, braided-stream terraces, ephemeral sluiceways, and low knolls. Local quarry sites, sources of chert for stone tools, were exploited on Crowley's Ridge and the adjacent Ozark Plateaus, representing frequent visits to the same localities. As population increased, foragers ranged less widely across the landscape.[12] Around 8100 BC the Mississippi River rapidly shifted from a braided-stream regime to a meandering system, changing the environmental setting of the LMV.[13]

Dalton sites have produced exotic, high-quality, and over-sized lanceolate blades, some measuring up to 38 cm (ca. 15 inches) in length.[14] Some blades, crafted from Burlington chert quarried just southwest of St. Louis, were chosen for their superior flint-knapping properties. Such blades have been found in more than thirty locations along a 700 km (434 mile) stretch of the Mississippi River. These unusually large bifaces were mortuary accompaniments in Dalton cemeteries, such as the Sloan site in northeastern Arkansas, which is the oldest documented cemetery in the Western Hemisphere.[15] Located on a low sand dune near the Cache River, the cemetery contained between twelve and twenty-five graves, each including individual caches or clusters of unused Dalton points. It is thought that production and exchange of these exotic bifaces integrated kin groups into regional alliances and fostered social interactions through rituals conducted in the context of feasting, festive gatherings, and gift exchange.

Early Archaic (7900–4000 BC)

As the Younger Dryas came to an end, a sharp rise in temperature characterized the onset of the Early Holocene era (10,000–5,000 BC), marking the beginning of the Early Archaic period. Modern flora and fauna replaced earlier Ice-Age biotic communities. Hunter-gatherer populations continued to increase after the Dalton period, and foraging territories slowly multiplied. Early Archaic territories were bound together by social networks that facilitated the flow of critical information, potential mates, and vital resources among family groups.

White-tailed deer continued to be the primary target prey of hunters, but small game and fish assumed increasing importance in the diet, in addition to a wide variety of plant foods. The abundance of stone mortars suggests that ground nuts and seeds had also become a major dietary focus. Dart point styles, including side-notched, corner-notched, and bifurcate-based bifaces, appear to have succeeded each other over time. These varying styles of hafting techniques aid archaeologists in charting cultural changes and determining the relative ages of sites. For example, Early, Middle, and Late Archaic temporal divisions, based on changing dart point styles, roughly coincide with broad climatic trends and changes in group mobility, hunting and gathering ranges, population levels, and raw material use.

Despite changes in the climate and environment, early post-Dalton foragers may have continued a lifestyle broadly similar to their late Paleoindian and Dalton ancestors. Although useful in marking temporal units, differences in projectile point styles may not signal significant changes in adaptation or behavior because groups of mobile hunter-gatherers with low population densities probably continued to maintain conservative subsistence strategies. Projectile point styles suggest that large territories were centered on floodplain bayous, lakes, and rivers, but until sites on Early Holocene land surfaces or in buried contexts are found and excavated, the Early Archaic period will remain poorly known.

Middle Archaic (4000–1700 BC)

Early Middle Archaic sites are rare and few have been excavated. As was the case with earlier Paleoindian and Early Archaic populations, short-term, seasonally occupied sites seem to be dispersed throughout the valley. These sites are located on late Pleistocene, braided stream surfaces, suggesting that that LMV may have been neither environmentally stable nor necessarily desirable for long-term habitation. Resources may have been patchy and unpredictable throughout the valley, resulting in a flexible and mobile lifestyle for hunter-gatherers. Moreover, the valley appears to have been avoided for much of the year, being visited only on a seasonal basis.[16]

Between 4000 and 3000 BC, significant cultural changes took place in the LMV with the establishment of an increasingly stable landscape, perhaps bringing about increasing dependence on more predictable foods.[17] Middle Archaic hunter-gatherers began to exploit riverine habitats through broad-spectrum and intensified foraging centering on deer, fish, and wild plants. As is the case with earlier sites, many Middle Archaic components may be obscured by more recent backswamp sediments. Some sites, such as Denton in the northern Yazoo Basin, are on Pleistocene terraces overlooking relict channels. Denton is noted for its distinctive lapidary industry, especially the manufacture of zoomorphic effigy beads.[18] The presence of well-crafted and exotic items is an indication of critical information being exchanged. The flow of such goods cemented communities within a matrix of reciprocal aid and support in times of environmental uncertainty, resource shortfalls, and social emergencies.[19]

The discovery of Middle Archaic mounds in the 1990s in northeastern Louisiana was an archaeological milestone of international significance. This monumental architecture in the LMV is the earliest in North America. The largest and most complex of the Middle Archaic mound centers, the Watson Brake site in northeastern Louisiana, includes eleven earthen mounds connected by a large, meter-high (ca. 3 feet) oval earthwork ridge about 280 meters (920 feet) in diameter. The largest mound stands 7.5 meters (25 feet) high.[20] Initial construction, based on a deliberate site design, began around 3400 BC. The periodically occupied mound and ridge surfaces were host to routine, domestic events, including processing and cooking white-tail deer, fish, and other game, in addition to plant foods. Ritual activities probably took place at the site, as mounds throughout the prehistory of the LMV had ritual, social, and symbolic value.[21] Bifaces and stone beads were manufactured at Watson Brake, but there is no evidence for the exchange of nonlocal materials.

Middle Archaic mound building in the LMV terminated around 2800 BC, coinciding with rapid climate change and locally devastating flood events.[22] During the mid-Holocene, unpredictable flooding was one component of LMV weather patterns.[23] Between 2800 and 1800 BC, a period of landscape transformation and instability—due to flooding and rapid sediment deposition—coincided with the abandonment of a large portion of the LMV.[24] Some archaeologists have suggested that mound construction may have been a communal response to stresses of climate changes, including droughts and flooding that caused food supplies to become less predictable.[25] On the other hand, LMV foragers were well adapted to floodplain subsistence and may have suffered little if any stress.[26] Much is yet to be learned concerning the Middle Archaic period in the LMV.

Late Archaic (1700–800 BC)

The Late Archaic Poverty Point culture emerged sometime after 1700 BC, when an era of stability characterized much of the floodplain, resulting in sedentary communities. Contemporary cultures in the St. Francis and Yazoo Basins had populated the valley several centuries earlier. Hunter-gatherers rapidly colonized the well-drained LMV natural levees.[27] They constructed a massive mound complex at the Poverty Point site and engaged in an interregional exchange network that reflects widespread cultural interaction.[28] The type site is a three-square-kilometer (1.2-square-mile) complex of mounded earth in six concentric, elliptical half-rings, one large mound, and a few smaller conical and flat-topped earthworks.[29] The ubiquitous baked-clay balls, characteristic of this period, were used in lieu of stones for roasting in earth ovens and boiling in skin bags and woven baskets. In addition, a well-developed lapidary and microlith industry emphasized the production of ground-stone gorgets and plummets. The latter were apparently used as weights for looms, suggesting a well-developed Late Archaic weaving technology.[30] The Poverty Point subsistence economy was based on intensified fishing, gathering, and hunting. Acorn and nut-bearing trees—such as hickory, oak, pecan, and walnut—may have been managed through periodic understory burning.[31]

The Poverty Point people exchanged a wide array of resources from throughout eastern North American: copper from the Great Lakes; flint from the southern Appalachian, Ouachita, and Ozark Mountains; galena, a native lead ore, from the Cumberland and Central Mississippi Valleys; and soapstone from the southern Appalachians. Many of these materials could have been reached by dugout canoe through the Mississippi's vast riverine network. The Poverty Point site was abandoned around 1200 BC. Shortly after 1000 BC, the long-distance exchange system abruptly ceased, perhaps as a result of cultural change and reorganization, landscape transformation, and widespread episodes of massive flooding that interrupted the valley-wide cultural and geological stability.[32] Public construction works were also abandoned, signaling that important changes were taking place for the Poverty Point people. The shift from Late Archaic to Early Woodland may have been due to a variety of factors, including climate change and exchange system collapse. The demise of the Late Archaic cultures is correlated with an episode of rapid global climate change that is hypothesized to be the cause of increased flooding in the LMV.[33]

Woodland Period (800 BC–AD 1000)

The original definition of Woodland beginnings included the appearance of cultivated plants, mound construction, pottery manufacture, and sedentary settlements, but recent research places these developments prior to the Woodland period.[34] Early markers for post–Archaic Woodland societies in the LMV are now defined principally by distinctive mound-based mortuary ceremonialism and ceramic designs that appear over a large area of the mid-South and adjacent Gulf Coast. LMV Woodland cultures include Tchula, Marksville, Baytown, Troyville, Coles Creek, and Plum Bayou. The Woodland period is also characterized by intense and long-lasting settlements that flourished and gave rise to considerable cultural elaboration. Long-term climate stability and the Mississippi River's confinement within a single channel brought about a period of widespread cultural fluorescence within the LMV.[35]

Early Woodland (800–200 BC)

The Tchula culture is marked by the widespread presence of distinctive pottery pastes and geometric surface treatments. Tchula in the southern Yazoo Basin and adjacent Gulf Coast is referred to as Tchefuncte, while sites in the northern Yazoo Basin are characterized as Lake Cormorant.[36] Tchula represents a cultural transition from the earlier Late Archaic period and is characterized by feasting, food production, midden deposits, mortuary activities, and mound building at scattered mound centers within the LMV and adjacent uplands.[37] These cultural characteristics continue into the Middle Woodland period. Sites are relatively rare in the LMV and few have been excavated, but known settlements are located on natural levees, in backswamps, and along adjacent upland bluffs. Interior sites are typically dispersed and small, perhaps reflecting seasonal occupations by a few families of mobile forager-gardeners, but coastal and delta sites are often large.

Native plant cultivation was widespread in eastern North America by 1200 BC, and, presumably, LMV populations were following suit with small, ephemeral gardens tended by partially sedentary hunter-gatherers. Tchula sites similar to LMV occupations are scattered throughout the uplands to the east, suggesting a widespread cultural pattern across the Gulf Coastal Plain. Occupations were connected and linked by exchange networks, mortuary rituals, and mound ceremonialism. The southern LMV may have been repopulated during Tchula times to some degree by people who returned from the adjacent uplands after the earlier flooding episodes.[38]

Middle Woodland (200 BC–AD 400)

Marksville culture (200 BC to AD 250) populations appear to have been semi-sedentary hunter-gatherers who harvested and managed wild subsistence resources and practiced limited gardening of indigenous, non-maize plants. They crafted distinctive iconographic motifs on utilitarian ritual ware, exchanged nonlocal goods, and raised burial mounds. Although Marksville material culture and community organization suggest cultural continuities with the earlier Tchula culture, they also reflect social interaction, especially with respect to iconography and religious beliefs, with societies outside the LMV, including contemporary Hopewellian groups in the Midwest.

Much of eastern North America participated in a vibrant exchange network through the movement of exotic goods: copper from the Great Lakes region, galena from the Ozark Plateaus, greenstone and mica from the southern Appalachians, and marine shell from the Gulf Coast. These raw materials were crafted into copper ear spools; marine shell beads, cups, and gorgets; panpipes; greenstone celts; stone and pearl beads; cut mica; and faceted galena cubes. Finished artifacts have been found with Marksville log tomb interments that were subsequently mounded over after the funeral rituals were completed. Populations resided in small, widely dispersed communities, each typically with a single mound; larger sites had clusters of mounds within earthen enclosures. The Marksville site in east-central Louisiana is the largest Middle Woodland site in the LMV.[39] Such large mound centers were not the communities of resident populations, but served local populations over a large area that collectively participated in mortuary rituals and world renewal ceremonies.

Between AD 250 and 400, exchange systems during the Middle Woodland period may have slowed down, with goods not being distributed as widely as in earlier times. Mortuary activities also show less flamboyance in comparison to earlier Marksville culture sites.[40] Few habitation sites have been excavated and most mound excavations lack solid dates. The best known of these "late" Middle Woodland expressions is Issaquena (AD 200–500), located in the southern Yazoo Basin and east-central Louisiana.[41]

Late Woodland (AD 400–700)

The Baytown culture (AD 300 to 650) witnessed important cultural changes differentiating it from the preceding late Middle Woodland cultures. These include a continued reduction in earlier inter-regional exchange networks and the emergence of new ceramic decorative ideas, community organization, and settlement types. Permanent villages, although still relatively small, become more abundant, suggesting an increase in population. Diet was still based primarily on fishing, gathering, and hunting, but cultivated local plants that produced starchy seeds (chenopodium, knotweed, and maygrass) and oily seeds (sumpweed and sunflower) may have become increasingly important in gardens or fields. The large Hayti Bypass site, located in southeastern Missouri, had a small-seed assemblage of chenopodium, knotweed, and maygrass, along with numerous deep, large storage pits and hearths, suggesting the site may have been occupied throughout the year.[42] Exotic cherts were obtained from southern Illinois, western Kentucky, and the Nashville Basin. Changes in pottery vessel manufacture, including large, grog-tempered jars and bowls, reflect innovations in food preparation.

Somewhat contemporary with Baytown cultures, Troyville cultures (AD 400–700) flourished between Vicksburg and Baton Rouge. A poorly defined Coastal Troyville culture found south of Baton Rouge had ties with Gulf Coastal Weeden Island cultures to the east. Mounds continued to be employed for communal burial programs. The monumental Troyville site, located in Catahoula Parish, Louisiana, consisted of nine mounds enclosed within an earthen embankment. The Troyville site may be instrumental in understanding changes that led to the later Coles Creek culture with its emphasis on monumental mound construction. Around AD 700, widespread changes took place in local cultures as new pan-LMV cultural patterns emerged that gave rise to Coles Creek cultures.[43]

Terminal Woodland–Early Mississippi (AD 700–1000)

The Coles Creek culture (AD 700 to 1200) is manifested in two areas in the LMV: Coastal Coles Creek, found south and west of Baton Rouge, and interior LMV Coles Creek.[44] Coles Creek marks an important transition from the relatively egalitarian Late Woodland–period, Baytown-Troyville cultures to increasingly hierarchical societies with their emphasis on hereditary chiefly privileges and social ranking. These political and social changes marked the gradual exclusion of the community from some aspects of ritual and the beginning of important and more exclusive roles for elites in ritual activities.[45] Coles Creek culture developments include increases in hierarchical settlement patterns, interregional exchange, greater diversity in material culture, larger mound groups, and reliance on maize, as well as distinctively decorated, grog-tempered pottery.[46] The earliest platform-mound-and-plaza precincts in eastern North America were constructed by Coles Creek groups, perhaps signaling the emergence of feasting and social elites.[47] These mound and plaza arrangements would have integrated members of the community through ritual performances that also displayed the aristocratic status of elites.[48] There are signs of increasing hierarchy at this time and perhaps some type of leadership positions based on feast providers and ritual specialists. Settlements include camps, hamlets, villages, and mound centers linked in a web of political affiliations, conflicts, and negotiations.[49] Corn agriculture was of minimal importance in the LMV until late in the Woodland period and in the southern LMV agriculture was even less important. Coles Creek populations did increase significantly, but subsistence was based on wild plant and animal foods, plus local domesticated seed crops. Mound sites are generally more numerous and larger than in earlier Late Woodland times.[50]

About AD 700, the bow and arrow, the latter tipped with small triangular and notched points, was introduced from the west, perhaps from present-day Texas or eastern Oklahoma.[51] The adoption of the bow and arrow apparently took place because of its increase in accuracy, range, and stealth over the atlatl. The bow and arrow may have given rise to increased conflict and changes in settlement patterns as populations grouped together for common defense. Intercommunity raiding may have become more lethal due to the increased effectiveness of the new weapon system.

The Toltec Mounds, located in the Arkansas-White River Basin southeast of Little Rock, Arkansas, is one of the largest Coles Creek sites, dating from AD 650 to 1050. Toltec is a Plum Bayou culture mound center. The mounds are aligned with solar events, suggesting the site served a calendrical and ritual function. As was the case with most Coles Creek sites, the main population lived in the surrounding countryside where they farmed cultivated indigenous seed crops, fished, gathered, and hunted—maize is virtually absent as a subsistence item. Social organization may have been based on ascribed elite leadership. Plum Bayou sites are found along the Arkansas River Lowland, the Lower White River, and the LMV's Western Lowlands.[52]

Changes leading to the emergence of Mississippian culture in the northern portion of the LMV and Plaquemine culture to the south become increasingly evident late in the Terminal Woodland period as innovations in sociopolitical organization and subsistence strategies took place. Interaction between Coles Creek and Mississippian polities is evident in some areas, especially in the Lower Yazoo and Tensas Basins.[53] The Coles Creek culture developed not only from within, but also from exchange connections far outside the LMV. Direct connections with the Mississippian urban centers at Cahokia and elsewhere in the American Bottom near St. Louis linked terminal Coles Creek cultures in the Yazoo Basin and the eastern edge of the Tensas Basin at the Lake Providence site with a pan-Mississippian world in which ideas, material culture, and people circulated widely.[54]

Mississippi Period (AD 1000–1541)

The shift from Late Woodland to Mississippian culture was significant, although many trends that characterize Mississippi period cultures had already developed by the end of the Late Woodland period, such as hierarchical societies, maize agriculture, and platform mounds. Mississippian patterns developed slowly in the region and resulted in two cultural expressions: Middle Mississippian in the northern LMV and Plaquemine south of the Arkansas River.[55] Differences between these southern and northern regions are expressed in distinctive ceramics, site plans, and social organization. The hunting, fishing, and gathering lifestyle that had been so successful was replaced with a reliance on domesticated crops and the growth of towns and villages. Settlements shifted from relict drainages and backswamps to natural levees adjacent to oxbow lakes near the Mississippi River. Social groups realigned the way in which they related to their natural, social, and spiritual worlds. Droughts and flooding played an important role in how people interacted with the landscape as field agriculture assumed increased importance in daily life.

Middle Mississippian was a farming culture found throughout much of the Midwest and Southeast from about AD 1000 to 1700. One of the most intensive Mississippian developments is found in the St. Francis and Upper Yazoo Basins. By AD 1000, population density began to increase and settlements grew larger and more diverse. Maize had been present in the LMV, but it became a dominant dietary staple only after AD 1200. Although squash had been important since 1200 BC, beans were not added to the diet until after AD 1350. Corn, beans, and squash, an eastern North American triumvirate, are often referred to as the "Three Sisters." Wild plant resources, such as acorns, berries, fruits, nuts, and tubers, and animals, including birds, fish, and mammals, remained important dietary items.

FIGURE 3.2. Mississippian ceramic vessel. *Photograph courtesy of David H. Dye.*

Shell-tempered pottery also became ubiquitous throughout the Mississippian world, giving rise to greater elaboration of vessel decoration and form. In some regions, especially the northern LMV, new vessel shapes—including bottles, bowls, jars, and pans—were appliqued, engraved, incised, painted, and punctuated with a variety of designs and motifs that reflect an ancient belief system centered on the "Above World" and the "Beneath World." Ceramic iconography embraced religious themes that emphasized life and death, reincarnation, and world renewal.

Mississippian settlement patterns and community organization are considerably diverse throughout the LMV. In southeastern Arkansas, for example, regional Mississippian settlement patterns consist of dispersed hamlets, farmsteads, and temporary foraging sites. These communities focused on a local ritual center with mounds and charnel houses, differing from the Mississippian towns to the north that were fortified and nucleated.[56] There was also variability in Mississippian houses, but generally they were rectangular in outline and constructed by placing wooden posts into excavated trenches, attaching woven, split-cane mats to the posts that supported and formed walls, and then plastering the walls with moist clay. Roofs were thatched with local prairie grasses, such as big blue stem, and other locally available materials. Houses were organized around plazas, which integrated and focused community activities. Politically important towns typically had two to eight mounds that supported buildings for aristocratic families as well as charnel houses that held ancestor corpses and statues and ritual paraphernalia.

FIGURE 3.3. Mississippian ceramic vessel. *Photograph courtesy of David H. Dye.]*

FIGURE 3.4. Mississippian ceramic vessel. *Photograph courtesy of David H. Dye.*

FIGURE 3.5. A reconstructed Mississippian house, Chucalissa Village, C. H. Nash Museum, University of Memphis. *Photograph courtesy of David H. Dye.*

Plaquemine (AD 1200–1700)

Plaquemine culture, generally found south of the Arkansas River, is characterized by the continuation of Coles Creek culture with greater emphasis on hereditary rights to leadership, hierarchical organization, social inequalities, and a robust program of mound construction, perhaps resulting from competition among elites and their constituent communities.[57] Strong continuities are evident between Coles Creek and Plaquemine cultures in belief systems, ritual, and sociopolitical organization.[58] A segment of late Coles Creek/emergent Plaquemine society began exercising its power more widely through external contacts, which empowered the rise of a Plaquemine elite through the manipulation and production of exotic exchange items and new ideologies.[59] Overall, Plaquemine culture was not a radically new way of life, but a continuation of earlier themes and traditions. Plaquemine culture, as was the case with Middle Mississippian culture, crossed ethnic and linguistic boundaries, and the various regional polities would have had differing political and social histories.[60]

Just as Coles Creek mound centers were often constructed over Baytown period mortuary and ritual centers, so did the Plaquemine elite appropriate Coles Creek mound complexes, suggesting that connections with ancestral ritual centers and their deities served as a basis for ritual authority. Multiple mounds constructed around plazas restricted the ritual boundaries of the mound-and-plaza precinct. Activities that took place at these precincts were more clearly distinguished from activities that took place in the neighboring hamlets and villages where the majority of the population lived.[61] The relative lack of occupation at Plaquemine mound centers is also reminiscent of Coles Creek mound sites, in contrast to Mississippian sites that have extensive midden deposits.[62] While Mississippian aristocratic families were resident in one continually dominant town to which smaller towns were subservient, the Plaquemine nobility were located in similarly structured towns; none of which catered to a higher authority.[63]

Except for a few large multi-mound sites, there is no evidence for palisades or defensive structures, although intersocietal conflict probably took place as communities tried to exert political authority over their neighbors.[64] Sites become larger and less numerous from Coles Creek times, suggesting increasing political centralization and larger territories. Unlike the Middle Mississippian sites north of the Arkansas River, Plaquemine towns are relatively uncommon. The community pattern is represented by dispersed settlements with site hierarchies reflecting multiple levels of political administration.[65] Plaquemine groups manifested considerable cultural variability throughout the LMV.[66] Archaeologists continue to discuss the degree to which contact with Middle Mississippian groups influenced the development of Plaquemine communities, but most researchers see Mississippian contacts as a minor factor in the rise of the Plaquemine elite.[67] The Winterville site, located in the Lower Yazoo Basin, is a major Plaquemine site that was occupied from AD 1200 to 1400.[68] The lack of cultural debris at the site suggests that such mound centers were occupied by the religious and social elite, except during ritual occasions when farming families from the hinterland temporarily resided at the center.

Access to and control of major water transportation routes was important in decisions concerning the location of these Plaquemine mound centers.[69] Overland trails, in combination with extensive waterways, created a matrix of exchange partnerships that connected great centers throughout the LMV, in addition to communities to the east and west of it.[70] Spiritually charged, Underwater Panther (Bellaire) pipes from the Vicksburg area were distributed east and west, while figurines, pipes, and pots came downriver from the American Bottom.[71] Exchange relations were crucial for ruling elites, who competed for positions of authority and sought to advance and legitimize their political interests and positions through genealogical claims and control of exchange items and ritual performances. The Middle Mississippian and Plaquemine worlds were transformed by climatic events and the effects of European presence in eastern North America in the sixteenth century.

Protohistoric Period (AD 1541–1700)

The Protohistoric period was a time of initial European contact, beginning with Hernando de Soto's entry into the LMV in the spring of 1541. After the mid-sixteenth century, events in the LMV are not well documented, but cultures there were deeply affected by events happening around them. Recent research has provided significant advances in knowledge of the Protohistoric period and the historic groups that emerged from the archaeological past.[72] Persisting lack of clarity and scholarly consensus about the Protohistoric period results from its relatively short time frame, at least in terms of archaeological conceptions of time, coupled with great cultural disturbances in the LMV and surrounding regions: climate change, intercommunity conflict, and polity collapse or transformation. Disruption of indigenous polities resulted in the decentralization of political authority, population movements, or social change, including community collapse and demise in some places. Consensus among archaeologists concerning the Protohistoric period has been difficult to build, largely because there are so few documentary sources for the post–de Soto era.

Demographic collapse in the post–de Soto, late sixteenth century is seen throughout much of the LMV. In addition to European diseases and slave raids, droughts may have played a role in triggering community migration, culture change, and population loss. Tree-ring reconstructions document two prolonged mega-droughts throughout the LMV between 1550 and 1600.[73] The most intense episode was between 1550 and 1575, more than a decade after the de Soto expedition.[74] Escalated warfare, population collapse, and settlement relocation may have been triggered by the continuing drought.[75] In the seventeenth century, devastating changes from European contact resulted from epidemics and slave raids when permanent European colonies began to take root along the Atlantic and Gulf coasts.

In archaeological terms, several broad cultural expressions are evident in the Protohistoric LMV, reflected in historically documented linguistic groups. Natchezan speakers may have resided in eastern Arkansas south of the Tunicans, in southeastern to south-central Arkansas, and in western

Mississippi. The Grand Village of the Natchez (1682–1729), located on St. Catherine Creek in present-day Natchez, Mississippi, was the main ritual center of the Natchez during the late seventeenth and early eighteenth centuries. The Grand Village (aka the Fatherland Site) was continuously occupied for over five centuries before its abandonment in 1730. French visitors to the village provide rare firsthand information about the Plaquemine belief system and political organization. In typical Plaquemine fashion, most of the population lived in nearby hamlets and villages, gathering periodically at the mound center for political, religious, and social events.[76] The Natchez were dispersed by the French following the Natchez Rebellion of 1729 and their attack on Fort Rosalie. The majority of Natchez today live among the Muscogee and Cherokee Nations in Oklahoma.

Sites possibly occupied by Tunican speakers include portions of eastern Arkansas, southeastern Missouri, and western Tennessee.[77] Some remnant Tunicans remained in southeastern Arkansas, but by 1700 virtually all had moved out of the LMV, relocating to east-central Louisiana or to Oklahoma. The distinctive Protohistoric Armorel material of northeastern Arkansas and southeastern Missouri may represent either Siouan or Tunican speakers.[78] Dhegihan Siouans may have resided in eastern Arkansas and southeastern Missouri as early as the fifteenth century.[79] There is no evidence for the existence of Siouan words in eastern Arkansas based on the de Soto expedition accounts, and Robert Rankin argues that the recorded words are Tunican.[80] The ceramic evidence does not appear to support a significant movement of people from eastern Arkansas/southeastern Missouri to the mouth of the Arkansas River in the seventeenth century.[81] Dhegihan speakers may have arrived at the mouth of the Arkansas River as late as 1660.[82]

The historic Chitimacha, located in the western Atchafalaya Basin, may also have direct connections to the Plaquemine culture.[83] Their Protohistoric villages are represented by a series of mound centers located between the Atchafalaya Basin and Teche Ridge. Chitimacha creation narratives place their origin in the vicinity of the Natchez, but over time they moved westward.[84] The remaining Chitimacha, although seriously reduced by epidemic disease and slave raids, reside in St. Mary Parish, Louisiana.

The ethnic assignment of LMV archaeological cultures is still an open question. Archaeologists as well as Native Americans continue to debate the archaeological recognition of historic linguistic groups. Continued refinement of regional chronologies and additional excavation will be required to evaluate connections among archaeological cultures, Protohistoric sites, and ethnohistoric sources.

Conclusion

The LMV was first peopled over 130 centuries ago by expert foragers who expanded out of Western Asia, across the Bering land bridge, and into the New

World, a hemisphere uninhabited by humans but teeming with abundant plants and game animals. The earliest, well-attested Paleoindians in the LMV are the Clovis people, who would have been widespread throughout the LMV. Unfortunately, the study of Paleoindian culture is made difficult by the scarcity of their material remains as a result of low populations and subsequent burial or erosion of archaeological sites.

As Archaic populations slowly expanded and multiplied, they adapted to new resources made available by the warming climate of the post-glacial Holocene world. With increased population densities, local groups began to maintain exclusive political and social boundaries and to store food against future needs. Monumental earthen construction signals the early development of mound rituals and perhaps world renewal ceremonies. Late in the Archaic period, domestication of native plants provided the groundwork for later cultural developments.

Burial mounds and utilitarian ritual ceramics characterize the Woodland period with its emphasis on earthwork construction, exchange in exotic raw materials, and mortuary programs, especially by Marksville populations. Although there is no evidence for high population levels, large and small religious centers were locales of religious rituals. By the end of the period, monumental earthwork construction shifted from mortuary functions to platform mounds that housed elite families and the bodies of their ancestors.

The addition of maize provided an easily stored surplus for Mississippian polities. The rise of Middle Mississippian in the upper LMV was followed by a related variant known as Plaquemine in the lower LMV. Members of de Soto's incursion into the upper LMV left vivid descriptions of flourishing Middle Mississippian polities, as did the French who recorded details of Natchez society. Their accounts add to the archaeological record by providing important details about Mississippian and Plaquemine cultures. With European discovery of the Western Hemisphere and the LMV, a clash of cultures ensued that resulted in the destruction and transformation of indigenous societies primarily through epidemic diseases and slave raids.

The LMV, for some thirteen thousand years, was home to societies ranging from Pleistocene foragers who discovered and created a new home along the Mississippi River to settled farmers who embraced a conservative belief system and social organization in their efforts to farm the land. Archaeology has documented many changes in Native American culture that took place over the millennia. Through adherence to scientific and social scientific procedures, archaeologists see the past not simply as artifacts and sites but also as human communities who solved problems through adherence to or changes in political strategies, religious beliefs, and social organization.

The land and cultures of the LMV continue to change, but archaeological interpretations depend on preserving the remaining vestiges of its earliest inhabitants. We cannot fully understand LMV prehistory without appreciating the LMV's archaeological record. Unfortunately, loss of archaeological sites is progressing at a staggering rate. Sites are being destroyed through agriculture and urban expansion. Although clandestine digging and looting has been limited with the passage of legislation that prohibits uncontrolled digging of Native American graves, looting still continues throughout the LMV. A fuller understanding of the LMV's first residents is possible only through the continued preservation of archaeological sites and the private and public support of professional archaeological organizations seeking to understand the past through scientific inquiry of a precious cultural resource.

Notes

I am grateful to Janelle Collins for her invitation to participate in *Defining the Delta* and for her comments that improved the chapter. Numerous colleagues contributed sage advice and corrected mistakes. My appreciation is extended to Ian Brown, Jeff Brain, Sam Brookes, Dorian Burnette, John Connaway, Randy Cox, Peggy Guccione, John House, Marvin Jeter, T. R. Kidder, Bob Mainfort, Charles McNutt, Ruth McWhirter, Dan and Phyllis Morse, Julie Morrow, Claudine Payne, Mark Rees, Lori Roe, Martha Rolingson, Vin Steponaitis, Roy Van Arsdale, Patty Jo Watson, Kit Wesler, and Rich Weinstein for their advice, comments, and corrections on earlier drafts of this chapter.

1. Roger T. Saucier, *Geomorphology and Quaternary Geologic History of the Lower Mississippi Valley* (Vicksburg: Waterways Experiment Station, Army Corps of Engineers, 1994).
2. Tristram R. Kidder, Katherine A. Adelsberger, Lee J. Arco, and Timothy M. Schilling, "Basin-scale Reconstruction of the Geological Context of Human Settlement: An example from the Lower Mississippi Valley, USA," *Quaternary Science Reviews* 27 (2008): 1255—70.
3. Harold N. Fisk, *Geological Investigation of the Alluvial Valley of the Lower Mississippi River* (Vicksburg: Mississippi River Commission, 1944); and Saucier, *Geomorphology and Quaternary Geologic History*.
4. Charles H. McNutt, "The Upper Yazoo Basin in Northwest Mississippi," in *Prehistory of the Central Mississippi Valley*, ed. Charles H. McNutt, 155–85 (Tuscaloosa: University of Alabama Press, 1996); Dan F. Morse and Phyllis A. Morse, *Archaeology of the Central Mississippi Valley* (New York: Academic Press, 1983); Michael J. O'Brien and Robert C. Dunnell, eds., *Changing Perspectives on the Archaeology of the Central Mississippi Valley* (Tuscaloosa: University of Alabama Press, 1998).
5. Arthur L. Blumm, *Geomorphology: A Systematic Analysis of Late Cenozoic Landforms*, 3rd ed. (Upper Saddle Creek, NJ: Prentice Hall, 1998), 390.
6. David G. Anderson and J. Christopher Gillam, "Paleoindian Colonization of the Americas: Implications from an Examination of Physiography, Demography, and Artifact Distribution," *American Antiquity* 65 (2000): 43–66.
7. David G. Anderson, "The Paleoindian Colonization of Eastern North America: A View from the Southeastern United States," in *Early Paleoindian Economies of Eastern North America*, Research in Economic Anthropology Supplement 5, ed. Kenneth B. Tankersley and Barry L. Isaac (Greenwich, CT: JAI Press, 1990), 163–216.
8. Kidder et al., "Basin-scale Reconstruction," 1255–70.
9. Martin Williams, David Dunkerley, Patrick de Decker, Peter Kershaw, and John Chappell, *Quaternary Environments*, 2nd edition (London: Arnold Publishing, 1998), 45.

10. Paul A. Delcourt and Hazel R. Delcourt, *Prehistoric Native Americans and Ecological Change: Human Ecosystems in Eastern North America Since the Pleistocene* (New York: Cambridge University Press, 2008).
11. Dan F. Morse, *Sloan: A Paleoindian Dalton Cemetery in Northeast Arkansas* (Washington, DC: Smithsonian Institution Press, 1997).
12. Kidder et al., "Basin-scale Reconstruction," 1255–70.
13. Tammy M. Rittenour, Michael D. Blum, and Ronald J. Goble, "Fluvial Evolution of the Lower Mississippi River Valley during the Last 100 k.y. Glacial Cycle: Response to Glaciation and Sea-level Change," *Geological Society of America Bulletin* 119 (2007): 586–608.
14. John A. Walthall and Brad Koldehoff, "Hunter-Gatherer Interaction and Alliance Formation: Dalton and the Cult of the Long Blade," *Plains Anthropologist* 43 (1998): 257–73.
15. Morse, *Sloan: A Paleoindian Dalton Cemetery.*
16. Kidder et al., "Basin-scale Reconstruction," 1255–70.
17. Andres Aslan and Whitney J. Autin, "Evolution of the Holocene Mississippi River Floodplain, Ferriday, Louisiana: Insights on the Origin of Fine-grained Floodplains," *Journal of Sedimentary Research* 69 (1999): 800–815.
18. John M. Connaway, *The Denton Site: A Middle Archaic Occupation in the Northern Yazoo Basin, Mississippi*, Archaeological Report 4 (Jackson: Mississippi Department of Archives and History, 1977).
19. Samuel O. Brookes, "Cultural Complexity in the Middle Archaic of Mississippi," in *Signs of Power: The Rise of Complexity in the Southeast*, ed. Jon L. Gibson and Philip J. Carr (Madison, WI: Prehistory Press, 2004), 97–113.
20. J. W. Saunders, R. D. Mandel, C. G. Sampson, C. M. Allen, E. T. Allen, D. A. Bush, J. K. Feathers, K. J. Gremillion, C. T. Hallmark, E. H. Jackson, J. K. Johnson, R. Jones, Roger T. Saucier, G. L. Stringer, and M. F. Vidrine, "Watson Brake: A Middle Archaic Mound Complex in Northeast Louisiana," *American Antiquity* 70 (2005): 631–68.
21. Sarah C. Sherwood and Tristram R. Kidder, "The DaVincis of Dirt: Geoarchaeological Perspectives on Native American Mound Building in the Mississippi River Basin," *Journal of Anthropological Archaeology* 30 (2011): 69–78.
22. Tristram R. Kidder, "Climate Change and the Archaic to Woodland Transition (3000–2500 cal B.P.) in the Mississippi River Basin," *American Antiquity* 71 (2006): 221.
23. Daniel H. Sandweiss, Kirk A. Maasch, and David G. Anderson, "Transitions in the Mid-Holocene," *Science* 283 (1999): 499–500.
24. McNutt, "Upper Yazoo Basin," 155–85.
25. Kidder, "Climate Change," 221; Saunders et al., "Watson Brake," 631–68.
26. Jon L. Gibson, '"Nothing but the River's Flood': Late Archaic Diaspora or Disengagement in the Lower Mississippi Valley and Southeastern North America," in *Trend, Tradition, and Turmoil: What Happened to the Southeastern Archaic?*, Anthropological Papers 93, ed. David H. Thomas and Matthew C. Sanger (New York: American Museum of Natural History, 2010), 33–42.
27. Clarence H. Webb, *The Poverty Point Culture*, 2nd ed., rev. (Baton Rouge: Louisiana State University, 1982).
28. Jon L. Gibson, *The Ancient Mounds of Poverty Point, Place of Rings* (Gainesville: University of Florida Press, 2000).

29. Tristram R. Kidder, Anthony L. Ortmann, and E. Thurman Allen, "Testing Mounds B and E at Poverty Point," *Southeastern Archaeology* 23 (2004): 98–113.
30. Carl P. Lipo, Timothy D. Hunt, and Robert C. Dunnell, "Formal Analyses and Functional Accounts of Groundstone 'Plummets' from Poverty Point, Louisiana," *Journal of Archaeological Science* 39 (2012): 84–91.
31. Gayle J. Fritz, "Native Farming Systems and Ecosystems in the Mississippi River Valley," in *Imperfect Balance: Landscape Transformations in the Precolumbian Americas*, ed. David L. Lentz (New York: Columbia University Press, 2000), 225–49.
32. Kidder et al., "Basin-scale Reconstruction," 1255–70.
33. P. A. Mayewski, E. E. Rohling, J. C. Stager, W. Karlén, K. A. Maasch, L. D. Meeker, E. A. Meyerson, F. Gasse, S. van Kreveld, K. Holmgren, J. Lee-Thorp, G. Rosqvist, F. Rack, M. Staubwasser, R. R. Schneider, and E. J. Steig, "Holocene Climate Variability," *Quaternary Research* 62 (2004): 243–55.
34. Martha A. Rolingson and Robert C. Mainfort Jr., "Woodland Period Archaeology of the Central Mississippi Valley," in *The Woodland Southeast*, ed. David G. Anderson and Robert C. Mainfort Jr. (Tuscaloosa: University of Alabama Press, 2002), 20–43.
35. Kidder et al., "Basin-scale Reconstruction," 1255–70.
36. Philip Phillips, *Archaeological Survey in the Lower Yazoo Basin, Mississippi, 1949–1955*, Papers of the Peabody Museum of Archaeology and Ethnology, vol. 60, pts. 1 and 2 (Cambridge: Harvard University Press, 1970).
37. Jay K. Johnson, Gena M. Aleo, Rodney T. Smart, and John Sullivan, *The 1996 Excavations at the Batesville Mounds: A Woodland Period Platform Mound Complex in Northwest Mississippi*, Archaeological Report 32 (Jackson: Mississippi Department of Archives and History, 2002).
38. Tristram R. Kidder, Lori Roe, and Timothy M. Schilling, "Early Woodland Settlement and Mound Building in the Upper Tensas Basin, Northeast Louisiana," *Southeastern Archaeology* 26 (2010): 121–45.
39. Edwin A. Toth, *Archaeology and Ceramics at the Marksville Site*, Museum of Anthropology, Anthropological Papers 56 (Ann Arbor: University of Michigan, 1974).
40. Tristram R. Kidder, "Woodland Period Archaeology of the Lower Mississippi Valley," in Anderson and Mainfort, *The Woodland Southeast*, 66–90.
41. Robert E. Greengo, *Issaquena: An Archaeological Phase in the Yazoo Basin of the Lower Mississippi Valley*, Memoir no. 18 (Salt Lake City: Society for American Archaeology, 1964).
42. Michael D. Conner, ed., *Woodland and Mississippian Occupations at the Hayti Bypass (23PM572), Pemiscot County, Missouri*, Special Publication 1 (Springfield: Center for Archaeological Research, Southwest Missouri State University, 1995).
43. Kidder, "Woodland Period Archaeology," 66–90.
44. For coastal LMV area, see Mark A. Rees, "Plaquemine Mounds of the Western Atchafalaya Basin," in *Plaquemine Archaeology*, ed. Mark A. Rees and Patrick C. Livingood (Tuscaloosa: University of Alabama Press, 2007), 66–93. For interior LMV area, see Martha A. Rolingson, "Prehistory of the Central Mississippi Valley and Ozarks after 500 B.C.," in *Handbook of the North American Indians: Southeast*, ed. Raymond D. Fogelson (Washington, DC: Smithsonian Institution, 2004), 534–44.
45. Douglas C. Wells and Richard A. Weinstein, "Extraregional Contact and Cultural Interaction at the Coles Creek–Plaquemine Transition: Recent Data from the Lake Providence Mounds, East Carroll Parish, Louisiana," in Rees and Livingood, *Plaquemine Archaeology*, 38–65.

46. Tristram R. Kidder, "Coles Creek Culture," in *Archaeology of Prehistoric Native America: An Encyclopedia*, ed. Guy Gibbon (New York: Garland, 1998), 171–73.
47. Vernon J. Knight Jr., "Feasting and the Emergence of Platform Mound Ceremonialism in Eastern North America," in *Feasts: Archaeological and Ethnographic Perspectives on Food, Politics, and Power*, ed. Michael Dietler and Brian Hayden (Washington, DC: Smithsonian Institution Press, 2001), 311–33.
48. Tristram R. Kidder, "Plazas as Architecture: An Example from the Raffman Site, Northeast Louisiana," *American Antiquity* 69 (2004): 514–32.
49. Douglas C. Wells, "Political Competition and Site Placement: Late Prehistoric Settlement in the Tensas Basin of Northwest Louisiana," *Louisiana Archaeology* 22 (1997): 71–91.
50. Kidder, "Woodland Period Archaeology," 86.
51. Michael S. Nassaney and Kendra Pyle, "The Adoption of the Bow and Arrow in Eastern North America: A View from Central Arkansas," *American Antiquity* 64 (1999): 243–63.
52. Martha A. Rolingson, *Toltec Mounds: Archaeology of the Mound-and-Plaza Complex*, Research Series 65 (Fayetteville: Arkansas Archeological Survey, 2012).
53. Jeffrey P. Brain, *Winterville: Late Prehistoric Culture Change in the Lower Mississippi Valley*, Archaeological Report 23 (Jackson: Mississippi Department of Archives and History, 1989); Wells and Weinstein, "Extraregional Contact and Cultural Interaction," 38–65.
54. Wells and Weinstein, "Extraregional Contact and Cultural Interaction," 38–65.
55. For Middle Mississippian, see Morse and Morse, *Archaeology of the Central Mississippi Valley*. For Plaquemine, see Mark A. Rees and Patrick C. Livingood, "Introduction and Historical Overview," in Rees and Livingood, *Plaquemine Archaeology*, 1–19.
56. Marvin D. Jeter, "The Outer Limits of Plaquemine Culture: A View from the Northerly Borderlands," in Rees and Livingood, *Plaquemine Archaeology*, 161–95.
57. Tristram R. Kidder, "Contemplating Plaquemine Culture," in Rees and Livingood, *Plaquemine Archaeology*, 196–205.
58. Lori Roe, "Coles Creek Antecedents of Plaquemine Mound Construction: Evidence from the Raffman Site," in Rees and Livingood, *Plaquemine Archaeology*, 20–37.
59. Wells and Weinstein, "Extraregional Contact and Cultural Interaction," 38–65.
60. Kidder, "Contemplating Plaquemine Culture," 196–205.
61. Tristram R. Kidder, "Mississippi Period Mound Groups and Communities in the Lower Mississippi Valley," in *Mississippian Towns and Sacred Spaces: Searching for an Architectural Grammar*, ed. R. Barry Lewis and C. Stout (Tuscaloosa: University of Alabama Press, 1998), 123–50.
62. Roe, "Coles Creek Antecedents," 20–37.
63. Ian W. Brown, "Plaquemine Culture in the Natchez Bluffs Region of Mississippi," in Rees and Livingood, *Plaquemine Archaeology*, 145–60.
64. Wells, "Political Competition and Site Placement," 71–91.
65. Kidder, "Mississippi Period Mound Groups," 123–150.
66. Kidder, "Contemplating Plaquemine Culture," 196–205.
67. Wells and Weinstein, "Extraregional Contact and Cultural Interaction," 38–65.
68. Brain, *Winterville: Late Prehistoric Culture Change*.
69. Jeffrey P. Brain, "Late Prehistoric Settlement Patterning in the Yazoo Basin and

Natchez Bluffs Region of the Lower Mississippi Valley," in *Mississippian Settlement Patterns*, ed. Bruce D. Smith (New York: Academic Press, 1978), 331–68.

70. Kidder, "Contemplating Plaquemine Culture," 196–205.
71. Vincas P. Steponaitis and David T. Dockery III, "Mississippian Effigy Pipes and the Glendon Limestone," *American Antiquity* 76 (2011): 345–54.
72. Robert C. Mainfort Jr., "The Late Prehistoric and Protohistoric Periods in the Central Mississippi Valley," in *Societies in Eclipse: Archaeology of the Eastern Woodlands Indians, A.D. 1400–1700*, ed. David S. Brose, C. Wesley Cowan, and Robert C. Mainfort Jr. (Washington, DC, Smithsonian Institution Press, 2001), 173–89; and Mark A. Rees, "Subsistence Economy and Political Culture in the Protohistoric Central Mississippi Valley," in *Between Contacts and Colonies: Archaeological Perspectives on the Protohistoric Southeast*, ed. Cameron B. Wesson and Mark A. Rees (Tuscaloosa: University of Alabama Press, 2002), 170–97.
73. David W. Stahle, F. K. Fye, Edward R. Cook, and R. D. Griffin, "Tree-ring Reconstructed Mega-droughts over North America since A.D. 1300," *Climate Change* 83 (2007): 133–49.
74. Edward R. Cook, Richard Seager, Mark A. Cane, and David W. Stahle, "North American Drought: Reconstructions, Causes, and Consequences," *Earth-Science Reviews* 81 (2007): 93–134.
75. Rita L. Fisher-Carroll, "Environmental Dynamics of Drought and Its Impact on Sixteenth Century Indigenous Populations in the Central Mississippi Valley" (PhD diss., University of Arkansas, 2001).
76. Brown, "Plaquemine Culture in the Natchez Bluffs," 145–60; Robert S. Neitzel, *The Grand Village of the Natchez Revisited: Excavations of the Fatherland Site, Adams, County, Mississippi*, Archaeological Report 12 (Jackson: Mississippi Department of Archives and History, 1983).
77. Marvin D. Jeter, "From Prehistory through Protohistory to Ethnohistory in and near the Northern Lower Mississippi Valley," in *The Transformation of the Southeastern Indians, 1540–1760*, ed. Robbie Ethridge and Charles Hudson (Jackson: University Press of Mississippi, 2002), 177–223.
78. Stephen Williams, "Armorel: A Very Late Phase in the Lower Mississippi Valley," *Southeastern Archaeological Conference Bulletin* 22 (1980): 105–10.
79. Dan F. Morse and Phyllis A. Morse, "Northeast Arkansas," in McNutt, *Prehistory of the Central Mississippi Valley*, 119–35.
80. Robert L. Rankin, "Language Affiliations of Some de Soto Place Names in Arkansas," in *The Expedition of Hernando de Soto West of the Mississippi, 1541–1543: Proceedings of the DeSoto Symposia 1988 and 1990*, ed. Gloria A. Young and Michael P. Hoffman (Fayetteville: University of Arkansas Press, 1993), 213–17.
81. Michael P. Hoffman, "Ethnic Identities and Cultural Change in the Protohistoric Period of Eastern Arkansas," in *Perspectives on the Southeast: Linguistics, Archaeology, and Ethnohistory*, ed. Patricia B. Kwachka (Athens: University of Georgia Press, 1994), 61–70.
82. Jeter, "From Prehistory through Protohistory to Ethnohistory," 177–223.
83. Rees, "Plaquemine Mounds of the Western Atchafalaya Basin," 66–93.
84. John R. Swanton, *Indian Tribes of the Lower Mississippi Valley and Adjacent Coast of the Gulf of Mexico*, Bureau of American Ethnology Bulletin 43 (Washington DC: Smithsonian Institution, 1911).

CHAPTER 4

An Environmental Historian's Perspective on the Mississippi Delta

MIKKO SAIKKU

The alluvial bottomlands of the Lower Mississippi Valley, commonly referred to as the Delta, have experienced enormous environmental change during the last centuries.[1] Agriculture, lumbering, and remaking of the vast floodplain's hydrological system have transformed the landscape, originally dominated by mature bottomland hardwood forest, beyond recognition and resulted in irrevocable alteration of local ecology. The long-term environmental history of the Delta, however, emerges as immensely more complicated. Significant human impact on the Delta's natural environment goes back much further than the nineteenth century and shows remarkable fluctuation even during the last decades.

A captivating perspective on the Delta from the viewpoint of social, economic, and political history was published in *Arkansas Review* more than a decade ago by Jeannie M. Whayne.[2] In that article, Whayne expertly outlined the classic themes for historiography of the region: plantation agriculture in myth and reality, slavery and sharecropping, unequal distribution of wealth, and troublesome race relations. This article strives to add yet another historical dimension to our understanding of the region. While the traditional subjects for the Delta's historiography must continue to be addressed, the viewpoint offered by environmental history can provide new perspectives to the core questions of regional history.

Environmental history may be described as an attempt to study the interaction between humans and nature in the past. Its aim is to deepen our understanding of how humans have been influenced by their natural environment through time and, conversely, how they have affected their surroundings and with what results. This relatively new field of historical study rejects the traditional assumption that human experience has been exempt from natural constraints or that the ecological consequences of past human activity can be ignored. In comparison with traditional historiography, environmental history emphasizes the role of humans as an integral part of their natural surroundings. Modern environmental history strives for a fuller understanding of today's environmental issues and, ideally, provides information for contemporary problem solving. Even as current environmental problems may differ from former ones, understanding of the past events may prove helpful.[3]

The most important questions within the field seem to be the different productive strategies of the human societies, their ideological backgrounds, and their consequences and comparisons across culture and place.[4] What kind of human society and natural environment emerge as a result of the interaction between these forces? In environmental historiography, the study of human-nature interaction often has to focus on long-term change. Many environmental changes, such as climatic transformations, are slow processes that may take centuries or millennia, while most ecological catastrophes, whether volcanic eruptions or explosions of nuclear power plants, are sudden occurrences that may have long-lasting repercussions for human societies. Thus environmental history often approaches what the French annalist historian Fernand Braudel called the history of long duration (*histoire de la longue durée*). [5]

Environmental history is also spatially more flexible than traditional historical research; natural entities, such as drainage basins or other geological formations, are often more important than administrative boundaries created by humans, such as the borders of nation-states or economic communities. With global industrialization and socioeconomic modernization, environmental problems such as industrial pollution have become international problems. Similarly, the integration of local economies into the world economy has affected the relationships between human societies and their natural surroundings. The fact that the exploitation of natural resources and the consumption of goods manufactured from these resources often take place in different parts of the world, combined with absentee ownership of production facilities, has caused ecological indifference that small and locally controlled economic systems could hardly accept. Due to their widespread temporal and spatial linkages, human-induced environmental changes of the past continue to affect contemporary life on earth. Soil impoverishment, erosion, deforestation, and pollution of air and water are among current environmental problems that have influenced human societies for a long time. Environmental history strives to reassess commonly accepted views of human history: the argument that humans have by means of religion, science, and technology liberated themselves from the limits set by the natural world seems absurd in the era of global climate change and nuclear and other industrial accidents.

However, any serious study on the history of human-nature interaction should counterbalance both technocratic fantasies and environmentalist utopias. Donald Worster has pointed out that one of the aims of environmental history is to "reject naive assumptions about a static, pristine, virgin world of unspoiled nature."[6] William Cronon has even suggested that one of the greatest contributions of environmental history to the conservation and environmental movements may be in providing counterbalance to ahistorical and antihistorical impulses within these movements.[7] Deeper understanding of contemporary environmental problems requires informed knowledge of past events, and the study of environmental history should provide some common ground in that quest.

Most environmental historians would probably agree with the claim that the greatest shift in human-nature relationship since the beginning of agriculture is the rise of industrial capitalism, connected to the economic and ecological expansion of Europe. The process gained momentum with the great explorations in the fifteenth century and, eventually, resulted in global colonization and a tremendous increase in the volume of trade and importance of the marketplace. Monoculture, or the extensive cultivation of certain agricultural products over vast areas solely for the emerging world market, began as early as the fifteenth century in the newly colonized areas. From the beginning, the new, interconnected network induced vast environmental changes in Europe and its economic fringes. Industrialization and economic development in the new peripheries of the European economy were built upon the export of natural resources in the form of raw or semi-refined products or staples. Local economic development soon became dictated by the supralocal needs of the world economy.

Environmental historiography has proved beyond doubt that the extraordinarily rapid downfall of many aboriginal nations in the areas of European colonization was largely caused by the smallpox and other germs imported by the European conquerors and not by the natives' perceived military or political inferiority. Aided by their biological allies, such as crops, weeds, domesticated animals, pests, and Eurasian microorganisms which created virgin soil epidemics in the Americas and elsewhere in the New World, the Europeans were able to create ecological representations of their homelands everywhere in the temperate zone of the globe. The European colonists were also able to adapt from the conquered lands the economically viable crops and other products of the aboriginals—such as the maize and the potato cultivated by Native Americans—and use these to their own advantage in creating new commercial empires.[8]

A series of transformations in the European economy and society, combined with technological innovation, resulted in the final breakthrough of industrial capitalism in the early nineteenth century and created the modern world with its environmental problems of unprecedented scale. In capitalist modernization, self-sustained economic growth was achieved for the first time in history. The old patterns of colonial trade were fortified, and the production of raw materials for export increased all over the world. Global economic trends and increasing integration of the world market contributed to escalating uniformity in the human use of natural resources. Human exploitation of nature under modern capitalism, however, is not omnipotent, since during recession phases production pressure on economic peripheries typically diminishes and natural systems may recover. Although maximization of profit, the basic aim of capitalist production, typically conflicts with the sustainable yield of natural resources, it can also offer incentives for rational management. It must furthermore be noted that traditional subsistence methods, as practiced under heavy population pressure, have often resulted in even more dramatic change in the natural environment.[9]

Ecological units characteristically maintain their structure not only despite but also because of external disturbances.[10] An important class of such events is often called a "disturbance regime," referring to a certain type of landscape that is created and maintained by a disturbance that occurs with characteristic frequency. Wildfires in northern coniferous forests or tall-grass prairies, hurricanes in tropical forests, and flooding on alluvial plains are examples of disturbances which are triggered by factors more or less external to the landscape itself, but essential to the maintenance of the whole complex in longer perspective. Disturbance regimes can provide a general norm for acceptable human intervention in nature: in order to preserve biological diversity and productivity, human modification should remain comparable to that caused by natural occurrences within disturbance regimes.

Natural catastrophes, as compared to disturbances, are unique events with spatial and temporal dimensions dramatically exceeding the resistance and resilience of the ecological complexes affected. All time-space scales are subject to catastrophes, but the definition of the term varies according to the size of the ecological complex in question. Mass extinctions of species, such as that of the dinosaurs at the end of the Cretaceous period, are catastrophes on the macro-level of evolution, whereas a normal forest fire affects an individual tree in the same way on the micro-level of organisms. A most significant difference between natural and human-induced catastrophes is that pertaining to their pace: the latest natural mass extinction some 70 million years ago took hundreds of thousands of years, while the current wave of extinctions proceeds on the temporal scale of centuries, if not decades.

It is the correspondence of natural and human scales that determines what kind of human-induced change is of consequence to a given ecological system. Human-induced environmental change results from the interaction between human-induced and natural dynamics on a shared scale. This correspondence also explains why intensive change on a limited scale is possibly insignificant (except for the individual organisms subjected to it), whereas a hardly observable change occurring uniformly on a large scale can prove truly disastrous, although its consequences might not be visible in the lifetime of individual organisms. Therefore, the most significant human-induced changes in history are results of systematic activities over vast regions and long time periods. Furthermore, as no single human being can through his/her activities cause relevant change on a larger scale, some sort of social formation is needed for creating human-induced environmental change of consequence. Natural resources are exploited and transformed into commodities within a social matrix.

As argued above, ecological systems and biological diversity typically ignore political and other manmade boundaries. By the 1970s, the US public land management agencies had realized the importance of land classification on the basis of regional variations in climate, vegetation, and landform as an aid to conservation problems. Regional differences were officially recognized in Robert G. Bailey's 1978 book, which divided the United States into sixty "ecoregions" and created a new taxonomy for American nature. Similarly, the World Wildlife Fund (WWF) today defines an ecoregion as "a large area of land or water that contains a geographically distinct assemblage of natural communities that share a large majority of their species and ecological dynamics, share similar environmental conditions, and interact ecologically in ways that are critical for their long-term persistence." Hundreds of scientists have contributed to the development of WWF's Conservation Science Program and identified over eight hundred distinct terrestrial ecoregions across the globe.[11]

During the last three decades, the concept of ecoregion (often referred to as "bioregion") has spurred a loose political and cultural movement that opposes the contemporary consumer culture and advocates local solutions in the search for sustainable development. Among contemporary environmental historians, Dan Flores has most vocally claimed that bioregion should also be recognized as a precise and useful term for historical study. Flores asserts that "the first step in writing environmental histories of place is recognition that natural geographic systems—ecoregions, biotic provinces, physiographic provinces, biomes, ecosystems, in short, larger and smaller representations of what we probably ought to call bioregions—are the appropriate settings for insightful environmental history." Bioregional histories should thus commence with geology, landform, and climate history. The second basis for bioregional history, beyond ecological parameters, is constituted by the diversity of human cultures across both space and time. Flores has criticized environmental historians for too often starting out with a single interpretive framework and forcing their studies of topographically, ecologically, and culturally diverse and geographically vast territories into some facsimile of that model. A better approach is to aim for "deep time, cross-cultural, environmental histories of places—and after a sufficient number of such case studies have been done, then to look for patterns." Bioregional history is, therefore, the story of different but successive cultures occupying the same space.[12]

Only some four centuries ago, the eastern half of the North American continent from the Mississippi River to the Atlantic Ocean, except for the northern extension of the Mississippi prairie, was generally covered with forest. The forest types changed from coniferous woods in the north through deciduous woodlands into tropical savanna in Florida. Eastern North America carries one of the most complicated and variable aggregations of vegetation in the temperate regions of the world. It has literally hundreds of species of trees, most of them deciduous. Hardwoods cover most of the forested area and largely determine the general characteristics of the forest. Not surprisingly, the eastern forest as a whole is usually known as the "Eastern Deciduous Forest."[13]

The major types of natural forest vegetation found within the Eastern Deciduous Forest have been arranged into various subdivisions by many authors. Nevertheless, the type of forest found in the Delta and on other southern floodplains is among the most distinctive of the subdivisions within the Eastern Deciduous Forest because of its unusual site conditions and sharply bounded topography. "Southeastern Bottomland Hardwood Forest" and "Southern Floodplain Forest" are some common terms that have been used to describe the vast bottomland and alluvial swamp forests that occur naturally on river floodplains of the southern United States. These forests are found wherever streams or rivers more or less regularly swell beyond their channels. Consequently, they are integral parts of delicate hydrological systems. The floodplain forests are dominated by woody species that show morphological and physiological adaptations enabling them to thrive in an environment where the soils within the root zone may be inundated or saturated for varying periods during the growing season.[14]

The largest single area of bottomland hardwood forest occurs on the alluvium of the southern Mississippi River Valley. In addition to the Mississippi, its southern tributaries, such as the Yazoo, White, Arkansas, Red, Ouachita, Tensas, and St. Francis, naturally support extensive areas of bottomland hardwood forest, as does the Atchafalaya, today a distributary. The Mississippi drains the world's third-largest basin. Its drainage area covers some 1.2 million square miles, or more than 40 percent of the coterminous United States, from New York to Wyoming. The main channel of the river, from Lake Itasca in northern Minnesota to the Gulf of Mexico, flows southward over a path almost 2,500 miles long. Near its midreach, the Mississippi is joined by the two largest tributaries, the Missouri and the Ohio, and, along its lower reaches, by several other major streams. Below the confluence of the Mississippi and the Ohio, the greatly increased flow results in the formation of an immense stream supporting a vast alluvial plain—the Delta.

The Delta more or less equals the "Lower Mississippi Riverine Forest" (Province 234) in Bailey's revised 1995 classification, covering approximately 44,000 square miles, or 1.2 percent of the United States. The ecoregion is a vast alluvial bed along the Lower Mississippi from the mouth of the Ohio River to the Gulf of Mexico, ranging from 30 to 40 miles in breadth at its northern end to 150 miles at the river's mouth in present-day Louisiana. It has been created by the meandering Mississippi River and attained its current form during the final cycle of the latest glaciation. The natural boundaries of the Delta ecoregion with Province 231, "Southeastern Mixed Forest," and Province 232, "Outer Coastal Plain Mixed Forest," are quite distinct. Similarly, WWF's ecoregions "Mississippi Lowland Forests" (NA0409), "Central US Hardwood Forests" (NA0404), and "Southeastern Mixed Forests" (NA0413) closely mirror Bailey's classification.[15] Thus the Delta offers a distinct natural setting for the study of human history while many of the historical processes taking place in the region can closely be tied to the land itself.

Water, soils, flora, and fauna were closely interdependent in the pre-Columbian Delta, together forming an identifiable ecoregion—the "Big Woods" described by William Faulkner as "the rich deep black alluvial soil which would grow cotton taller than the head of a man on a horse, already one jungle one brake one impassable density of brier and cane and vine interlocking the soar of gum and cypress and hickory and pinoak and ash, printed now by tracks of unalien shapes—bear and deer and panthers and bison and wolves and alligators and the myriad smaller beasts."[16]

Biotic communities of the Delta, as known during historical times, assembled only since the last major glaciation, the past 20,000 years. Like other southeastern vegetational assemblages, the bottomland hardwood forests of the Delta have not been stable either in composition or in location through long periods of time. Still, pollen data suggests that elements of an oak-hickory forest have been present at the Delta for the past 16,000 years, providing a relatively established zone for faunal—and human—habitation. The pre-Columbian biota of the Delta was a rich mosaic of plant and animal communities pursuing compositional equilibrium but still responding dynamically to impacts of climate, geomorphic processes, fire, and Native American populations.

Vastly differing subsistence systems and human modes of production evolved in the region since the arrival of the first Native Americans.[17] Originally based on hunting and gathering, the Native American subsistence systems in the Delta changed over time to include various horticultural practices. The amount of horticulture and trade practiced by the Native American societies did not show a continuous, linear increase, but fluctuated considerably. For example, during the late Mississippian era, Native American populations of the region were probably larger and more dependent on agriculture as their subsistence base than ever before or after. During this time period, the environmental impact of Native Americans on the bottomland forests accordingly attained significant levels in certain locations. Consequently, true wilderness, meaning nature uninfluenced by human activities, hardly existed in the region at the commencement of Euro-American settlement.

In the Delta, Native American settlements and agricultural fields often concentrated on natural levees in the proximity of rivers. Fields in the vicinity of permanent villages could become quite extensive and, combined with other land use practices, resulted in localized deforestation. Still, the typical

Native American use of the bottomland hardwood forest presumably did not endanger the natural variety of the land or its capacity for self-renewal. Much of the Native American use of timber remained, furthermore, insignificant in relation to deforestation in the bottomlands. It was mostly the vicinity of scattered Indian villages on natural levees that became a patchy landscape of fields, grasslands, and young woods, and actions by the Indian generally resulted in local modification of the bottomland hardwood forest rather than its large-scale eradication. In the absence of flood control devices, the human-induced change of the Delta environment was largely restricted to the highest elevations within the floodplain, affecting directly only a very small percentage of the total land area.

The early European accounts customarily claimed that the Indians encountered in the South lived in harmony with their natural surroundings. Whether or not this was true in regard to the utilization of natural resources by the Indian populations, the escalating clearing of forest for timber harvest and cultivation by the European colonists, combined with expanding alteration of the floodplain hydrological regime, was to represent a major shift in the degree to which humans affected their natural environment in the Delta.[18]

Large-scale Euro-American settlement of the Delta commenced with the rapid expansion of cotton economy to the region at the turn of the nineteenth century. In many places, the Delta landscape still possessed many features of pristine wilderness because the elimination of aboriginal populations by virgin soil epidemics had enabled forest vegetation to reclaim the areas affected by Native American agriculture. With the arrival of Europeans, Native American consciousness, which presumably viewed all life forms as equal subjects, was supplanted by a vision in which humans were largely considered separate from the rest of the world, which was viewed as a pool of resource objects.

The Euro-American period of subsistence-oriented agriculture in the Delta was typically brief, as the development of the market and transportation systems during the early nineteenth century geared the local economy toward production for the capitalist marketplace from the very beginning. The analytical and quantitative consciousness of Euro-Americans emphasized efficient management and increasing control over nature from the establishment of the first cotton field on the Delta bottomlands. Consequently, major human-induced environmental change on the floodplain was unavoidable. During the nineteenth century, agricultural and lumbering activities expanded enormously compared with Native American practices. The first commercial products of the Delta in the industrial age, cotton and cypress, were later supplemented by other crops and tree species, while land use on the fertile floodplain progressively intensified.

During the nineteenth century, a marked increase in agricultural development and population took place on the rich Delta bottomlands. The explanation for the rise of a New South cotton kingdom on the Mississippi floodplain can largely be summed up in two words—flood control. Land clearing by commercial logging for cotton production and other purposes was closely connected to the growing human control of the hydrological processes of the floodplain, a development gradually attained during the late nineteenth and early twentieth centuries. There was a close relationship between the agricultural expansion, amount of land cleared, and growth of flood control and drainage systems in the Delta.

By the turn of the twentieth century, the bottomlands of the Lower Mississippi Valley had become a leading production center for cotton and hardwood lumber while the system of protective levees on the floodplain expanded enormously. The human exploitation of the Delta's natural resources had turned into what Faulkner described as "a mad and pointless merry-go-round . . . : the timber which had to be logged and sold in order to deforest the land in order to convert the soil to raising cotton in order to sell the cotton in order to make the land valuable enough to be worth spending money raising dykes to keep the River off of it."[19]

As more dependable levees for the protection of croplands and other property were established, it became evident that maximum yields could be obtained from the flat Delta lands only by providing adequate drainage. These developments furthermore facilitated the removal of the original forest cover. For example, by 1932 only 2.1% of the Yazoo basin supported old-growth forest and numerous species had become extinct.[20] In the following decades, land clearing continued and the hydrological system of the whole Lower Mississippi floodplain was remade in a way that offered little chance for the regeneration of the bottomland hardwood forest in the future.

The immense ecological transformation in the Mississippi Delta was not limited to the reduction of forested acreage or removal of mature trees by planters and loggers: the native fauna of the bottomlands similarly underwent significant changes. The Delta had formed an important part of the original range of several animals, which more or less disappeared by the mid-twentieth century. The most striking examples of local—and sometimes even global—faunal extinctions were provided by the red wolf, cougar, Carolina parakeet, and ivory-billed woodpecker.[21]

Probably the most important question in assessing human-induced change in ecological systems is whether the systems affected retain their resilience. Human modification of much of the Delta environment up to the late nineteenth century generally remained comparable to that caused by natural occurrences within disturbance regimes, and the change stayed within bounds that enabled the recovery of the natural systems after disturbance. As evidenced by the successful forest regeneration on the agricultural fields abandoned by the Mississippian people and by the Civil War–era planters, Delta forests retained their resilience well into the modern era. Similarly, deer herds decimated by the deerskin trade during the eighteenth century made a comeback within decades.

The massive human efforts to restrain the rivers for the first time endangered the ability for recovery of the whole natural complex. From the late nineteenth century on, the system of protective levees on the floodplain expanded enormously. The immense human-induced change in the hydrology of the Mississippi and its tributaries, especially after the disastrous flood of 1927, has aimed at making floodplains safe for agriculture and human habitation. After centuries of hard work and enormous investments by local, state, and federal governments, it has been customarily assumed that almost any amount of high water can be safely transported through the Lower Mississippi Valley. Still, the collapse of New Orleans levees in 2005 after a storm surge and the record heights of the Mississippi in 2011 have raised serious doubts about the reliability of the whole flood control system. Human inhabitants in the Delta have not succeeded in liberating themselves from the limits set by the natural world. In many places along the Lower Mississippi River and its tributaries, it might have been wiser—also from an economic point of view—to regulate development on the floodplain rather than restructure the bioregion at enormous cost and with considerable loss to the biological diversity of the region.

Even before the arrival of the European settlers, the shift from hunting and gathering to agriculture and trade among the Delta's native peoples had led to a general increase of human influence on the natural environment. Still, it was the region's inclusion in the greater European economy that revolutionized the way the floodplain was exploited by humans. For the first time, local economic development became dictated by the supralocal needs of the world economy. The basic productivity of the bioregion remained the same, but now it was the capitalist world system that ultimately mandated how the Delta was utilized as a pool of natural resources. The first commercial products of the Delta in the industrial age, cotton and cypress, were later replaced by other crops and tree species, while land use on the fertile floodplain progressively intensified. As a result, great personal fortunes were made, always based on the biological productivity of the land. For most of the people involved in the transformation of the Delta bottomlands, especially black slaves, sharecroppers, and agricultural workers, economic gain and social mobility remained severely limited.

The factors that contributed to the successful utilization of the natural resources in the Delta are not unique to southern or American history: exploitation of disadvantaged people and the natural environment in a globalizing culture geared at continuous economic growth is an unavoidable theme in modern history. The whole planet seems to be fast realizing Faulkner's 1942 Delta prophecy that, despite given our chance and "warning and foreknowledge, too," the natural world we devastate will be "the consequence and signature of [our] crime, and [our] punishment."[22]

Notes

1. The term "Delta" has other, more localized connotations. For example, the Yazoo-Mississippi basin, which forms a part of the greater Mississippi alluvium, is in Mississippi usually referred to as the Delta. Much of the material in this essay is based on Mikko Saikku's *This Delta, This Land: An Environmental History of the Yazoo-Mississippi Floodplain* (Athens: University of Georgia Press, 2005), which includes complete documentation.
2. Jeannie M. Whayne, "What Is the Mississippi Delta? A Historian's Perspective," *Arkansas Review: A Journal of Delta Studies* 30, no. 1 (1999): 3–9.
3. The following discussion on environmental history and its methodology is based on Saikku, *This Delta, This Land*, 5–24, and the sources quoted therein.
4. Good introductions to the field include John R. McNeill, "Observations on the Nature and Culture of Environmental History," *History and Theory* 42 (2003): 5–43; Donald Worster, "Doing Environmental History," in *The Ends of the Earth: Perspectives on Modern Environmental History*, ed. Donald Worster (Cambridge: Cambridge University Press, 1988), 289–307; and the round table panel in the *Journal of American History* 76 (March 1990): 1087–1147.
5. Timo Myllyntaus, "Environment in Explaining History," in *Encountering the Past in Nature: Essays in Environmental History*, ed. Timo Myllyntaus and Mikko Saikku (Athens: Ohio University Press, 2001), 145–51; Fernand Braudel, *Écrits sur l'histoire* (Paris: Flammarion, 1969), 11–13.
6. Donald Worster, "Nature and the Disorder of History," *Environmental History Review* 18, no. 2 (summer 1994): 2.
7. William Cronon, "The Uses of Environmental History," *Environmental History Review* 17, no. 3 (fall 1993): 10–12; and "The Trouble with Wilderness; or, Getting Back to the Wrong Nature," *Environmental History* 1, no. 1 (January 1996): 7–25.
8. Two classic studies on the subject are Alfred W. Crosby's *The Columbian Exchange: Biological and Cultural Consequences of 1492* (Westport, CT: Greenwood, 1972); and *Ecological Imperialism: The Biological Expansion of Europe, 900–1900* (Cambridge: Cambridge University Press, 1986).
9. Yrjö Haila and Richard Levins, *Humanity and Nature: Ecology, Science, and Society* (London: Pluto, 1992), 208–11. The large-scale application of slash-and-burn agriculture in societies as dissimilar as preindustrial Finland and contemporary Brazil offers an example of the potential destructiveness of traditional subsistence methods.
10. The following discussion on change in ecological units and scaling is based on Haila and Levins, *Humanity and Nature*, 72–74, 184–90.
11. Robert G. Bailey, *Description of the Ecoregions of the United States* (Ogden, UT: USDA Forest Service, 1978); WWF (World Wildlife Fund), "Ecoregions," http://www.worldwildlife.org/science/ecoregions/item1847.html, accessed March 28, 2012; David M. Olson et al., "Terrestrial Ecoregions of the World: A New Map of Life on Earth," *Bioscience* 511, no. 1 (November 2001): 933–38.
12. Dan Flores, "Place: An Argument for Bioregional History," *Environmental History Review* 18, no. 4 (winter 1994): 6, 14.
13. Thomas R. Cox, Robert S. Maxwell, Philip Drennon Thomas, and Joseph J. Malone, *This Well-Wooded Land: Americans and Their Forests from Colonial Times to the Present* (Lincoln: University of Nebraska Press, 1985), 2; Michael Williams, *Americans and Their Forests: A Historical Geography* (New York: Cambridge University Press, 1989), 3–4.

14. On southeastern bottomland hardwood forests, see Saikku, *This Delta, This Land,* 26–51, and the sources quoted therein.
15. Robert G. Bailey, *Description of the Ecoregions of the United States,* 2nd rev. ed. (Washington, DC: USDA Forest Service, 1995), 22–24; WWF, "Ecoregions."
16. William Faulkner, *Big Woods: The Hunting Stories* (New York: Random House, 1955), 166.
17. On Native Americans and the Delta environment, see Saikku, *This Delta, This Land,* 52–86, and the sources quoted therein.
18. The following discussion on environmental change in the Delta is based on Saikku, *This Delta, This Land,* especially 247–55, and the sources quoted therein.
19. Faulkner, *Big Woods,* 201.
20. I. F. Eldredge, *Preliminary Report on the Forest Survey of the Bottom-Land Hardwood Unit in Mississippi* (New Orleans: USDA Forest Service, Southern Forest Experiment Station, 1934), 2–5.
21. On the Delta's fauna and population changes, see Saikku, *This Delta, This Land,* 45–50, 226–36, and the sources quoted therein.
22. Faulkner, *Go Down, Moses* (New York: Random House, 1942), 349.

CHAPTER 5

Landscape, Environment, and Geography of the Mississippi River Delta Region

FIONA DAVIDSON AND TOM PARADISE

In geography, the concept of a *region* has focused on often poorly defined, ephemeral, and ever-changing boundaries, which are based on definable variables and characteristics but rarely on fixed borders. Over time, spaces and places may change, as many of the features that define them change through natural or anthropogenic influences. Regional archetypes and boundaries are based on various features, factors, and processes that may be broadly classified as physically defined (i.e., mountain ranges, river basins), human-impacted (i.e., demographic, commercial), or related to the interaction of humans and the environment (i.e., hazards, natural resources). However, in some political landscapes, regions may be defined by legal jurisdiction and administration (i.e., county, state, and national borders).

In addressing the determining factors of regions and regionalism, the aspect of *scale* is fundamental as well. On global scales, regions are often defined by physical attributes, which may include atmospheric, climatic, hydrospheric, and oceanic characteristics, and are often measured in thousands of miles. However, at continental scales, regions may be delineated and explained through human, natural, and environmental factors—individually or in combination—and measured in tens or hundreds of miles. For example, regions of natural-hazard occurrences and impacts are often demarcated through the variables of population density, vulnerability, and phenomenal frequency and magnitude (i.e., tornadoes, earthquakes).

So in describing and explaining the Mississippi Delta, geographers examine the various factors and features that may define the space and place. In human geography, we assess the spatial and temporal extent of the area's demographics, ethnography, politics, and socioeconomic activity. In physical geography, we scrutinize the role and effect of natural features and processes on the landscape; these may include topography, hydrology, climate, geology, and biotic distribution. Increasingly, however, geographers explain regions through the interactions, connections, and intersections of various characteristics of the landscape. Hence, fields that represent the combination of human and natural landscapes can define agricultural regions, raw material availability, natural resource access, and natural-hazard occurrences and risks. This is indeed the case with the Mississippi Delta, where it is the melding of natural and human-based traits and activities that may define its extent today. Nowadays, analysis of these spatial intersections is best facilitated through the use of geomatic tools such as cartographic, spatial statistics, and geographic information modeling (GIS).

In defining regions it is also critically important to understand the terminology and classification of types of regions that are recognized by geographers (National Geography Standards). Formal, functional, and vernacular (or perceptual) regions all provide different theoretical ways of examining the characteristics and activities that bind places together as distinct geographic entities. Of these three, the Mississippi Delta, as traditionally conceived, is a vernacular or perceptual region, albeit a rather unconventional one. Conventional vernacular regions have no universally agreed upon boundaries or characteristics; they exist as ill-defined regional identities whose extent varies depending on the perception of different individuals. The traditional Mississippi Delta clearly does have well-defined boundaries. It is that slice of land between the Mississippi and Yazoo Rivers (sometimes referred to as the Yazoo basin); however, as will become apparent in this chapter, in the twenty-first century there is no geographic rationale for this region to be distinguished from the surrounding areas, and as such the only reason for it to be discussed as an area distinct from its surroundings is the perception that it is a unique and distinct region.

As this chapter demonstrates, the Delta is included in much broader geographically valid formal and functional regions that extend from southern Illinois to the Gulf of Mexico and across both sides of the Mississippi. Formal regions are defined by human and physical characteristics such as a shared language, religion, and political or ethnic identity or by homogeneous physical characteristics such as pedology, topography, climate, or inclusion in a watershed. Thus, the Delta is part of the Mississippi River watershed and also part of a region in which a majority of the population is African American; it is also part of the majority Baptist-affiliation region and part of the southern extension of the Corn Belt.

Finally, functional regions exist around activity nodes, primarily cities and other aggregations of economic and political activity. Bounded by lines of communication, the Delta exists as part of the greater Memphis economic region, the political region identified as the State of Mississippi, and a transportation region centered on river traffic that stretches from New Orleans to the Great Lakes.

Using the theoretical construct of formal and functional regions, this chapter creates a geographical definition of the Delta that provides an alternative to the traditional vernacular vision of a geographically constrained and historically static region bounded by the Mississippi and Yazoo Rivers. The layering of formal and functional regional characteristics creates a palimpsest of physical, environmental, and human features and activities that define a north-south-trending series of counties that stretch two to three deep on both sides of the Mississippi from Cairo, Illinois, to just north of Baton Rouge.

Physical Landscapes

In defining the Delta as a region it is important to first note that a strictly hydrological definition would set the northernmost extent of the Mississippi Delta at the confluence of the Mississippi River and the first distributary that begins the process of separating the river flow into a multitude of downstream channels that eventually reach the Gulf of Mexico. This point is currently marked by the Old River Control structure in Concordia Parish, Louisiana / Wilkinson County, Mississippi, where 30 percent of the flow of the Mississippi River diverts into the Atchafalaya River, a flow that is controlled by the Corps of Engineers using a structure designed to prevent the eventual, and some would say inevitable, shift of the main channel to the Wax Lake outlet west of Morgan City, Louisiana.

Located within this physiographic region, the traditional Mississippi Delta is a north-south-trending extent of land bounded by Memphis, Vicksburg, and the Mississippi and the Yazoo Rivers: a region that is physiographically indistinguishable from the neighboring landscape. However, when we examine the physical nature of its landscape, it becomes apparent that the extent of the Delta represents a prehistoric embayment of the Lower Mississippi River. In modern terms, the alluvial plain of the river can be defined as the lowlands from Monroe, Louisiana, to Vicksburg, Mississippi; from Little Rock, Arkansas, to Jackson, Tennessee; and from Cape Girardeau, Missouri, to Paducah, Kentucky, all draining to the south from its confluence at Cairo, Illinois, to become the bird-foot-style delta of the Mississippi River in southern Louisiana.

In geologic terms, the Delta is not a true delta, but a wide river valley filled with riverine deposits since sea level rose 40 to 60 million years ago during the early Tertiary period. As sea level dropped during the late Tertiary, the coastline retreated to the south and exposed much of the floodplain of the Mississippi we see today. So, unlike the iconic river delta of the Mississippi that extends beyond New Orleans into the Gulf of Mexico, the Delta addressed in this book can be described simply as a relict deep river valley filled with layers of sediments that have produced a broad, sinuous, flat floodplain. Moreover, this broad plain now cradles a much smaller river that once flooded these lowlands from bank to bank when it was larger from the melting glacial packs to the north. Since the last Ice Age, when the river was higher, with a greater discharge (order of magnitude greater), the modern river has a decreased discharge with a lower level and can rarely, if ever, fully flood these broad lowlands today.

A more generous physiographic definition of the Delta can include the flat terrain that borders the Mississippi River south of Cairo, Illinois, at the point of confluence of the Mississippi and Ohio Rivers. Enhanced by the volume of the Ohio River, some five times that of the Mississippi at this point, the river then flows south into the New Madrid Seismic Zone, where it spreads out

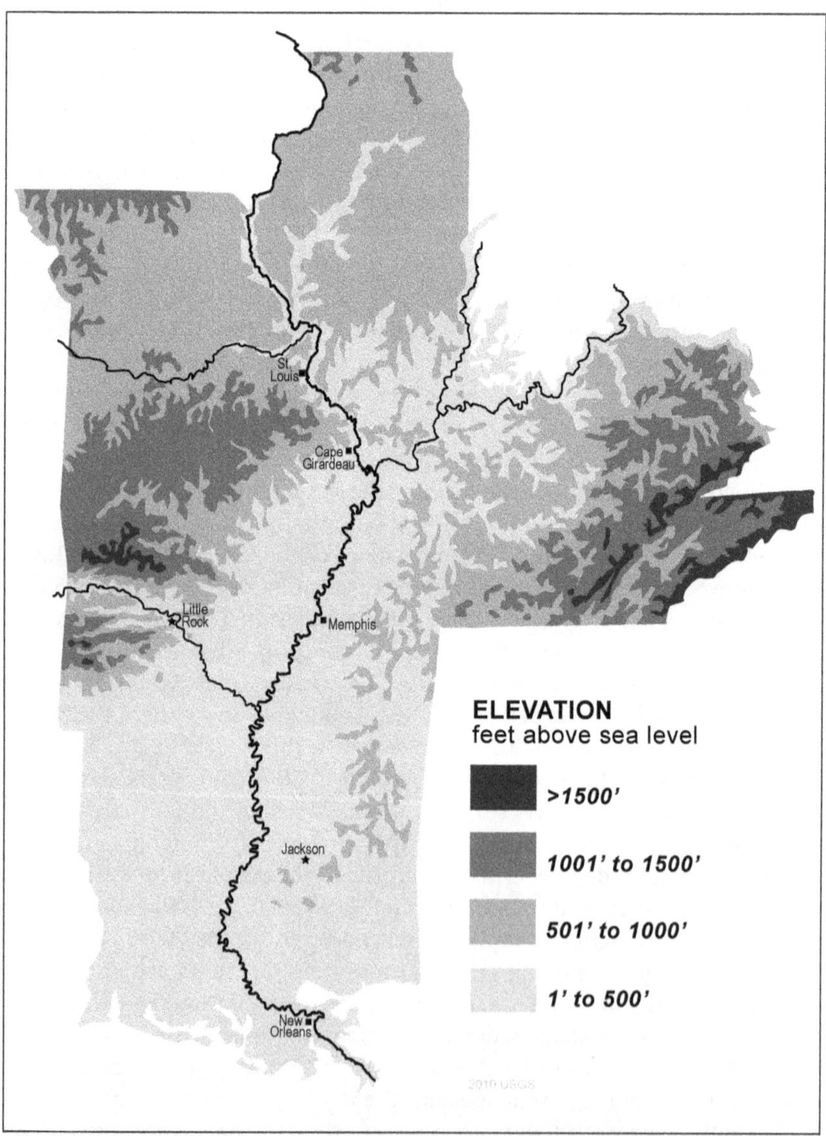

FIGURE 5.1. *Cartography by Tom Paradise.*

across an extensive alluvial plain created over thousands of years of annual flooding, sedimentary deposition, and river channel migration (fig. 5.1). Also, it is this deep alluvial plain that has buried and concealed mineral resources that may have facilitated the region's early non-agricultural economic development. The regional coalfields, for instance (fig. 5.2), have only been found

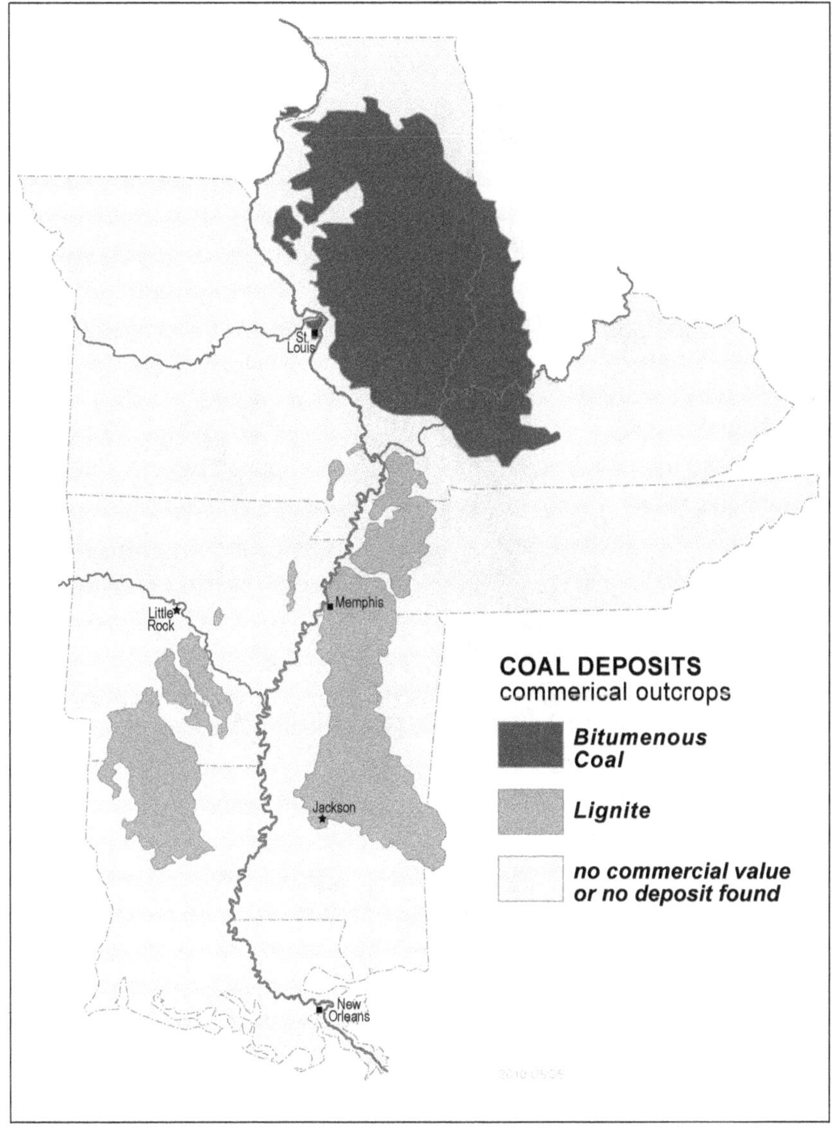

FIGURE 5.2. *Cartography by Tom Paradise.*

outside and above these floodplains, where their overburden has not obscured coal seam discovery and recovery. However, it is that same alluvium which has created an abundance of biomass—the total mass of living matter within an environmental area (fig. 5.3). It is this fertile substrate, in tandem with an ideal climate, which has spawned a wealth of renewable organic material, which in turn has produced its prolific agricultural legacy. It is these alluvial plains capped with fertile soils and crisscrossed with accessible irrigation that have

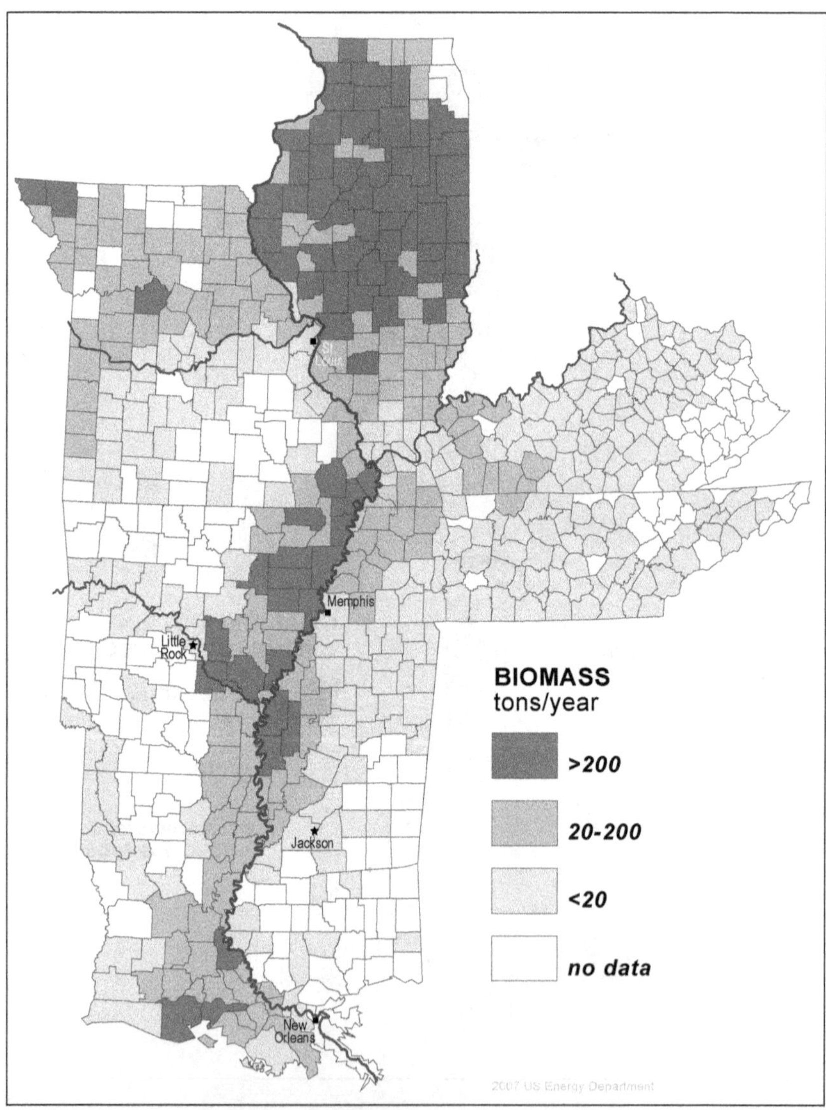

Figure 5.3. *Cartography by Tom Paradise.*

attracted inhabitants since the earliest days. Indigenous peoples, European settlers, migrant workers, and resource and transport-based industries have all sought these prime lowlands for agricultural production, raising cotton, grains, soybeans, and sugar.

Economic Landscapes

Once we have established that the traditional Mississippi Delta is physiographically indistinguishable from the areas surrounding it, then we must rely on socioeconomic, demographic, and political characteristics to create a workable geographic definition of the Delta.

Agriculture

Historically, the traditional Delta was a region that relied almost exclusively on agriculture, in particular, the production of upland cotton, as its economic base. With an absence of recoverable industrial minerals, particularly coal, a lack of serviceable industrial infrastructure, and minimal indigenous investment capital in the wake of the Civil War, the region, along with most of the rest of the South, largely failed to keep pace with US industrialization rates in the late nineteenth and early twentieth centuries. As a result, the capital that did exist was turned to the service of cotton production. From the 1830s, when native claims on the land were finally dismissed, through the opening up of the Mississippi bottomlands in the wake of the Civil War, the expansion of upland cotton farming provided income and employment for a growing population of tenant farmers and sharecroppers from Reconstruction through the early twentieth centuries. However, mechanization, price depression through the globalization of cotton production, and the internal shift of cotton production to the western United States (particularly West Texas, with the postwar expansion of gas-powered irrigation in the Great Plains and Southwest) all led to the decline of cotton-related employment, while leaving much of the cotton production in place (fig. 5.4).

In the early twenty-first century, cotton still makes up between 5 to 50 percent of all the harvested cropland in the Delta counties; however, it is equally important on both sides of the Mississippi in a series of counties that stretches in an unbroken line from the boot heel of Missouri to central Louisiana. So while the Delta can still be defined as a cotton region, the production of cotton actually defines a much larger region than the traditional Delta.

The production of several other crops, notably rice, soybeans, and corn, also all overlap in the alluvial flatlands on either side of the Mississippi, and it quickly becomes apparent that the river serves as a conduit through a region rather than as the border of a region. In terms of topography, geology, climate, and pedology, the flat alluvial lands on either side of the river form a single agricultural region in which a variety of crops that, like cotton, require long growing seasons, high summer temperatures, and regular, abundant rainfall can flourish in the absence of irrigation (figs. 5.5, 5.6, 5.7).

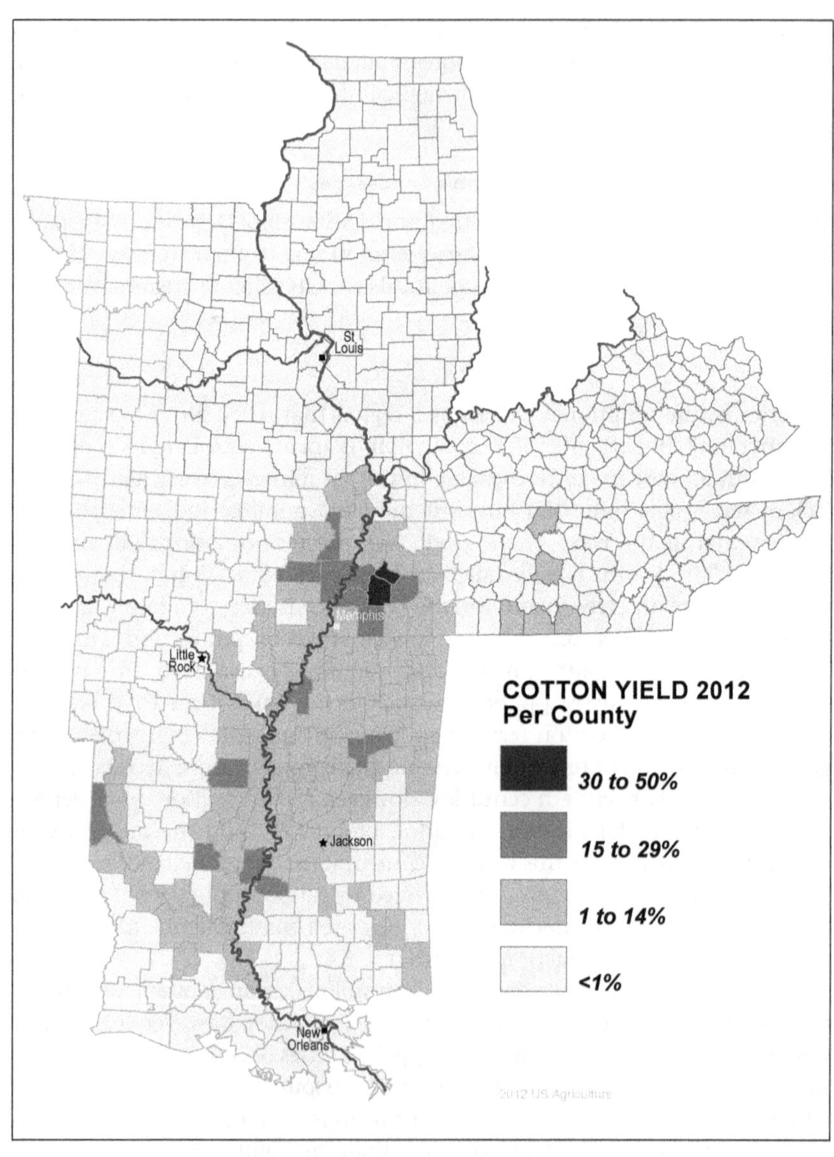

FIGURE 5.4. *Cartography by Tom Paradise.*

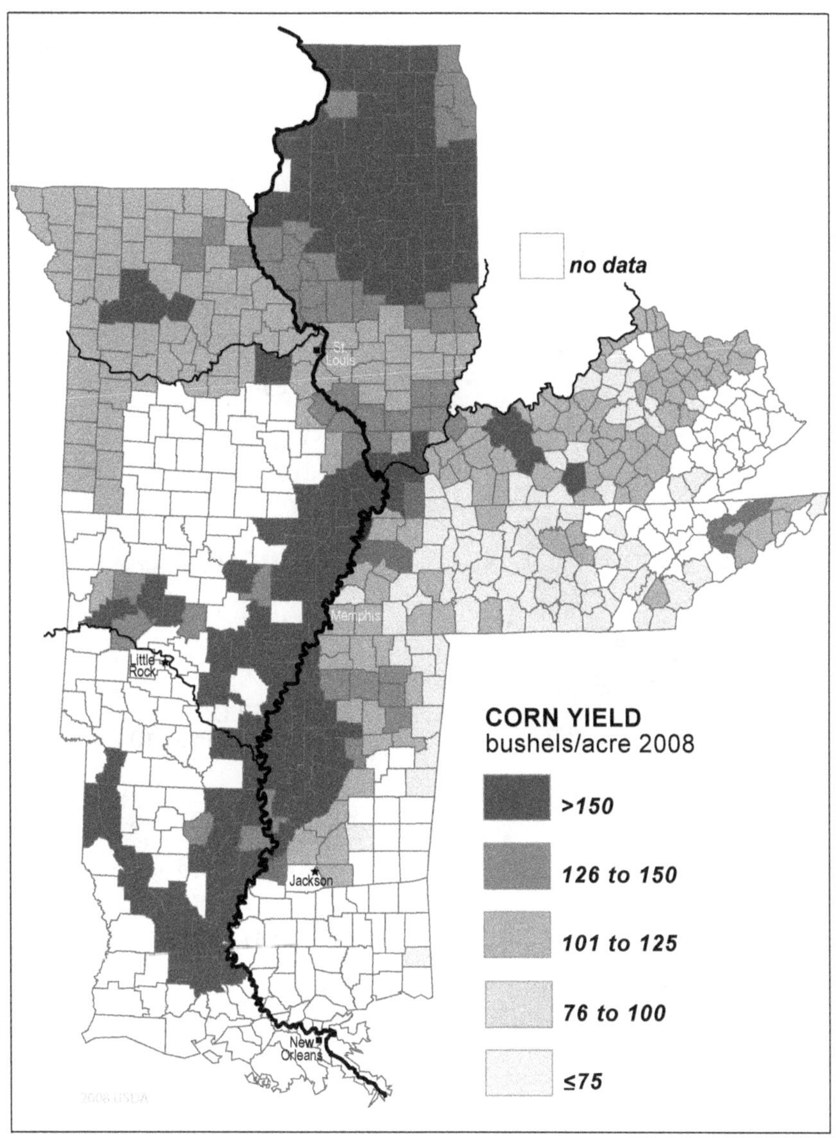

FIGURE 5.5. *Cartography by Tom Paradise.*

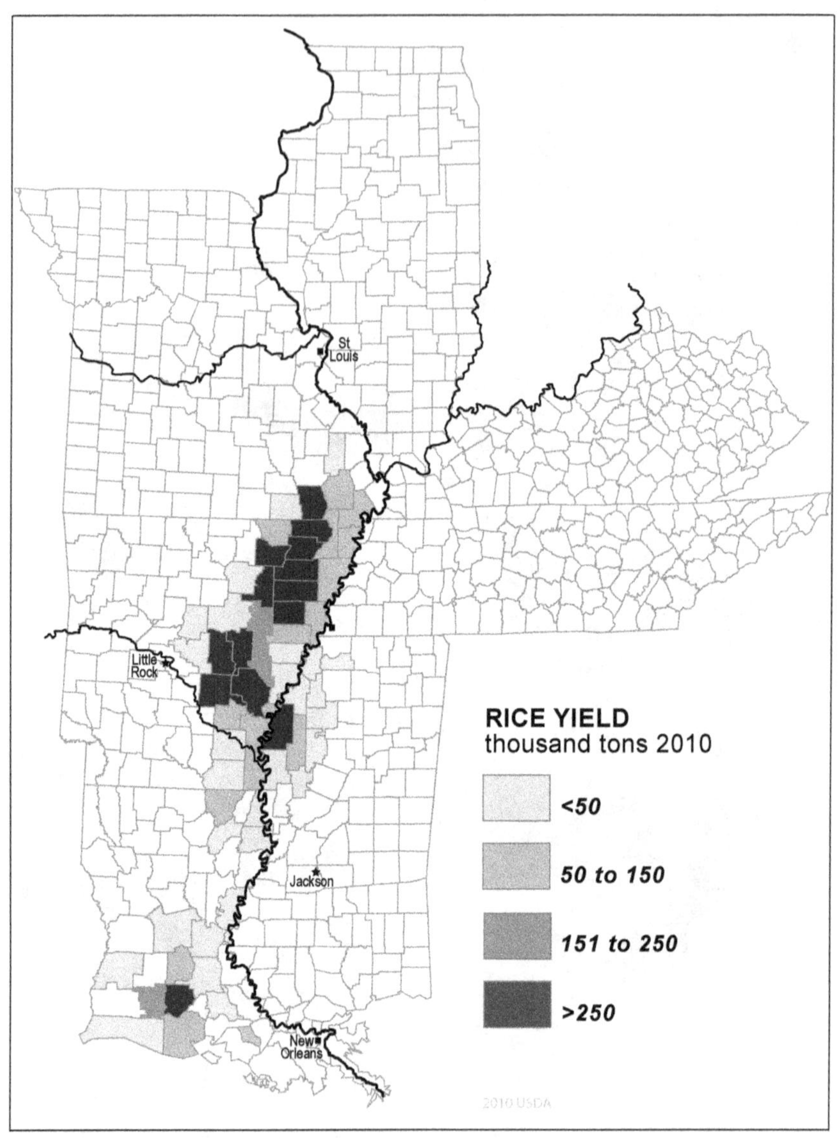

Figure 5.6. *Cartography by Tom Paradise.*

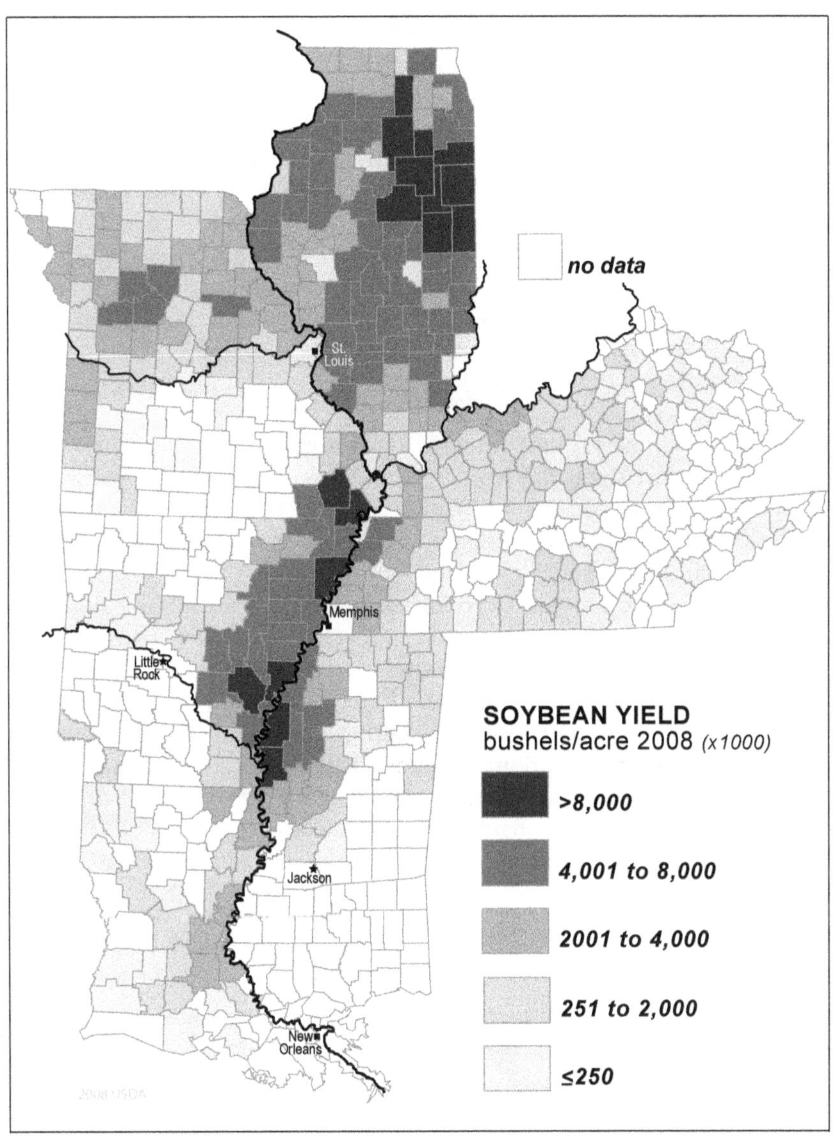

Figure 5.7. *Cartography by Tom Paradise.*

The area of the Mississippi Valley in which the soybean crop produces significant yields stretches, like that of cotton, from the Missouri boot heel to central Mississippi, with the highest concentrations in northern and central Arkansas and northern Mississippi. Rice is concentrated largely in east-central Arkansas but is also found throughout the Delta and down into south-central Louisiana. And counties in which corn production is concentrated again follow the river from Cairo, Illinois, to east-central Louisiana. Corn best demonstrates the relationship between the riverine environment and agriculture as counties with high corn production also follow the line of the Red River through Louisiana (a pattern that also exists, although less noticeably, with cotton and soybeans).

Transportation

The maps of primary transportation routes through the Mississippi River Valley are, at first sight, the best superficial indication that the Mississippi Delta exists as a defined, isolated region. Bounded on the west by the river itself, with only two major crossing points—at Greenville and West Helena—between Memphis and Vicksburg, and by Interstate 55 on the east, the map of major roads in the region provides an apparent sketch of the boundaries of the Yazoo basin, otherwise the Mississippi Delta. However, it is important to understand that transportation routes operate as elements of functional regions, not as formal regions, and as such they act as conduits of activity, not as regional barriers (fig. 5.8). When we examine the road network in particular, it is clear that the area traditionally defined as the Delta is actually embedded in a larger region that is connected across the river by US Highways 49 and 82 and is connected to the regional transportation nodes of Memphis, Vicksburg, and Jackson by Interstate 55 and US Highways 62, 82, and 49E and W.

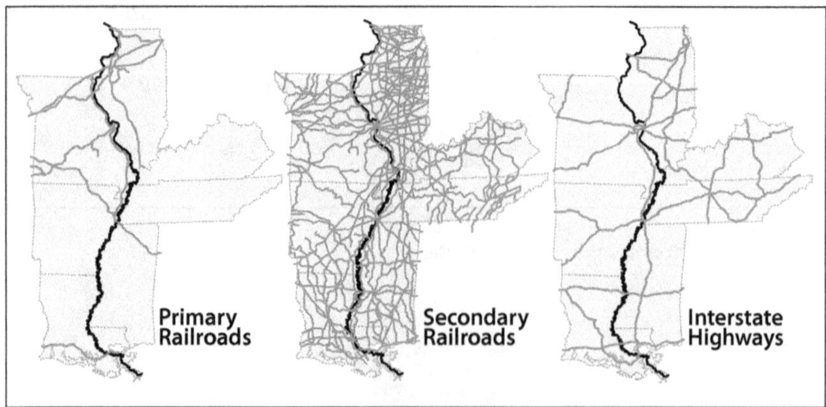

FIGURE 5.8. *Cartography by Tom Paradise.*

This is the functional region of the Mississippi River Valley, an area that is defined by commuters, regional news coverage (both print and radio/television), and support for professional sports teams: a region that is tied together by a road network that interlaces the population of the Delta with the surrounding regions on both sides of the river.

On a commercial level, the Delta is similarly interconnected through the secondary rail network with half a dozen minor freight lines that connect local industrial and agricultural activity to the regional commercial nodes of Jackson and Memphis and facilitate the connection of this region to the national and global economies.

Thus, in both the road and rail networks the Delta is just one area within a much larger functional transportation region that operates around Memphis and the Jackson-Vicksburg axis on Interstate 20: a region that provides commerce, transportation, employment, and urban services for the entire mid–Mississippi Valley area. Operating largely on a north-south axis, the functional area is bounded on the south by Baton Rouge and to the north by St. Louis, although the rapid transportation made possible by interstate corridors extends the influence of Memphis and Jackson east and west as far as Little Rock, Shreveport, and Nashville. Given its location between Memphis and Jackson/Vicksburg and the relatively dense network of roads and rail lines that intersect the region, the Delta is actually a very well connected part of a much larger functional region that occupies much of the Lower Mississippi Valley.

Income Measures

Historically heavily reliant on undercapitalized and labor-intensive agriculture, the Delta has a long history of poverty and below average household earnings. A series of county-level maps displaying income, poverty, unemployment, and reliance on federal assistance indicates that the characteristic poverty of the Delta is actually widespread throughout the river valley.

Counties with a per capita income of less than $10,000 a year line the Mississippi River on both sides from southern Illinois to southern Louisiana. The traditional Delta is clearly distinguishable as a group of counties that stretch from the southern suburbs of Memphis to just north of Jackson; however, low income is a feature of the entire Mississippi River Valley, particularly south and west of the Delta and in a non-contiguous belt of counties through central Mississippi (fig. 5.9).

Complementing the spatial distribution of low-income counties, the spatial extent of poverty, as defined by households with incomes below the federal poverty level, is again delimited by counties on either side of the river, with the Delta forming a visible, but not discrete, region of deprivation in central Mississippi (fig. 5.10).

Counties with the highest percentage of households dependent on federal assistance again follow the river, but the extent of this region is greater

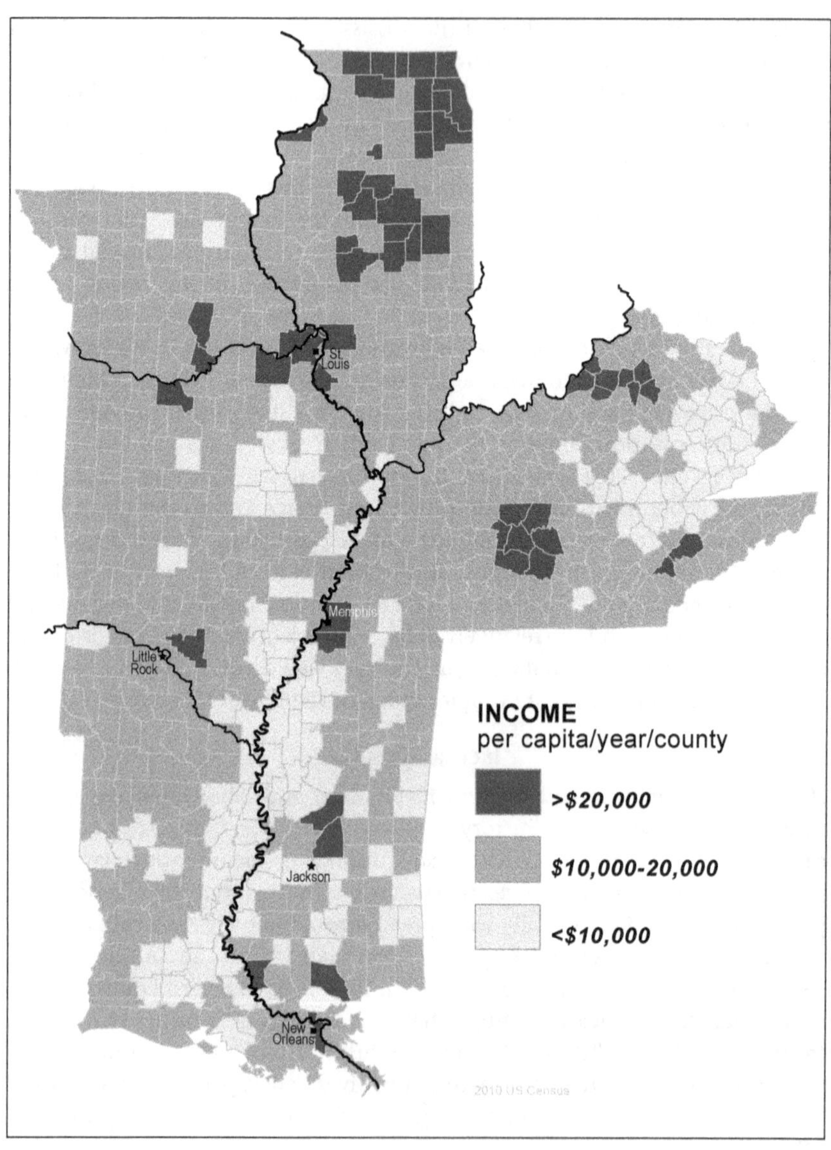

FIGURE 5.9. *Cartography by Tom Paradise.*

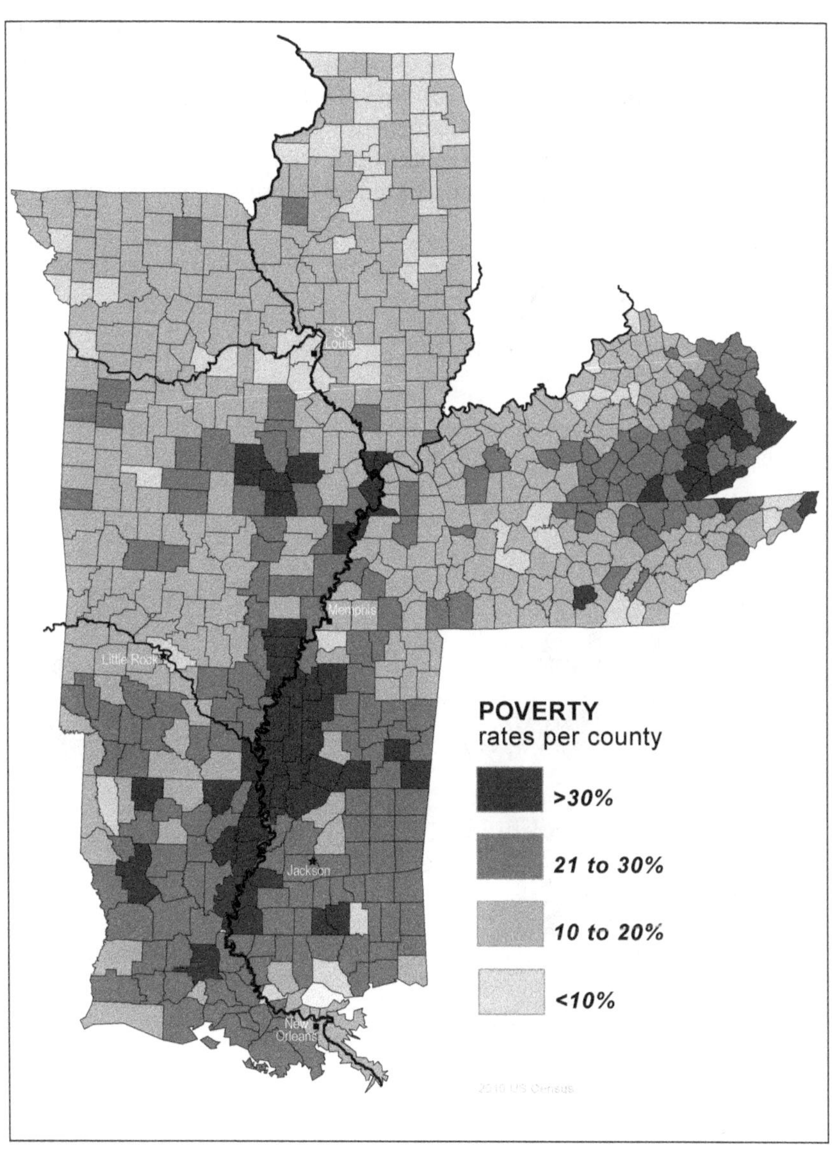

FIGURE 5.10. *Cartography by Tom Paradise.*

than that for low income and poverty and the Delta is much less clearly defined, subsumed within a large area of eastern Arkansas, Louisiana, western Tennessee, and more than half the counties of Mississippi (fig. 5.11).

High rates of unemployment, the underlying cause of much of the deprivation in the Mississippi River Valley, likewise are found in counties along the river, although in a few notable cases, including Bolivar County, Mississippi, at

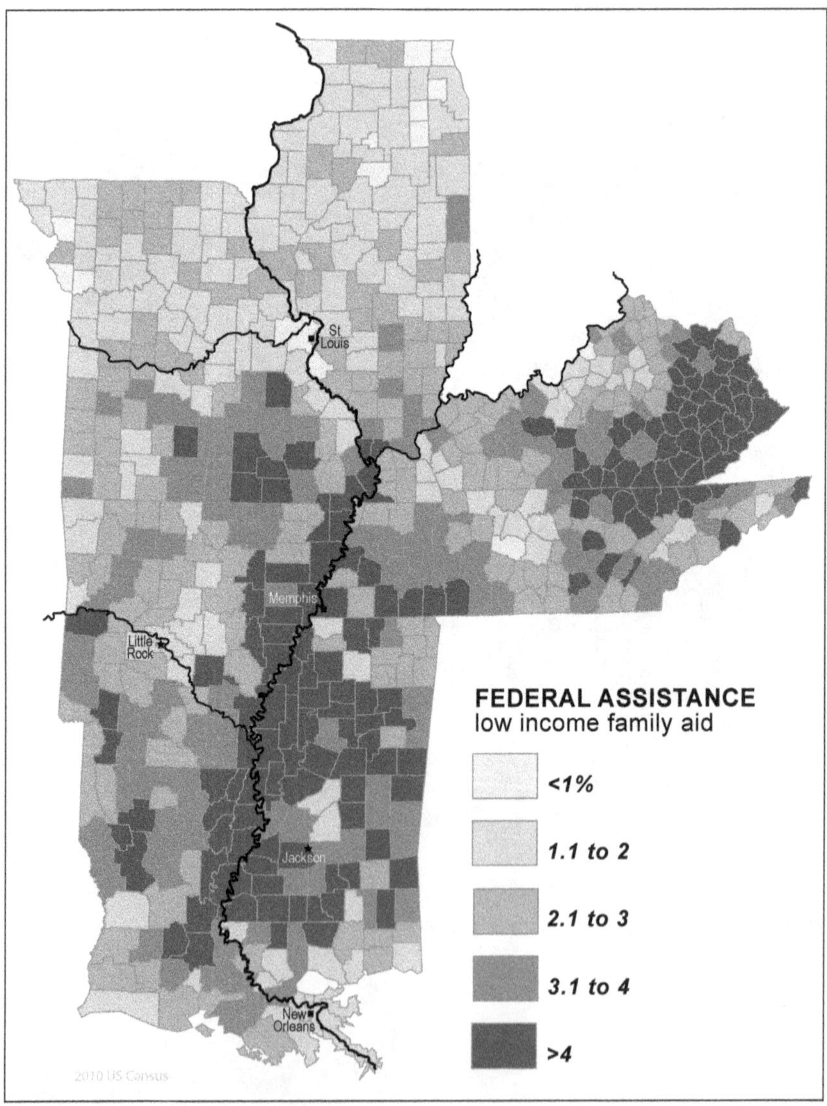

FIGURE 5.11. *Cartography by Tom Paradise.*

the heart of the Delta, unemployment is not as high as in neighboring counties. The region of high unemployment is more geographically contained than the regions of poverty and low income, stretching from western Tennessee to the northern counties of eastern Louisiana. The mismatch between poverty or income and unemployment indicates a higher than average presence of low- and minimum-wage jobs in these counties of the Mississippi River Valley (fig. 5.12).

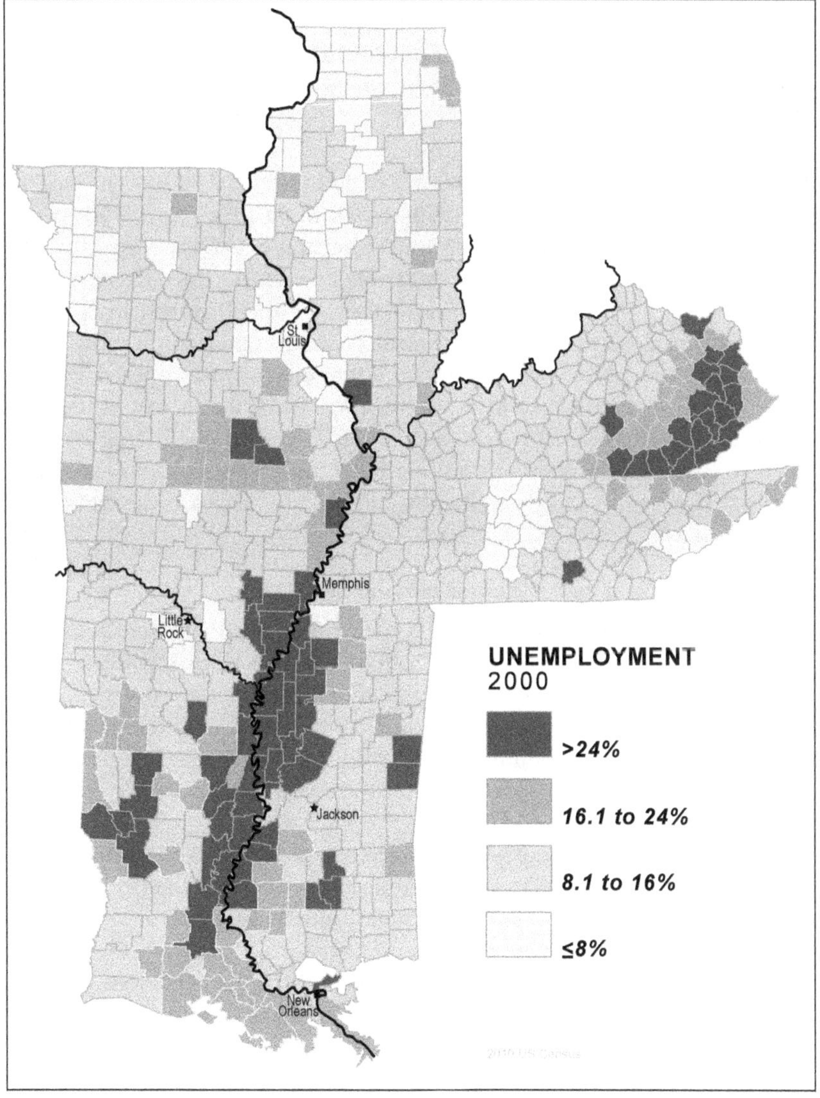

FIGURE 5.12. *Cartography by Tom Paradise.*

The income and employment indicators clearly show that the economic characteristics traditionally ascribed to the Delta are in fact more broadly distributed throughout the Mississippi River Valley. When the formal regions created by sociocultural indicators are added to the analysis, the rationale for extending the definition of the Delta beyond the Mississippi/Yazoo basin becomes even clearer.

One of the most definitive delineations of the Delta has historically been through the presence of a high percentage of African Americans in the Delta counties. While the Great Migrations of the early and mid-twentieth century greatly reduced the number of counties in which African Americans constituted a majority of the population, there are still numerous counties along the Mississippi where the black population exceeds 50 percent. Unsurprisingly, many, but not all, of the Delta counties are in that group. However, as with physical and economic indicators, a characteristic that has been used in the past to define the Delta is actually much more widespread, with counties that are 50 percent or more African American extending north to West Memphis (Crittenden County) in Arkansas and south, through northeastern Louisiana, to the Mississippi-Louisiana state line. Unsurprisingly, these figures reflect, to some degree, the legacy of slaveholding in the Mississippi River Valley. Most of the counties with a majority African American population today were either counties where slaves made up at least 60 percent of the total population in 1862 or counties that contained large areas of wooded Mississippi bottomland (along both the Mississippi and the Yazoo Rivers) that were intensively settled by emancipated slaves at the end of the Civil War (figs. 5.13, 5.14).

Sociodemographics

In keeping with the low income and high unemployment rates for the region, the entire Mississippi Valley south of Memphis is plagued by low high school graduation rates. Again, the data clearly show the boundaries of the Yazoo basin in which every county has a graduation rate below 75 percent, but poor education performance extends northwest into a trio of Arkansas counties and southwest into Louisiana, in a line of counties that follows the river all the way to New Orleans (fig. 5.15).

Poor health outcomes are similarly related to poverty and poor educational attainment; and layers of data representing the presence of obesity, diabetes, and sexually transmitted diseases in Delta counties bear out these relationships. However, high rates of obesity and diabetes are prevalent throughout the South, and the counties of the Delta are indistinguishable from those of the rest of Mississippi, southern and eastern Arkansas, southern and eastern Louisiana, and western Tennessee on both measures. Nutrition-related health issues are intimately related to poverty and low educational attainment, as

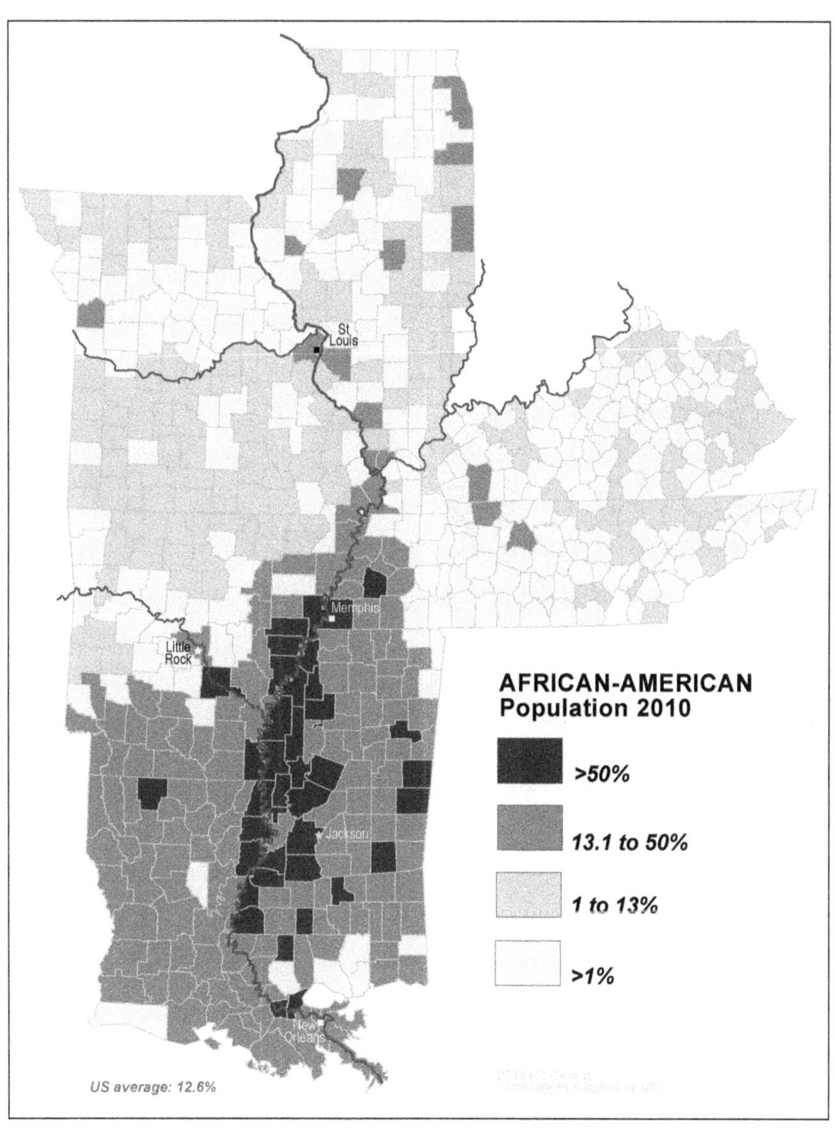

FIGURE 5.13. *Cartography by Tom Paradise.*

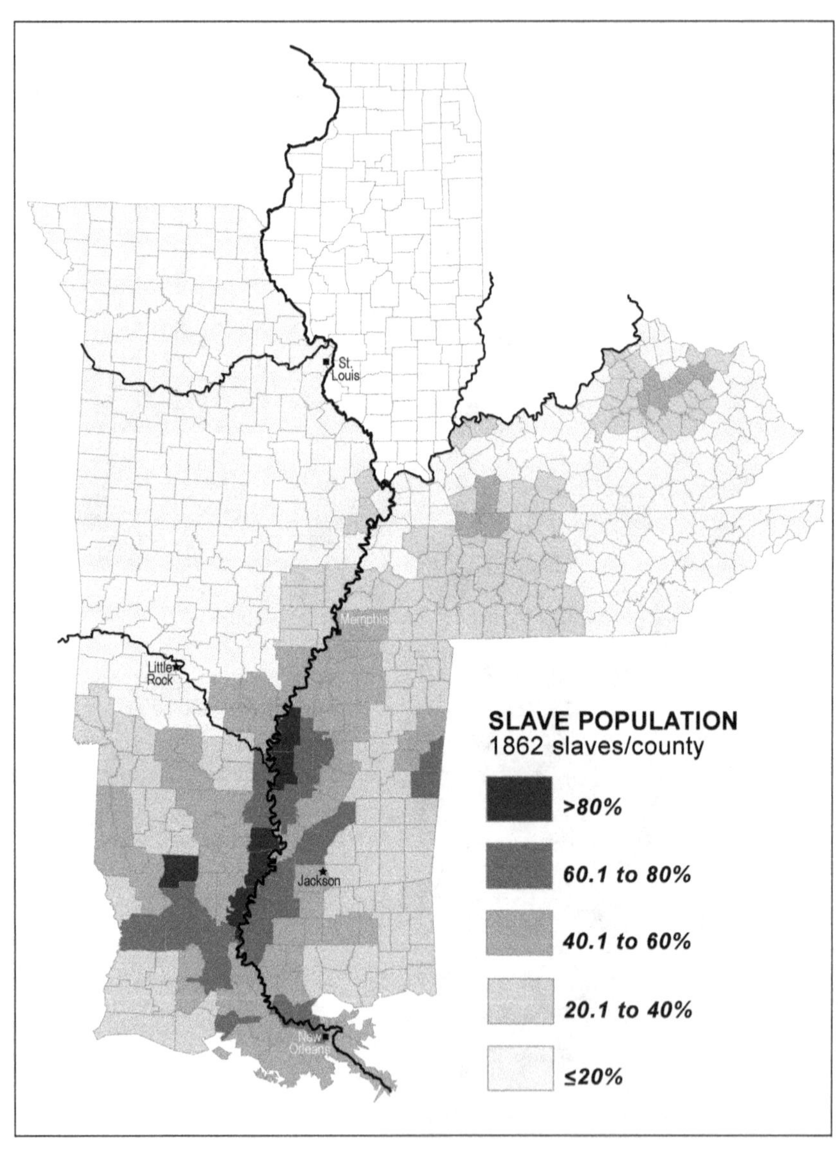

FIGURE 5.14. *Cartography by Tom Paradise.*

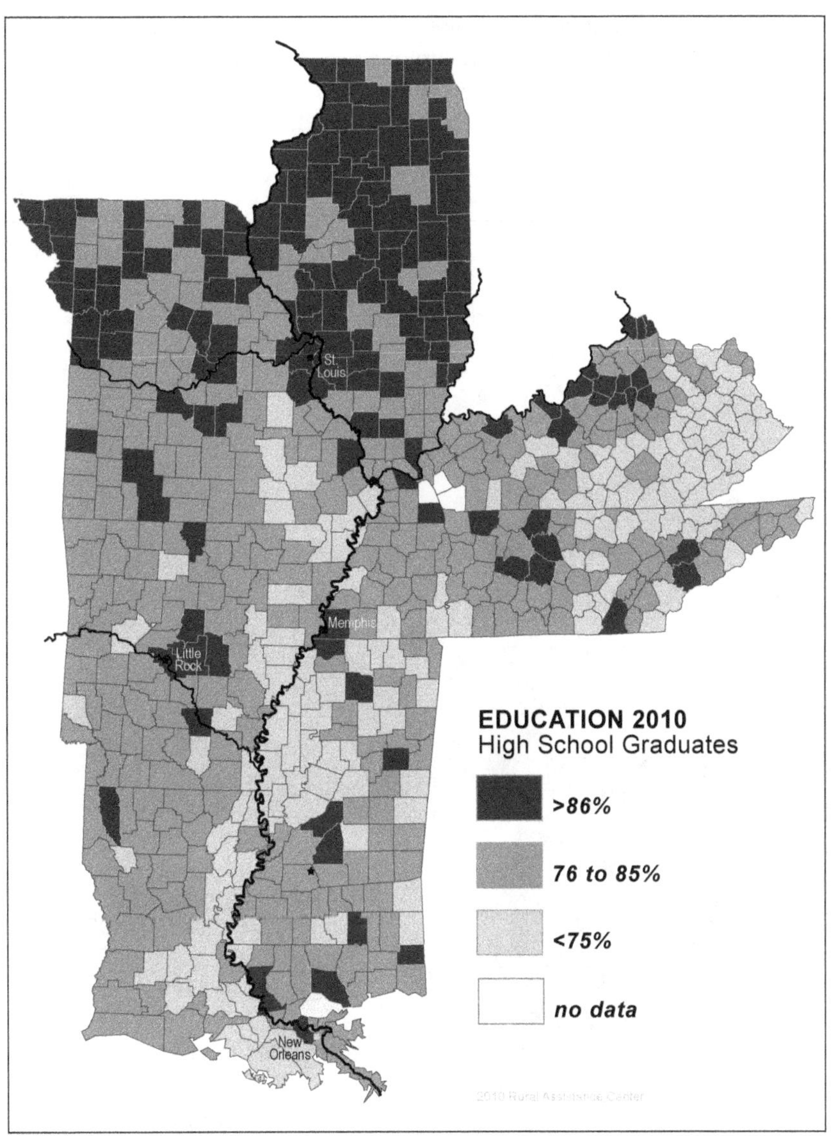

FIGURE 5.15. *Cartography by Tom Paradise.*

are rates of sexually transmitted disease. The incidence of chlamydia in the Mississippi River Valley suggests the unique regional characteristic of the Delta within Mississippi, but, as the map shows, high rates of the disease are also prevalent on the Arkansas bank of the river, in a line that stretches all the way north to the Missouri border (fig. 5.16).

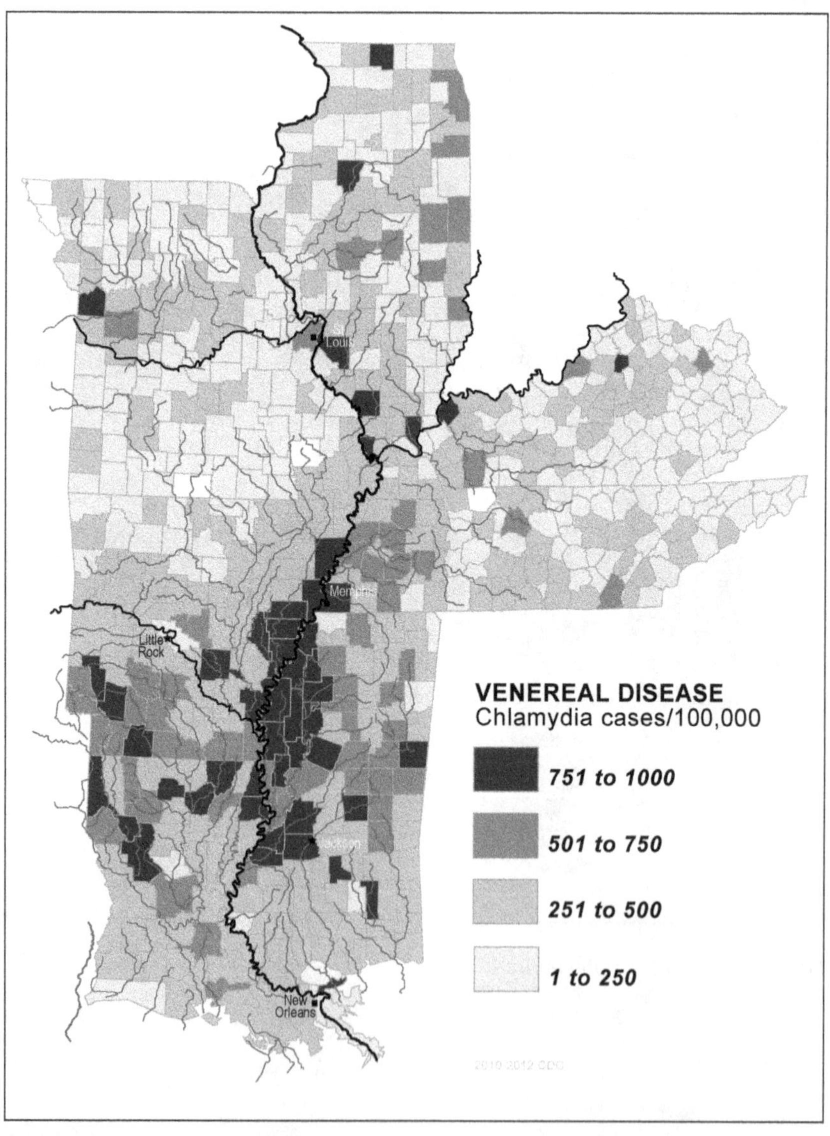

FIGURE 5.16. *Cartography by Tom Paradise.*

Politics

The final indicator of possible Delta uniqueness is a modern artifact of the changing political allegiances of the South. Historically, the Delta was part of an indistinguishable Democratic bloc that stretched from Texas through Georgia and north to Virginia. This Democratic stranglehold on the South was fractured in both 1948 and 1968 by the third-party southern responses to the increasing importance of civil rights platforms in the national Democratic Party. In the wake of George Wallace's failed presidential run in 1968, support for the Republican Party grew dramatically in presidential, congressional, state, and local elections throughout the South, support that was almost exclusively driven by white flight from the national Democrats, who were forcing desegregation and civil rights on the south. Forty-five years later, political affiliations in the state of Mississippi are almost exclusively racially aligned, with 95 percent of black voters selecting Democrats at the polls and an almost equivalent percentage of whites favoring Republicans. Consequently, the map of counties that voted for a Democratic presidential candidate in 2012 almost exactly maps onto the counties that are majority African American. Once again, however, this phenomenon is not unique to the Mississippi Delta, throughout the Mississippi River Valley, in a line of counties on both sides of the river from southern Louisiana to central Arkansas, the Democratic vote lines up with majority African American counties (fig. 5.17).

Summary

There is, in fact, not a single significant economic, political, or demographic measure that is unique to the Delta; nor is it possible to layer different indicators to create a composite of factors that allows the Mississippi Delta to emerge as a unique region that can be defined by particular physiographic, sociodemographic, economic, or political criteria. If the definition of the *Delta* region is extended to encompass the Arkansas Delta, a formal region on the west side of the Mississippi that is defined by the extent of the alluvial plain in Arkansas, the region then becomes under-bounded, as many of the characteristics ascribed to the Delta (poverty, slave heritage, percentage African American) do not currently extend over this entire area.

Consequently, from a geographic perspective, the Delta can be said to exist only as a perceptual or vernacular region. Historically considered the area bounded by the Mississippi and Yazoo Rivers, it was a *region* with clearly demarked boundaries, and yet the counties of the region share physiographic, social, economic, and political traits with counties to the north and south and on the western bank along the Mississippi. Rivers are as much corridors and pathways as they are barriers and margins. In this chapter, when the maps of the region are examined, it is evident that the Yazoo and Mississippi Rivers do not determine the edges of the region but act as conduits of growth, change, movement, accessibility, and commonality. The rivers do not confine or restrict it.

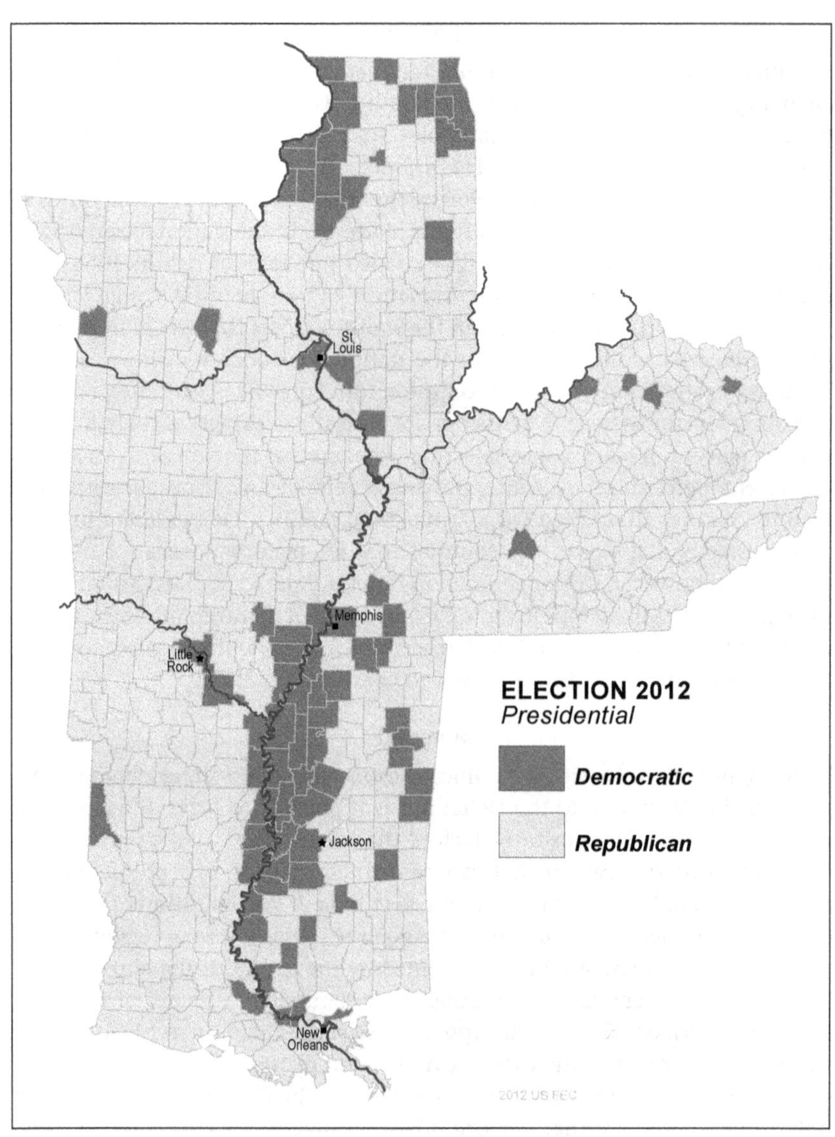

FIGURE 5.17. *Cartography by Tom Paradise.*

In essence, the *Delta* region is delineated by the variables used to define it. Regions of unemployment and poverty, health and wellness, agricultural and resource productivity, income, education, transportation, politics, and demographics all identify, distinguish, and demarcate the multifaceted Mississippi Delta region. So, when we look at the complex palimpsest of the region's demographic, physiographic, commercial, political, and economic parameters, in tandem with its resources, transportation, and vernacular spaces and places, there is not one Mississippi Delta but many, defined by the characteristics being used to describe or influence the Delta.

CHAPTER 6

The Death-Dealing Delta

A CLIMATOLOGICAL VIEWPOINT

MARY SUE PASSE-SMITH

Combining the word "Delta" with the word "climate," at least in the mind of this physical geographer who lives within the region and specializes in climate and hazards, conjures up much the same image referenced by Jeannie M. Whayne, a historian reviewing the Delta in 1999: hoop-skirted belles fanning themselves while lounging in the shade of graceful trees near a stately plantation mansion with nearby gentlemen conferring over cold mint juleps. But as Whayne points out, this grand plantation tradition is more suited to areas outside of the Delta as defined herein, and perhaps the climate is also not specific to the Delta. As can be seen from the maps (figs. 6.1 and 6.2) of typically defined "climate," temperature, and precipitation averages over time, the Delta shares many of its climatological characteristics with much of what is called "Dixie" by geographers.[1] Like most of the southeastern one-third of the United States, the Delta, save for a small segment of Missouri, lies in the humid subtropical or transitional humid subtropical climate zone. Due to its proximity to a fetch of maritime tropical air from the Gulf of Mexico, summer days can be oppressively humid, even during the driest months of the year. Nocturnal temperatures also do not lower appreciably due to the greenhouse effect of water vapor, with or without cloudiness. A fluttering, lacy fan, for most modern Delta dwellers, will barely cut the air itself on many a summer day. In fact, Delta towns such as New Orleans, Shreveport, and Memphis proudly stand on the *Old Spice* All-Time Top 20 Sweatiest Cities. In the past, press releases show that Little Rock and Baton Rouge have also made the cut, with Jackson, Mississippi, never far behind.[2]

Figure 6.1 shows that the Delta as a climate region is not very cohesive when considering average annual temperatures. Mean temperatures range from the 70s (darkest tones) to the low 50s over the period 1970–2000. The average daily temperature across the entire year in the Delta is a pleasant 61.6 degrees, ranging from an average of 40.7 in January to 80.5 in July. Figure 6.2 also shows little continuity when considering average annual precipitation. Precipitation values range from about 40 inches to over 60 (darker grays) per

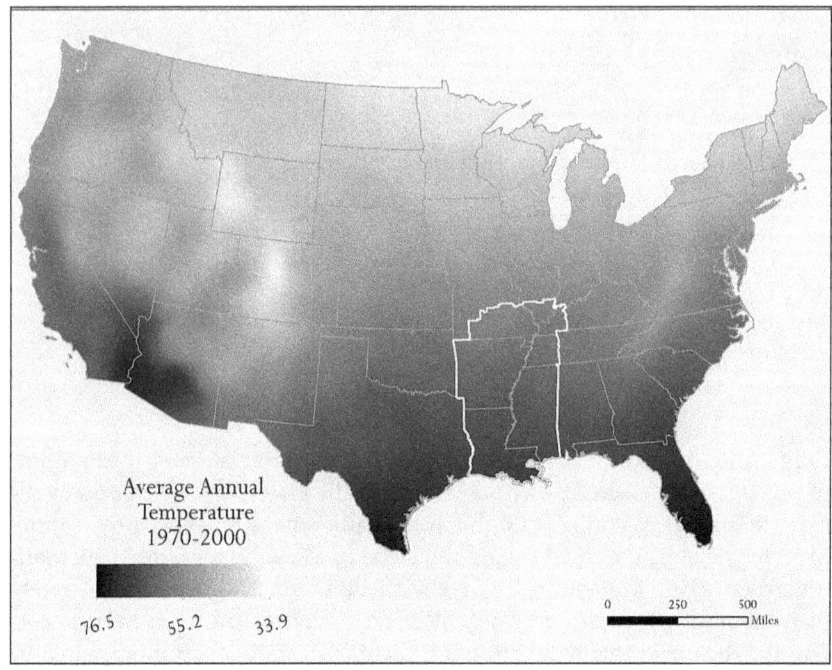

FIGURE 6.1. Thirty-year average annual temperature in contiguous United States, 1970–2000. Delta region is outlined in white. *Temperature data from the National Climate Data Center; state boundaries from US Census Bureau.*

year. Regionally, the average is 54 inches per year; precipitation is fairly evenly distributed throughout the year, with most months averaging between 4 and 5 inches. March is the wettest month at 5.26 inches average across the Delta, while August is driest, at 3.51 inches. Looking at these two figures, it is clear that the goal of defining the Delta as a cohesive climate region requires some deeper digging.

If twenty-four maps of monthly variation in mean temperature and precipitation could be fit on the page instead, there would still be a great deal of variation across the area. For while it is 40.7 degrees in January across the whole Delta, averages range from around 53 degrees on the Gulf coast to below freezing in Missouri and Illinois. July, the hottest month, shows the least range in mean value: 76.5 degrees in the northern reaches of the Delta to about 83 on the Gulf. With regard to precipitation, May's precipitation varies the least, ranging from 4.3 inches in the north to 6.4 inches, not along the Gulf, as might be expected, but in far western Arkansas. So, while the Delta can be defined as primarily "humid subtropical" climate, the most consistent thing seen thus far is a lack of consistency.

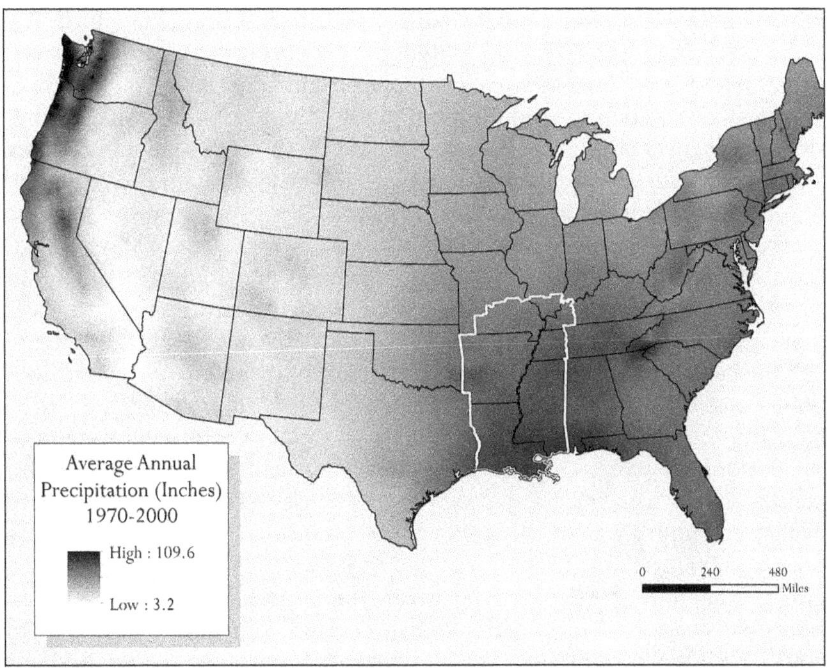

FIGURE 6.2. Thirty-year average annual precipitation for contiguous United States, 1970–2000. Delta region outlined in white. *Precipitation data from National Climate Data Center; state boundaries from US Census Bureau.*

Certainly snow is not a cohesive factor. As is clear from the temperature figures above, few Delta cities make the coldest places lists; no Delta city even breaks the top one hundred snowiest. But while winters in the Delta do tend to be somewhat mild, those less-frequent incursions of polar air often give rise to the Delta's more interesting and devastating climatological exhibits, depending on which air is on top: severe thunderstorms, tornadoes, ice storms, flash and riverine floods. Perhaps a common thread can be found looking not at the climate norms, but at meteorological hazard norms.

Days of Thunder

That occasional blast of cold, dry air shoving aside warm, moist air often means thunderstorms, and many Delta storms become severe. Most of the region experiences between fifty and sixty days with thunderstorms on average per year. Southern Louisiana and a slice of the southern quarter of Mississippi are more thunder prone, dodging storms from sixty to eighty days per year right on the coast. Lightning is a definite hazard in the three states included in their entirety in the Delta: from 1959 to 1994, Arkansas was ranked third in per-capita lightning fatalities; Mississippi was fifth; and Louisiana, ninth. The National Lightning Safety Institute website cites encouraging news, however: from 1990 to 2003, only Louisiana remained in the top ten.

A less-talked-about hazard from thunderstorms, *derechos* (long-lived windstorms that cover large areas, usually associated with bands of rapidly moving storms) affect the area frequently; much of the Delta suffers one derecho each year. The northern sections (most of Arkansas, western Tennessee, extreme southern Illinois, and southeast Missouri) are affected most from May through August, while the southern Delta (all of Mississippi and Louisiana, the southeastern two-thirds of Arkansas, and, again, western Tennessee) experiences more from September through April. On May 8, 2009, one of the most intense and unusual derechos ever observed struck southeast Missouri, southern Illinois, western Kentucky, and northern Arkansas with hurricane-force winds, hail, and flash flooding, as well as tornadoes. Gusts over 90 mph were measured along its path (Carbondale, Illinois, measured a gust of 106 mph), and the complex traveled over 1,000 miles and caused millions of dollars in damage. Memphis was struck in July of 2003, causing 750,000 people to lose power; 47,000 still were in the dark ten days later. Although this storm caused only two fatalities, in general, derechos are responsible for *more* fatalities than nearly 90 percent of their better-known cousins, tornadoes, from 1986 to 2003. The little-known Delta "Derecho Alley" is thus definitely risky business that should garner respect from residents within.

Tornadoes and the "Dixie" Tornado Alley

Those above-mentioned incursions of cold air combined with winds streaming from the south at one level and from the southwest to northwest at other levels of the atmosphere are but a few of the ingredients that have moved much of the Delta (along with much of Alabama) squarely within a new definition of "Tornado Alley": the "Dixie Alley," which has "the highest frequency of [high intensity] tornadoes in the country."[3] This is evident in Figure 6.3, which shows a true hotspot across a large portion of the Delta for strong to violent tornadoes, which are those rated at or over EF2. "EF" refers to the Enhanced Fujita scale, the measure of intensity for tornadoes, ranging from EF0 (very weak) to EF5 (with winds over 200 miles per hour). So while some places in the United States may have more tornadoes, the Delta keeps pace with what is thought of as the traditional Tornado Alley in the Great Plains when it comes to strong storms. The highest map value of 0.26 means that within 1,000 square miles of any cell in the darkest shade (and note that the mean size of a US county is 1,100 square miles), twenty-six EF2 or greater tornadoes have occurred. Thus, it is no surprise that the mean intensity for tornadoes inside the Delta is over 1.4, while for all other tornadoes it is but 0.88.

Delta tornadoes are most numerous in April, while outside of the region the peak month is May; but, as will be discussed below, Delta tornadoes also

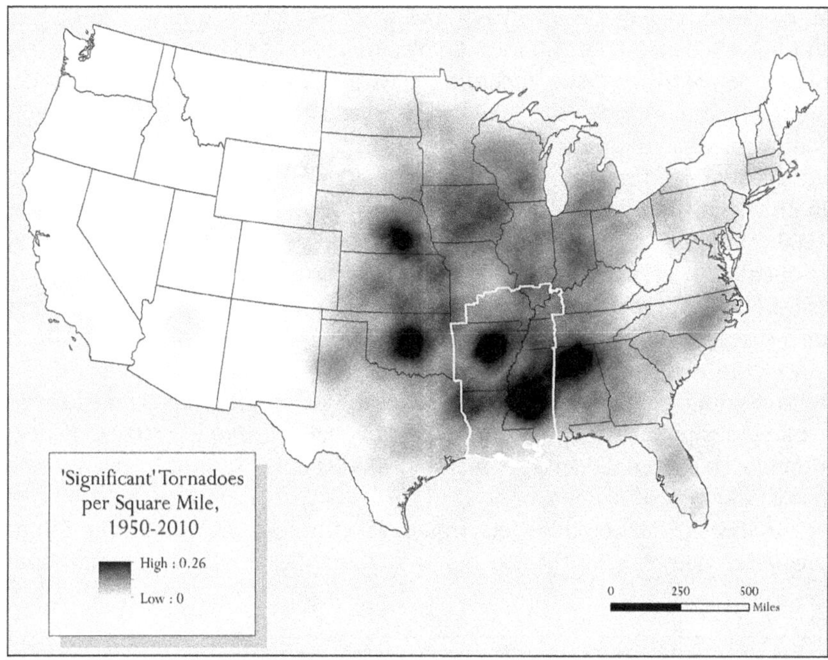

FIGURE 6.3. "Significant" tornadoes (with an EF scale of 2 or above on the Enhanced Fujita scale) per square mile, 1950–2010. A value of 0.26 significant tornadoes per square mile equates to 26 EF2 or higher-intensity tornadoes every 100 square miles. *Tornado tracks from Storm Prediction Center; state boundaries from US Census Bureau.*

tend to happen during the "off" months of November through February more often than in other regions. As is true of tornadoes in general, they also are most frequent between 4:00 and 9:00 p.m.; they are more distributed across the day, however, than are tornadoes outside of the Delta. Delta tornadoes have a mean length of 10.8 miles and a mean width of 197 yards and affect more area than the average tornado elsewhere, which has a path length of 5.9 miles and 131 yards of width.

Figure 6.4 shows the distribution of fatalities in tornadoes (along tracks) for the period 1950–2010. While there are other areas with high fatality rates (the darkest shades here represent 2 fatalities *per square mile* over the area), much of the Delta is in the darkest, highest fatality area. This is perhaps the most climatologically unique aspect of the Delta: its dubious distinction as the area suffering the highest tornado fatality figures in the country before and since the advent of tornado warnings, whether the measure is per capita, per tornado, or per square mile. This modern-day (post-1950) map does not show the carnage quite as well as a review of the history of violent tornadoes in the United States. Of 200 tornadoes causing over 18 fatalities since records began, 56 are in the Delta, including most of the path of the daddy of them all, the infamous Tri-State Tornado of 1925, which killed nearly 700 people in southeast Missouri and southern Illinois. In fact, three others in the top ten are in the Delta: the Natchez tornado of 1840 (ranked 2), the Tupelo tornado in 1936 (ranked 4), and Amite, Louisiana, to Purvis, Mississippi, tornado of 1908 (ranked 8). Of the nearly 2 million square miles of US territory that is frequented by tornadoes (i.e., the area east of the Rocky Mountains), only a tenth (195,445 square miles) makes up the Delta region, yet nearly 30 percent of the deadliest tornadoes in US history have taken place there. And this in a far more rural region than much of the rest of the country, making direct hits on populated areas even less likely. Kevin M. Simmons and Daniel Sutter rank Mississippi first and Arkansas second in fatality *and* injury rate from 1950 through 2007, and Arkansas is first and Mississippi is second in fatality rate from 1900 to 2007.[4]

Analysis of all tornado data from 1950 through 2010 from the Storm Prediction Center shows that the top twenty counties with the highest fatalities

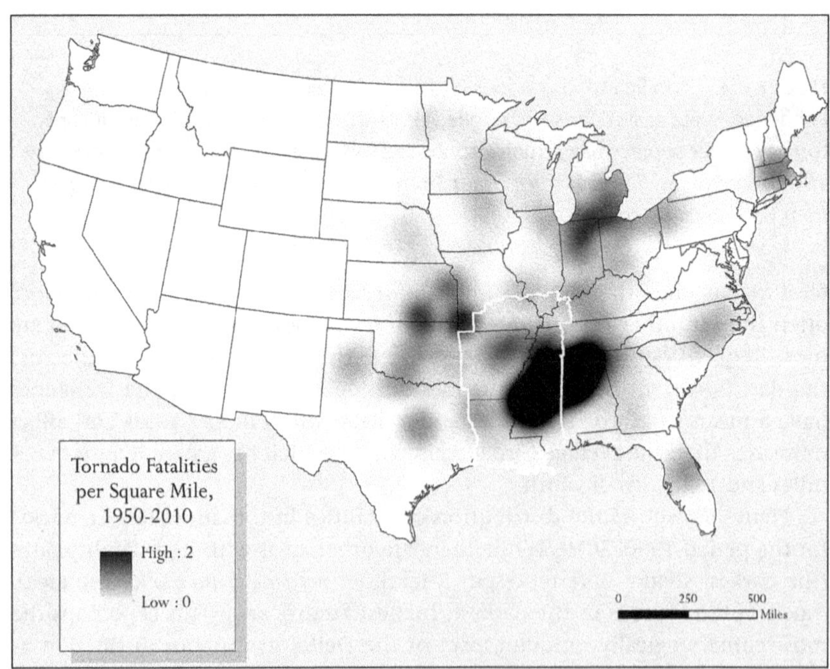

FIGURE 6.4. Sixty-year tornado fatalities per square mile (1950–2010). *Tornado tracks from Storm Prediction Center; state boundaries from US Census Bureau.*

per square mile are in the Delta (nineteen are in Mississippi); the top twenty highest tornado density counties are in the Delta (sixteen are in Mississippi). Three of the top four counties suffering the most tornadoes are in the Delta (in Arkansas); six of the top ten counties experiencing strong to violent tornadoes are in the Delta (all but one are in Arkansas). Woodruff County, Arkansas (population of 14.0 per square mile), East Carroll Parish in Louisiana (21.0 per square mile), and Jefferson County, Mississippi (18.2 per square mile), are also in the top ten for tornado fatalities per capita over the period 1950–2010. But more telling is the fact that per-tornado fatalities across Delta counties are 35 per 100 tornadoes as compared with 14 per 100 tornadoes for the rest of the United States. Even in the "traditional" Tornado Alley, an area roughly the same size as the Delta to its north and west, only 19 fatalities per 100 tornadoes occur. Per capita measures also tell the same tale: over 8 people out of every 100,000 have died in a tornado, on average, across Delta counties from 1950 to 2010. That is over six times the number for all other counties in the United States outside of the Delta.[5]

What might explain this anomaly: a relatively rural, low-population-density area with such high death tolls? Explanations run the gamut from logical to the ridiculous. Logically, a strong-to-violent tornado alley would likely produce more fatalities. But since there are, as the map shows, other areas suffering high numbers of strong twisters with less carnage, one explanation, offered in the 1970s, was that the benighted southerner "face[d] the whirlwind alone with his God" while tech-savvy Illinoisans (and thus all northerners) were prepared by virtue of their use of media and implicit trust in Weather Service warnings to take more appropriate action.[6] From interviews with a few people in Alabama and Illinois, these findings were extrapolated to the entire South and North, showing fatality maps that *did* reflect similarities to those herein. An interview with an Itawamba County resident who survived the April 27, 2011, EF5 tornado near Smithville, Mississippi, however, lends little credence to this notion in present times:

> We were aware that the weather news was not good. I look at the NWS Memphis web site daily and we had a weather radio. . . . We were also tuned to WTVA, channel 9, in Tupelo. . . . Before our power went out we knew there was a possible tornado, in northern Monroe County, headed in our general direction. . . . Our vision, in all directions, is limited due to the terrain where we live. The local ridges block our view and the timber we owned was still standing. . . . We had about 32 acres of mature mixed upland hardwood.[7]

Here it is also notable that the geography of the Delta can contribute to higher death tolls. As noted above, many parts are heavily treed and hilly away from the Mississippi River floodplain areas; these factors lead to far less distance visibility than in the Great Plains. As well, with the exception of a small part of the Arkansas and Missouri Ozark region, the Delta is shown by the EPA to be defined as "wetland," a place where "the water table is usually at or near the surface." Underground storm shelters and basements are not viable in areas of high water tables. Thus, the stock of safe havens during a tornado is less than in many other parts of the country.

The poverty seemingly endemic to the area, cited by other Delta authors in this collection, in conjunction with the very climatology that produces the tornadoes themselves is also a likely part of the explanation. Both milder climate and lower incomes mean more mobile homes and other types of less-sturdy housing. Mississippi ranks eighth, Arkansas ranks tenth, and Louisiana ranks thirteen in percentage of housing units that are mobile. Multifamily and multistory rental units also are more dangerous. A Paducah, Kentucky, National Weather Service review of a 1982 tornado that killed ten in Marion, Illinois, notes, "Ten persons lost their lives, all in Marion. Seven of the ten died in the Shawnee Village Apartments, a low cost housing project. . . . It appeared that the two-story Shawnee Village Apartment complex was not as well constructed as some privately owned businesses."[8] Simmons and Sutter show that mobile home fatality rates are ten times greater than those for permanent homes, and low education is strongly correlated with fatalities; both also correlate with low income.[9]

Simmons and Sutter also address the fact that nocturnal and off-season tornadoes are also more deadly, and they note that Louisiana and Mississippi have a higher percentage of nocturnal tornadoes than other states.[10] The Delta region is squarely in the high off-season tornado area; memorable outbreaks occurred in Arkansas on December 2 and 24, 1982, in January 1999 (an all-time record for January tornadoes in any state on any day as well as the largest outbreak in Arkansas history), and on February 5, 2008 (Arkansas and western Tennessee). In February 1971 in Jackson, Mississippi, the NWS site described how the "Arklamiss" was crushed by strong and violent tornadoes; three separate, violent (F4 or F5) tornadoes occurred, killing a total of 117 people and injuring over 1,200. November 2005 produced fifty tornadoes, many strong or violent, and most in the Delta; November 23–24, 2001, spawned more storms across the Arklamiss and parts of Tennessee, mostly in the middle of the night, killing 9 people.[11]

Tropical Storms

In the Delta, the blues are a tradition. One blues artist seems to have weather not only in his soul, with songs like "That Mean Old Twister," but even in his name, Lightnin' Hopkins. In "Hurricane Betsy," Hopkins lamented the losses from the terrifying passage of a major hurricane through New Orleans:

> Betsy passed through Louisiana today; she had people runnin', they was tryin' to hide; killed so many folks that the rest that were left, they couldn't

> be satisfied. . . . [B]ackwater rise; comin' all in them windows and doors. . . .
> Betsy come through there; Lord, it swept them buildings down.[12]

Fifty-eight people lost their lives in Louisiana during Hurricane Betsy in 1965. Fast-forward forty years and replace the name "Betsy" with "Katrina" in the song. The sadly obvious place to start this section is with that incredible devastation and heartbreaking fatality count unfortunately *revisited* upon the city of New Orleans and, less remembered at times, coastal Mississippi in August 2005. Attending a conference in March of that same year at the former Fairmont Hotel in New Orleans and sharing the table with a panel of Louisiana State University hurricane researchers, I was treated to a discussion, in no uncertain terms, of what was to come. They predicted that where we were sitting—on the second floor of the hotel—would be under several feet of water. The hotel was indeed flooded and was four years in renovation.[13] Katrina was the costliest hurricane in US history and among the five deadliest—and it must be remembered that this death toll occurred in an era of nearly universal mass communications when approximately 80 percent of people actually heeded evacuation warnings.

Louisiana State University and the Weather Channel produced an episode of *It Could Happen Tomorrow* about a major hurricane hitting New Orleans in April of the same year.[14] And seemingly everyone else knew what was coming, with the hindsight provided by Hurricane Betsy, so why did an estimated nearly 2,000 people die as a direct result of the storm in the two Delta states of Louisiana and Mississippi?[15] If this chapter is beginning to peg the Delta as the fatality-at-the-hands-of-nature capital of the United States, indeed, there does seem to be a stream of evidence coming to light. The death toll from Katrina may never be complete, and not just because of the people unaccounted for who were washed out to sea or drowned in hurricane storm surges and water from failed levees. As noted by the Social Science Research Council, which has been studying effects of the storm since it hit, the emotional and psychological devastation continued after the storm: "The true death toll changes by the moment. Suicide rates are at an epidemic level in New Orleans and elsewhere, but those grim statistics do not include those who die from too much alcohol or too many drugs, those who die because they no longer care enough to take proper care of themselves, those who die because the levels of stress or despair or pain have become intolerable."[16]

Worse, much of the death and devastation was evidently caused by humans; the BBC reported in November 2009 that the US Army Corps of Engineers was found guilty of negligence in the flooding of parts of New Orleans. The specter of the Delta as a region of poverty reared its head. As explained by the *San Francisco Chronicle*, "Hurricane Katrina's winds ripped away barriers that kept one city's poor out of sight and, for most people, out of mind. As the world watched, the deadly storm thrust the nation's enormous economic disparities into plain view."[17] A term used repeatedly to describe Katrina in its entirety—result, response, inability to release a death count—is "Third World" and is expressed with surprise and shock that such a thing could happen in the United States Delta.[18] Perhaps this offers some insight into why the climatology of the Delta is so deadly—what citizens here choose *not* to see on a day-to-day basis.

Katrina was no fluke; Mississippi has been hit by nineteen hurricanes since 1851, nine of which were major (over Saffir-Simpson Scale Category 3), one of which was one of three Category 5 storms to ever strike the US coast: Camille, in 1969, which claimed the lives of between 130 and 150 people (accounts vary) on the Mississippi gulf. Fortunately—if one can state this about a Category 5 hurricane—Camille's eye-wall was very small (only 12 miles across), with hurricane-force winds extending out only about 75 miles, while Katrina, "only" a Category 3 on impact, had hurricane-force winds extending out 125 miles or more. Katrina's central pressure was the third lowest ever recorded—right behind Camille. Nonetheless, Pass Christian and Long Beach in Harrison County were under about 22 feet of water during Camille's storm's surge, and winds at landfall were estimated to be over 190 miles per hour. Over 5,600 homes were destroyed in Louisiana and Mississippi; another nearly 15,000 suffered major damage.[19]

Again, was the extent of Camille's destruction a tragedy that could have been avoided? It seems that residents in Pass Christian were listening to New Orleans for storm reports, perhaps feeling they had dodged a bullet because the New Orleans reports were described as low-key and not alarming. Those to the east, listening to Biloxi stations, heard a "highly emotional plea to get out."[20] For whatever reason, only about one-half of residents who should have evacuated did so (81,000 out of 150,000) before low-lying bridges became swamped, blocking escape, well before landfall. The low evacuation numbers helped perpetuate the "hurricane party" myth. One iconic image from Camille is a photo of the destroyed Richelieu Apartments in Pass Christian, the site of a party supposedly held by twenty-three residents and whose deaths were reported on national news. According to Ben Duckworth, a resident who survived the destruction of the apartment building by hanging onto a tree limb in 200 mph winds, part of the time under water—a memory he would still rather forget—of those twenty-three people verified to have been at the apartment complex during the storm, only eight were identified as dead or missing. Even Mary Ann Gerlach, the woman who purportedly hosted the party and herself perpetuated the popular myth, had actually taken a nap and woke to find her apartment disintegrating around her.[21]

Louisiana, too, has suffered mightily, aside from Katrina. Audrey in 1957 was the deadliest event in Southwest Louisiana history (and seventh all-time in the United States) as an estimated 500 persons lost their lives in that storm. Taking almost the same path as Audrey, and comparable or worse in many ways, Rita in 2005 killed 7 people in Louisiana, a relatively happy note in an otherwise not-too-pleasant litany of lessons not seemingly learned. Gustav in 2008 also resulted in 7 fatalities, 2 in a hurricane-spawned tornado. The deadliest, however, took place long ago: the 1893 Chenier Caminada hurricane killed over 2,000. Perhaps the worst Delta place of all to spend hurricane season is Isle Dernière (now Isles Dernières, after the island was split into many islands by repeated hurricane hits). "Last Island" was hit three times in the 1800s: 45 people perished in 1812 and 1,500 perished in 1831. Finally, in 1856, when another 200 persons lost their lives, the island was abandoned. It is interesting to note that the 1856 storm demolished a luxury resort and most of those killed were wealthy planters.[22]

It may come as somewhat of a surprise that 181 Atlantic or Gulf of Mexico tropical cyclones have crossed the Delta in some form since 1852, when recordkeeping began (see fig. 6.5). Of those, sixty-three crossed through the totally landlocked Delta counties of Arkansas, Illinois, Kentucky, Missouri, and Tennessee. Many of these retained at least tropical-storm-force winds and some hurricane-force gusts. In recent times—between 1970 and 2007—80 percent of hurricane-related fatalities took place *outside* of the landfall county. These fatality figures exclude Katrina, although most of those fatalities also took place outside of the landfall county.[23] As recently as September 2008, Hurricane Gustav inundated the entire Delta region, from the Louisiana and Mississippi coasts through Arkansas and Missouri, bringing much of Arkansas 5 to 12 inches of rain and knocking out power to many, including the author. Barely two weeks later, Hurricane Ike produced several tornadoes and more heavy rain in Arkansas. The National Weather Service (NWS) in Springfield, Missouri, reported winds near West Plains of over 65 mph, while the Paducah NWS reported wind gusts of 78 mph at Grand Rivers, Kentucky; 75 mph at Calvert City, Kentucky; and 67 mph at Cairo, Illinois; all near or over hurricane force. The center of circulation of both storms remained well defined for an unusually long period.[24]

Inland hurricane hazards include some astonishing rainfall totals. Allison in 1989 inundated Arkansas with 13.91 inches of rain, and 29.92 inches fell in Louisiana. Georges produced 32.2 inches in Mississippi in 1998, causing worse damage than its landfall did at Pascagoula, Mississippi. Rita, which occurred a scant three weeks after Katrina, spawned tornadoes across the Delta states; and, as if Jackson, Mississippi, did not have enough "natural" tornadoes, land-falling hurricanes bring them during the less active months of August through October. The severe weather outbreak spawned from the outer edges of Hurricane Rita lasted about thirty-six hours; fifty-five tornadoes occurred across the Jackson NWS county warning area (CWA), making it the largest tornado outbreak in the CWA in recorded history. Of the fifty-five total tornadoes, one was rated an F3 and seven were rated F2. Tornado outbreaks are, unfortunately, not uncommon in the Delta during land-falling tropical

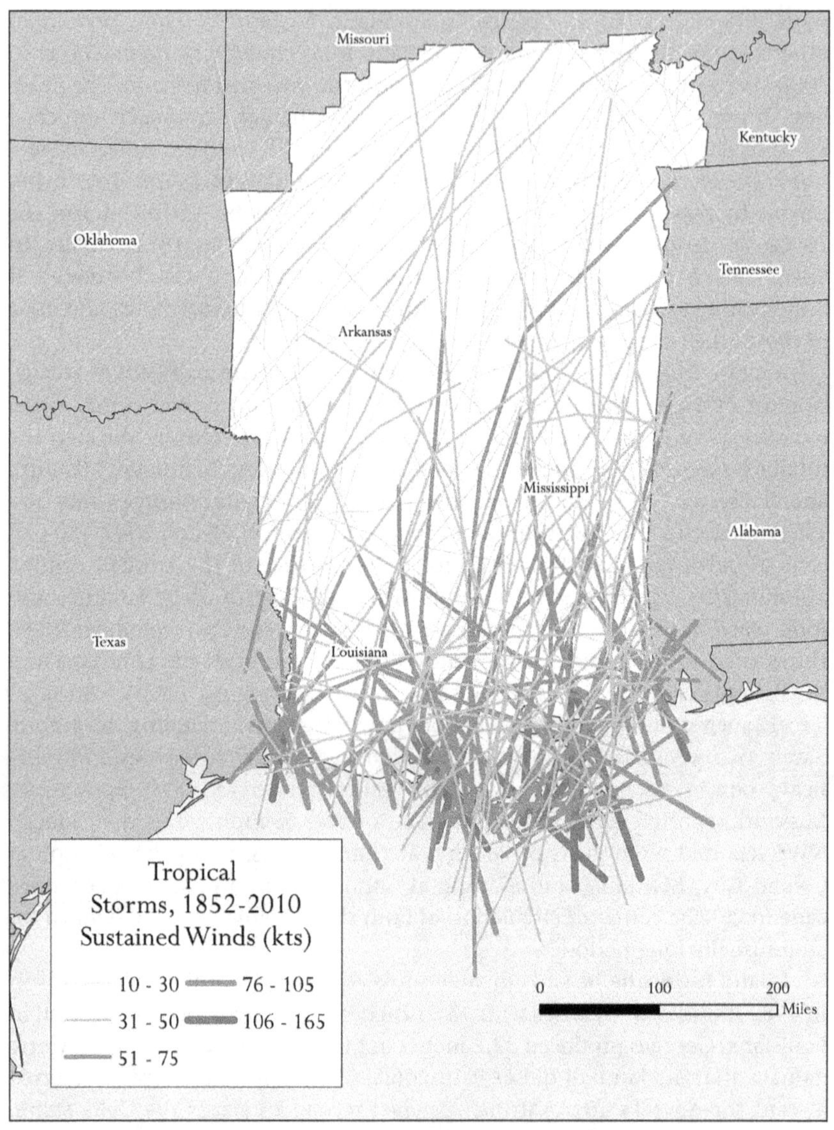

FIGURE 6.5. Tropical storm tracks affecting the Delta region, 1852–2010, graded by sustained winds along track in knots. *Hurricane data from National Hurricane Center; boundaries from US Census Bureau.*

cyclones; Hurricane Andrew produced twenty-six tornadoes across Mississippi in late August 1992.

Of ironic interest is also that in 2005 much of northern Louisiana and southwestern Arkansas was suffering from a severe drought (NWS Shreveport). Even Rita and Katrina could not alter the placement of cities such as El Dorado, Arkansas, and Monroe, Louisiana, in the top ten driest years in history.

Feast and Famine: Flood and Drought

Even more ironic is the fact that in 2011 Louisiana experienced the second driest year in history while at the same time the Mississippi River was experiencing a historic flood. The *Wall Street Journal* reported May 23 of that year that farmers having to move their cattle to higher ground ahead of floodwaters diverted into long-unused channels could find nothing there for them to eat; the earth was parched and dead. In a year filled with horrifying tornado disasters, *USA Today* reported that the drought of 2011, which was extreme to exceptional over one-half of Louisiana and Southwest Arkansas as well as other parts of the United States, was ostensibly the most costly.[25] Droughts and associated heat waves are often the least covered, most costly disasters befalling human beings; their effects are hard to summarize, unlike the discernible path of a tornado or hurricane center. It is estimated that the two costliest disasters in US history are droughts (1980 and 1988), and there is some indication that these droughts may be responsible for 10,000 heat- or drought-related deaths *each*.[26]

As noted, 2011 saw historic flooding on the Mississippi River. Riverine flooding is normally less associated with fatalities than is flash flooding due to its predictability and slow onset, but its economic costs can be terrible. In 2011, much of the Delta experienced near record March through May rainfall, following record snowfalls near the headwaters and along tributaries to the Mississippi. Delta towns experiencing amazing totals during April 1–May 5 include Cape Girardeau (24.89 inches); Poplar Bluff (24.04 inches), and West Plains, Missouri (18.68 inches). The NWS in Memphis shows that downtown Memphis, Dyersburg, and Jackson, Tennessee, all experienced totals over 15 inches, while their facility approached 18 inches. Paducah, Kentucky, was at nearly 22 inches, and Harrison and Mountain Home, Arkansas, collected totals of nearly 18 inches each. Shelby County, Tennessee, home to Memphis, experienced over $2 billion in damage from the flood; and Tunica, Mississippi, $1 billion alone. Many other counties along the river suffered damage as well. Natchez and Vicksburg, Mississippi, also set a new all-time flood crest on May 19; serious disruption of boat and car traffic due to closed bridges and dangerously low wiring occurred. There were 350 residences destroyed and 1,500 damaged in the Jackson warning area (which includes most of the "Arklamiss"); but, as noted above, due to lead times, only one person was killed in this historic flood.

But the flood that will always be the most memorable in the Delta is the 1927 Mississippi River flood, the "Great Flood." It was never supposed to happen, as massive levee systems were "guaranteed" to hold back the river. But five massive storms drenched the entire watershed in the spring of 1927. Within just ten days, one million acres were flooded with water ten feet deep, and water continued to flow through the gap for months. The flood lasted for two months and covered nearly seventeen million acres in seven states. At its widest point, just north of Vicksburg, Mississippi, the swollen river formed an inland sea nearly a hundred miles across.[27]

At its peak, the area under water equaled the size of the states of Massachusetts, Connecticut, New Hampshire, and Vermont combined. The "official" death toll was 500 persons; Christine Klein and Sandra Zellmer state that others believe over 1,000 perished in Mississippi alone. In Arkansas, thirty-six of seventy-five counties were under water, and nearly 100 people perished. Here again, it is likely not only that African Americans, made at gunpoint to work on the levees, were not counted but also that they were denied aid in many cases, due to fears that they would get "soft" or, worse, leave the area.[28]

On Easter 1979, the worst flood of record struck Jackson, Mississippi, breaking its previous record crest by over six feet following three months of above-average precipitation. The flood destroyed at least a thousand homes and displaced 17,000 people. Sadly, the 1979 Easter flood proved twice as costly as the previous flood of record in 1961, partially because developers, citing the safety of the new levees, built in the floodplain; this flood, too, was never supposed to happen.

Although flash flooding kills more people on average in the United States than any other weather-related phenomenon (many statistics are kept at the state level), it is difficult to extract information that is specific to the portions of the several states making up the northern Delta. From 1959 to 2005 (excluding the fatalities from Katrina), there were 85 fatalities in floods identified as flash floods in the states of Louisiana, Arkansas, and Mississippi, and 52 fatalities in those identified as river floods in the same area.[29] Only Mississippi ranks in the top ten for flood fatalities of all types (flash, river, tropical systems, and "unknown"). As is the case with almost all other climate-related fatalities, per-capita flood fatalities are high in Mississippi.

Ice Storms

When the rare blast of frigid air holds its ground at the surface and warm, moist Gulf air rains into it from just above, the result can be inches of ice coating trees, power lines, and roads. While not quite as dramatic as tornadoes and hurricanes, and not quite as frequent as in the northeastern United States (see fig. 6.6), there have been some notable ice storms in the Delta over the last sixty years. David Call notes that "ice storms in the south, while less common than elsewhere, are more likely to cause ice accumulations sufficient for severe damage."[30] Figure 6.6 shows "catastrophic" icing events, defined as those with over $1 million in property loss. Arkansas suffered two back-to-back in December of 2000, arguably the worst disaster in the state in recorded history, according to Little Rock NWS. Then-governor Mike Huckabee said of the storm, "We've never had a storm like this. . . . I'm using words like apocalyptic and cataclysmic."[31] At one point during the Christmas Day storm, 315,000 Arkansans were without power; as of New Year's Day, about 46,000 homes and businesses still remained without electricity.

Not to be outdone, Mississippi (and many other Delta areas such as southeastern Arkansas, western Tennessee, and northern Louisiana) suffered a catastrophic ice storm in February 1994, damage from which was compared to that of Category 5 Hurricane Camille in 1969. Curiously, the nation heard little about this storm, perhaps a sad commentary on, as noted earlier, an area more known for its poverty than its plantations. Paducah NWS described the January 2009 ice storm: "Life abruptly changed as a growing layer of ice

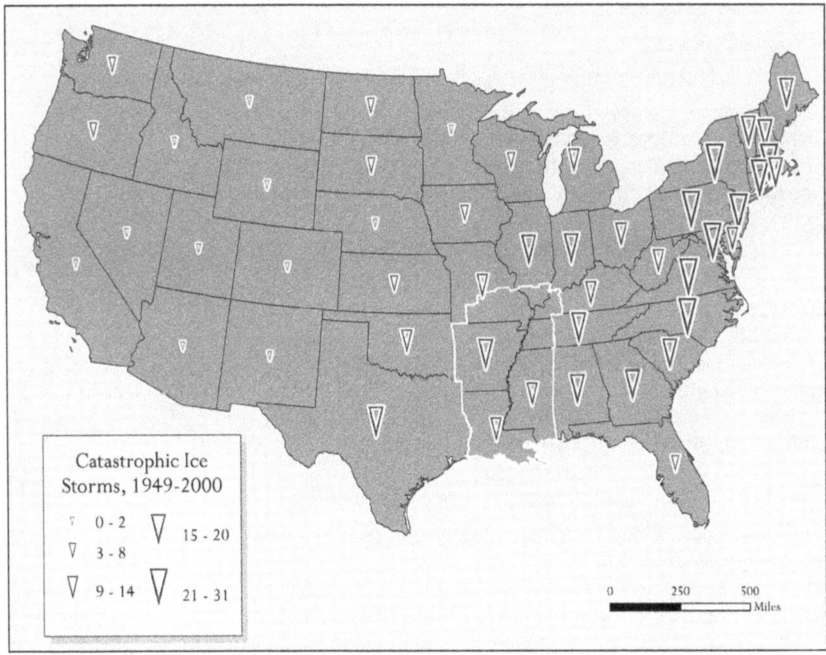

FIGURE 6.6. Catastrophic ice storm events, 1949–2000. "Catastrophic" is here defined as those ice storms with over $1 million in loss to property. *Ice storm information from David A. Call, "An Assessment of National Weather Service Warning Procedures for Ice Storms," Weather and Forecasting 24, no. 1 (2009): 104–20, map p. 105; boundaries from US Census Bureau.*

dragged trees and power lines to the ground. Lights and heat went out for days and even weeks. Such basics as food, water, and gas were difficult to obtain." Over 90 percent of southwestern Kentucky was without power from the storm on January 27; although by February 4 power was restored to most cities, some rural areas in western Kentucky and extreme southeastern Missouri were still facing weeks without power. There were at least fourteen fatalities either directly or indirectly related to the storm.

The Common Thread

Sadly, then, it seems what molds the Delta's climatology into a cohesive whole is weather-related fatality figures out of line with its population density, and often storm strength, across most of history. Proximity to the Gulf of Mexico and its warm, humid air and tropical storms, generally temperate winters that leave the Delta vulnerable and ill prepared when polar air does make its rare appearance, and, apparently, its inheritance of poverty collectively paint the Delta as a difficult and often deadly place to face what the climate has to offer.

Website References

NOAA (National Oceanic and Atmospheric Administration)

http://www.srh.noaa.gov/jetstream/tstorms/tstorms_intro.htm
http://www.srh.noaa.gov/jetstream/derechos/derecho_intro.htm
http://www.spc.noaa.gov/misc/AbtDerechos/papers/Ashley_2005.pdf
ftp://ftp.ncdc.noaa.gov/pub/data/papers/200686ams1.2nlfree.pdf

National Weather Service Sites (Historic Events Databases)

Jackson, Mississippi: http://www.srh.noaa.gov/jan/
Lake Charles, Louisiana: http://www.srh.noaa.gov/lch/?n=events
Little Rock, Arkansas: http://www.srh.weather.gov/lzk/
Memphis, Tennessee: http://www.srh.noaa.gov/meg/
Paducah, Kentucky: http://www.crh.noaa.gov/pah/
Shreveport, Louisiana: http://www.srh.noaa.gov/shv/events/
Springfield, Missouri: http://www.crh.noaa.gov/sgf/

Data Sources for Mapping and GIS Analysis

Tornado tracks: Storm Prediction Center Severe GIS Database, http://www.spc.noaa.gov/gis/svrgis/
Hurricane tracks: NOAA Coastal Services Center, http://www.csc.noaa.gov/hurricanes/#
Temperature and precipitation data: National Climate Data Center, US Climate Normals, http://cdo.ncdc.noaa.gov/cgi-bin/climatenormals/climatenormals.pl?directive=prod_select2&prodtype=CLIM81&subrnum
State boundaries courtesy of US Census Bureau (TIGER line files).

Notes

1. All maps were designed and produced by Mary Sue Passe-Smith using Esri's ArcGIS software. See D. H. Aldermanand and R. M. Beavers, "Heart of Dixie Revisited:

An Update on the Geography of Naming in the American South," *Southeastern Geographer* 39, no. 2 (1999): 190–205.

2. See Cara McCoy, "Vegas Lands in No. 3 Spot on Sweatiest Cities List," *Las Vegas Weekly*, July 2, 2009; "Old Spice Names Phoenix Sweatiest City in America in Sixth Annual Ranking of Summertime Perspiration," *Proctor and Gamble*, June 18, 2007; and "El Paso Named the Sweatiest City in America in Third Annual Old Spice Sweatiest City Study," *PR Newswire*, June 16, 2004.
3. Steve Tracton, "Tornado Alley? What's That?," *Washington Post*, April 26, 2010, http://voices.washingtonpost.com/capitalweathergang/2010/04/tornado_alley_wheres_that.html.
4. Kevin M. Simmons and Daniel Sutter, *Economic and Societal Impacts of Tornadoes* (Boston: American Meteorologist Society, 2011).
5. The per-capita figures here, in the interest of ease of calculation, use 2000 population figures for counties to divide by number of tornadoes. It would be preferable to divide each decade's tornadoes by each decade's population figures, but past research has shown that the results come out much the same. See, for example, Mary Sue Passe-Smith (master's thesis, University of Arkansas, 2004).
6. John Sims and Duane Baumann, "The Tornado Threat: Coping Styles of the North and South," *Science* 176 (1972): 1386–92.
7. See http://www.srh.noaa.gov/meg/?n=apr2011stormaccount, accessed April 5, 2015.
8. See http://www.crh.noaa.gov/pah/1982/report.php/, accessed April 5, 2015.
9. Simmons and Sutter, *Economic and Societal Impacts of Tornadoes*.
10. Ibid.
11. Storm data is available on the Mississippi National Weather Service website under the "Past Weather Events" links: http://www.srh.noaa.gov/jan/, accessed April 5, 2015.
12. Sam "Lightnin'" Hopkins, "Hurricane Betsy," from the album *Lightnin' Strikes* on Tradition Records, recorded October 4 and 5, 1965, in Los Angeles, California. Lyric from *Youtube* video, http://www.youtube.com/watch?v=8idFEcXIQ5Y, accessed September 30, 2014.
13. Ashley Nanco, "Katrina-Flooded Hotel Reopens in New Orleans," *Preservation*, June 18, 2009, http://www.preservationnation.org/magazine/2009/todays-news/katrina-flooded-hotel-reopens.html.
14. "The Weather Channel Achieves High Rating for 'Lost Episode' of It Could Happen Tomorrow," *The Weather Channel*, 2007, http://press.weather.com/press_archive_detail.asp?id=116.
15. The first estimate of fatalities of 1,335 came from a service assessment of Hurricane Katrina performed by NOAA, published in June of 2006, nearly ten months after the storm. The *Houston Chronicle* notes a figure of 1,464 in Louisiana alone in an article published five years after the event, in August 2010 (an estimate for all deaths of about 1,800) but also states that the true fatality count from the storm will never be known; there are still many missing and unaccounted for, unnamed, and unclaimed bodies, etc. See http://www.nws.noaa.gov/om/assessments/pdfs/Katrina.pdf, accessed April 5, 2015.
16. Kai T. Erikson and William R. Kenan, "Katrina, the Mighty—and Unending—Storm," *Social Science Research Council*, August 28, 2007, http://www.ssrc.org/features/view/katrina-the-mighty-and-unending-storm/.

17. Marc Sandalow,"Katrina Thrusts Race and Poverty onto National Stage/Bush and Congress under Pressure to Act," *San Francisco Chronicle,* September 23, 2005, http://www.sfgate.com/cgibin/article.cgi?f=/c/a/2005/09/23/MNG64ESMPQ1.DTL.
18. See Virginia R. Dominguez, "Seeing and Not Seeing: Complicity in Surprise," *Understanding Katrina: Perspectives from the Social Sciences,* June 11, 2006, http://understandingkatrina.ssrc.org/Dominguez/; and Lise Olsen, "5 Years after Katrina, Storm's Death Toll Remains a Mystery," *Houston Chronicle,* August 30, 2010, http://www.chron.com/news/nation-world/article/5–years-after-Katrina-storm-s-death-toll-remains-1589464.php.
19. Roger Pielke, Chantal Simonpietri, and Jennifer Oxelson, *Thirty Years after Hurricane Camille: Lessons Learned, Lessons Lost* (Boulder, CO: National Center for Atmospheric Research, 1999).
20. Dan Ellis, *All about Camille: The Great Storm of '69* (North Charleston, SC: CreateSpace, 2010).
21. Ibid.
22. David Roth, *Louisiana Hurricane History* (Camp Springs, MD: National Weather Service, 2010).
23. Jeffrey Czajkowski, Kevin Simmons, and Daniel Sutter, "An Analysis of Coastal and Inland Fatalities in Landfalling US Hurricanes," *Natural Hazards* 59, no. 3 (2011): 1513–31.
24. See http://www.crh.noaa.gov/news/display_cmsarchive.php?wfo=sgf, accessed April 5, 2015.
25. Doyle Rice, "USA's Drought Costs Exceed $10 Billion," *USA Today,* November 11, 2011, http://www.usatoday.com/weather/drought/story/2011-11-10/drought-south-midwest/51159668/1.
26. Neal Lott and Tom Ross, "Tracking and Evaluating Billion Dollar Weather Disasters, 1980–2005," February 2006, ftp://ftp.ncdc.noaa.gov/pub/data/papers/200686ams1.2nlfree.pdf.
27. Christine A. Klein and Sandra B. Zellmer, "Mississippi River Stories: Lessons from a Century of Unnatural Disasters," *Southern Methodist University Law Review* 60, no. 4 (2007): 1471–1538.
28. Stephen Ambrose, "Great Flood," *National Geographic,* May 1, 2001, http://news.nationalgeographic.com/news/2001/05/0501_river4.html.
29. Sharon T. Ashley, and Walker S. Ashley, "Flood Fatalities in the United States," *Journal of Applied Meteorology and Climatology* 47 (2008): 805–18.
30. David A. Call, "An Assessment of National Weather Service Warning Procedures for Ice Storms," *Weather and Forecasting* 24, no. 1 (2009): 104–20.
31. "Ice Storm Victims 'Tough It Out,'" *Houston Chronicle,* December 30, 2000, http://www.chron.com/CDA/archives/archive.mpl/2000_3270276/ice-storm-victims-tough-it-out.html.

CHAPTER 7

Sociodemographic Snapshots of the Mississippi Delta

JOHN J. GREEN, TRACY GREEVER-RICE, AND GARY D. GLASS JR.

Framing the Mississippi Delta

The region of the United States known as the Mississippi Delta represents the intersection of people and space.[1] This intersection encompasses an overlay of cultural, social, demographic, and economic characteristics within a set of geographic and political boundaries. Over time, what marks the Delta from the non-Delta has been fluid—affected by the legacy of the plantation economy, federal and state economic and social policy, the evolution of cultural norms, and engineered change in the river system itself. Although residents up and down the Delta share similarities when compared to citizens of non-Delta counties in their respective states, differences within the Delta exist as well. Both of these dimensions—similarity and variation—are important in defining the Delta. And, when trying to better comprehend the forces shaping development in the region, it is necessary to use both of these vantage points.

To delineate the region for this sociodemographic exploration of Delta people, we focus on the combination of population and spatial characteristics. Broadly defined, the Delta consists of the relatively flat and highly fertile lands bordering the Mississippi River. This alluvial plain includes a wide range of tributaries to the broader Mississippi River system. With opportunities for agricultural production and river transport, the region was transformed for intensive crop production, especially cotton and, later, rice, primarily producing for export markets.[2] This trajectory influenced the people who came to reside in the region and the socioeconomic institutions that shaped their lives.

As a way of framing the snapshots presented here, we operationalize the Delta as those 231 counties and parishes located in the vicinity of the Mississippi River across seven states. Specifically, we include those counties that are part of the Delta as defined by the Delta Regional Authority—a federal- and state-level development partnership—in Illinois, Kentucky, and Missouri (the upper Delta), and Arkansas, Louisiana, Mississippi, and Tennessee (the lower Delta).[3] In doing so, we focus attention on population counts and change, racial composition, educational attainment, economic position, and health factors. Additionally, we draw upon insights derived from various case studies conducted within a particular subregion of the Delta, the so-called Core Delta counties in the state of Mississippi.

Choosing a Conceptual Lens

For this review and analysis, we use the "livelihoods framework" to explore and interpret the development trajectory of the Delta and the influences this has had on people and spaces, thus providing a conceptual lens for the snapshots. Focused on intentional development pathways and both their anticipated and unanticipated outcomes, the livelihoods framework directs attention to opportunities and barriers in accessing and utilizing material (natural and built), cultural, social, and economic forms of capital, especially in relation to security and well-being at the individual, household, and community levels.[4] This is similar to the community capitals framework, although the latter focuses specific attention toward these forms of capital at a more collective level and with consideration of the interaction between different actors.[5] Furthermore, informed by ecological and sustainability models, the sustainable livelihoods variant to the framework considers the ways in which particular arrangements lead to greater vulnerability or, conversely, resiliency in the face of short-term shocks and long-term stressors. We add to this by attending to the overlay and intersection of sociodemographic characteristics, economic conditions, and health outcomes, helping us to consider the ways in which population characteristics vary across space.[6] Addressed in the community capitals literature, potential exists for different development arrangements for places to "spiral up" toward greater assets or "spiral down" to fewer assets.[7]

As articulated by Leo de Haan, the livelihoods framework is ideally an action-oriented approach, focusing on the interplay between agency and structure at the local and global levels.[8] In this chapter, however, we focus primarily on the structural and institutional forces influencing broad regional development that help to shape, and which are shaped by, people's actions. We use extant literature to inform this discussion combined with sociodemographic data to illustrate the patterns resulting from these development trends.

Historical Sociodemographic Snapshots

The Delta as a whole was developed largely for the purposes of agricultural production. With the history of rivers flooding and the deep, rich soil deposits left behind, coupled with high rainfall, the lands bordering the Mississippi River and its tributaries were productive for agriculture, especially intensive cotton production. Once cleared of their thick forests and marshes, crops thrived on the relatively flat lands.[9]

The Mississippi River system was both a blessing and a curse to the agricultural economy. It provided the basis for fertility and access to shipping channels, yet flooding was a perpetual challenge. Controlling rivers and their impacts dominated much of the political economy of the Delta, and those experiences had far reaching impacts on broader development policy throughout the country.[10] In many ways, flood control to maximize agricultural production became the predominant focus for development, and this framed people's understanding and measurement of development, most notably after the great flood of 1927. Although numerous levies and other control mechanisms have been put in place, flooding along the Mississippi River continues to impact the region, as seen in the early 1990s and the 2010s.

Within the United States, there have been three dominant models of agricultural production—family farms, plantation-style sharecropping and tenant farming, and corporate farming.[11] Based on proximity, farming in the upper Delta reflected the Midwestern model of family farming typical of that broader region, whereby family members owned and controlled the business and provided the bulk of the labor. However, with wide tracts of land and the possibility of growing southern crops (such as cotton), some areas adopted features of agricultural production more similar to the lower Delta. That agricultural system was heavily influenced by the broader southern region and the dominance of plantation-style production, first rooted in slavery and later incorporating sharecropping and tenancy systems. Plantation systems are organized hierarchically with most of the decision-making power resting with the plantation owner and his managers, highly specialized with a few number of commodity crops, labor (and later technology) intensive, and require high capitalization.[12] In *The Cotton Plantation South since the Civil War,* Charles Aiken provides the following comparison between agricultural systems and regions:

> Primarily on the basis of the differences in agriculture, a Lowland South with its emphasis on plantation crops evolved distinct from an Upland South dominated by white yeoman farmers. The Lowland South had larger landholdings and a larger black population and emphasized the great staples—rice, tobacco, sugarcane, and cotton. Not only did the Upland South have smaller landholdings and fewer blacks, but grains, especially corn and wheat, and livestock, primarily hogs, cattle, horses, and mules, were the basis of its husbandry.[13]

As noted in the previous quote, blacks were brought into the region to provide labor for the plantation system, thus setting the basis for African Americans' presence throughout much of the region today.[14] The overlap of race and class position led to protracted racial exploitation and fueled the struggle for civil rights. Although a majority of blacks provided labor through tenant-farming sharecropping, some went on to lead family farms.[15]

As with other regions and nations with economies rooted in plantation systems, reliance on this exploitative system shaped much of all forms of subsequent development—sociodemographic, economic, and cultural. From his international analysis, including parts of the American South, George Beckford notes,

> Plantation economy never gets beyond the stage of underdevelopment. For within the system itself there are structural factors which impede further economic progress for plantation society as a whole. . . . The institutional arrangements that exist contribute to perpetuation of a continued state of underdevelopment.[16]

In the face of these experiences, a culture was formed with a sense of place identity, as identified by James Cobb in *The Most Southern Place on Earth*.[17] This included strong attachment to place, unique foodways, and special music in the form of the blues. Interestingly, these cultural attributes arose from the tensions caused by the plantation system and racial stratification.

Consistent with the evolution of agricultural production globally, the Delta experienced waves of increased mechanization resulting in an agricultural production shift from being labor intensive to being capital and technology intensive. With these changes, labor became redundant, pushing people from the region and the agricultural sector.[18] Beginning in the early twentieth century, the pull of better opportunities in northern industrial cities led to many residents leaving the Delta in the so-called Great Migration.[19]

In contemporary times, other than crop specification, upper and lower Delta farms operate similarly. They tend to consist of large acreages, produce commodity crops, and rely on complex technologies and capital arrangements, with farm managers playing important roles. Smaller-scale farmers also exist, including those producing vegetables, fruits, and livestock.[20] Still, these are fewer and farther between than in the non-Delta sections of the southern states.

Manufacturing did take hold in some rural areas of the Delta region, but it did not evolve to dominate the economic system, and most manufacturing industries that emerged continued to be modeled on a low-skill, low-wage framework. The early stages of this path are outlined by James Cobb in his description of industrial development in the first half of the twentieth century across the broader southern region, encompassing but not limited to the Delta: "With industry still growing slowly, merchants and professionals in the early twentieth-century South retained strong ties to agriculture, and thus offered no significant challenge to the region's planter class."[21]

Although there are exceptions to the generalization, especially in urban areas (specifically, Memphis, Tennessee, but ranging from St. Louis, Missouri, to New Orleans, Louisiana), the agricultural base of the Delta region created a culture of local governance that eschewed investment in basic education, health, and other social infrastructure. This lack of investment resulted in structural underdevelopment whereby the economy lacked diversity and became stagnant, low levels of household income and both high inequality and poverty rates prevailed, and the region was increasingly isolated.[22] These challenges led to lower quality of life and poorer health outcomes, setting the region apart on a variety of measures, including disparities in life expectancy between Delta and non-Delta places.[23]

Taking the long view, since 1910, Louisiana is the only state to experience greater population growth in Delta than in non-Delta counties. Most Mississippi and Arkansas counties have experienced a similar pattern of population change, regardless of their Delta status, while Illinois, Kentucky, Missouri, and Tennessee have experienced far greater population growth in their respective non-Delta counties (figs. 7.1 and 7.2).

Furthermore, the limited educational and economic opportunities in the region served to push people in search of a better life, and they have been pulled by larger cities in the region, but even more so, by cities outside of the Delta region. As can be seen in figures 7.3 and 7.4, the Delta counties experienced greater overall population declines between the years 2000 and 2010 than did non-Delta counties. Investigation of county-level net migration

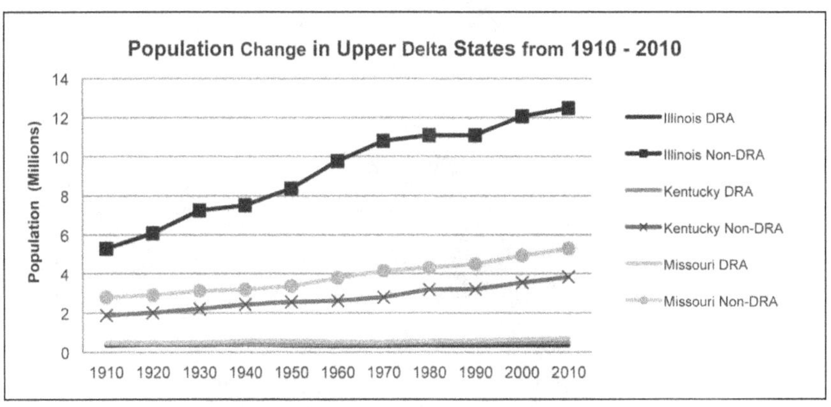

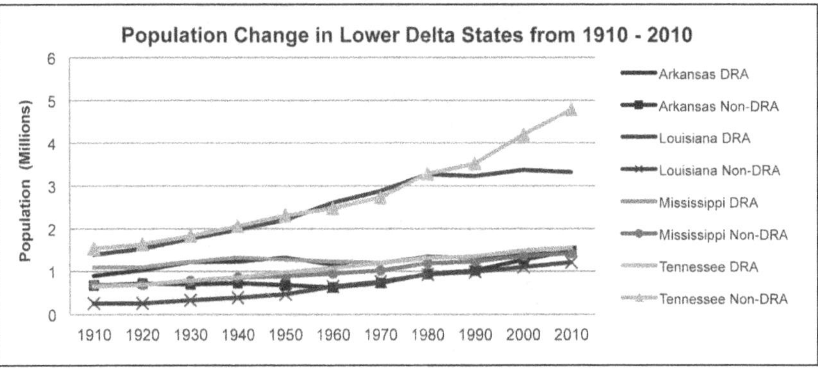

FIGURES 7.1 AND 7.2. Population change by Delta status and state (1910–2010). *US Census Bureau, 1910 through 2010 Decennial Censuses. Graphs created by the University of Missouri Office of Social and Economic Data Analysis.*

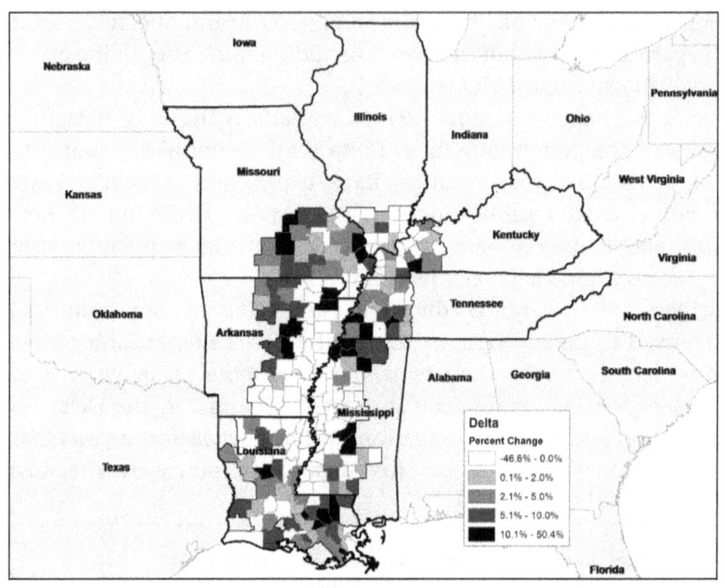

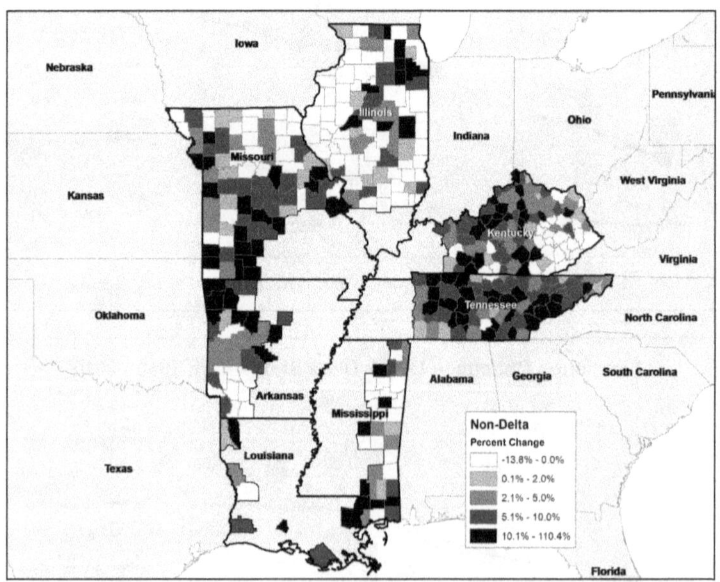

FIGURES 7.3 AND 7.4. Percentage population change in the Delta and non-Delta regions. *US Census Bureau, 2000 and 2010 Decennial Censuses. Map created by the University of Missouri Office of Social and Economic Data Analysis.*

patterns (i.e., population change due specifically to migration, controlling for births and deaths) by other researchers shows that a majority of the Delta counties faced outmigration from 2000 to 2010, and the rates were particularly high for people between twenty-five and twenty-nine years of age.[24] Throughout the Delta states, outmigration patterns continue to reflect the century-long depopulation of rural America.

Contemporary Sociodemographic Snapshots

With the historical narrative presented above serving as a backdrop, here we focus attention on the contemporary sociodemographic characteristics of Delta people. This exploration is primarily based on secondary data from the US Decennial Census (2000 and 2010), American Community Survey (ACS, five-year estimates, 2008–2012), and National Center for Health Statistics (2005–2011) data reported through the County Health Rankings and Roadmaps program. Comparisons are made at several different levels, including Delta counties (n=231) to non-Delta counties (n=424) aggregated in each state. In addition to description, the sociodemographic characteristics presented here were selected based on their predictive value as indicators of personal and community well-being.[25]

Today, Louisiana has the greatest number of people living in its Delta region, followed by Tennessee, Mississippi, and Arkansas (table 7.1), all in the lower Delta subregion. As a percentage of the total state populations, nearly three-quarters of Louisiana's population live in Delta parishes, while approximately half of the populations of Mississippi and Arkansas live in Delta counties.

TABLE 7.1: Population count and percentage of total population by Delta status and state (2010)

	DELTA COUNTIES		NON-DELTA COUNTIES		STATE
	% of state population	Population count	% of state population	Population count	Population count
UPPER DELTA					
Illinois	2.69	344,594	97.31	12,486,038	12,830,632
Kentucky	11.52	500,107	88.48	3,839,260	4,339,367
Missouri	11.49	688,122	88.51	5,300,805	5,988,927
LOWER DELTA					
Arkansas	49.09	1,431,528	50.91	1,484,390	2,915,918
Louisiana	73.34	3,324,911	26.66	1,208,461	4,533,372
Mississippi	52.11	1,546,272	47.89	1,421,025	2,967,297
Tennessee	24.62	1,562,650	75.38	4,783,455	6,346,105

Source: US Census Bureau, 2010 Decennial Census. Table created by the University of Missouri Office of Social and Economic Data Analysis.

Delta counties are home to a greater percentage of racial and ethnic minorities in four of the seven states considered in this analysis, when compared with their non-Delta counterparts (table 7.2). The minority population primarily consists of African Americans.

TABLE 7.2: Minority and non-minority (white, non-Hispanic) population by Delta status and state (2010)

	DELTA COUNTIES		NON-DELTA COUNTIES		STATE	
	Minority (%)	Non-minority (%)	Minority (%)	Non-minority (%)	Minority (%)	Non-minority (%)
UPPER DELTA						
Illinois	10.07	89.93	24.28	75.72	23.89	76.11
Kentucky	11.60	88.40	11.01	88.99	11.08	88.92
Missouri	7.36	92.64	16.63	83.37	15.56	84.44
LOWER DELTA						
Arkansas	28.69	71.31	15.08	84.92	21.76	78.24
Louisiana	38.02	61.98	30.04	69.96	35.89	64.11
Mississippi	48.02	51.98	31.27	68.73	40.00	60.00
Tennessee	43.14	56.86	13.77	86.23	21.00	79.00

Source: US Census Bureau, 2010 Decennial Census. Table created by the University of Missouri Office of Social and Economic Data Analysis.

For the most part, the Delta and non-Delta county populations in Delta states achieve comparable levels of educational attainment. Table 7.3 presents the percentage of adults twenty-five years of age and older with and without a baccalaureate degree. The Delta counties in Missouri are an exception to the general pattern, with the Delta region having substantially lower educational attainment. Attainment of a baccalaureate at the personal level is a strong predictor of lifetime earnings potential. Generally, at the population level, communities with higher levels of educational attainment are more successful in creating and attracting higher wages and higher-skilled jobs.[26]

Income (table 7.4) and poverty (table 7.5) are particularly important measures of a region's well-being. Here we see that families in the Delta are less likely to have incomes at or above the medians for their respective states than those in non-Delta spaces (five of seven states); however, the differences are not pronounced. Much of the Delta has higher rates of poverty among families with young children, with differences of five percentage points or more in five of the seven states. Additionally, the Delta counties have higher rates than do their non-Delta counterparts of low-birth-weight infants (table 7.6), an important predictor of the risk of developmental delays and school readiness.[27]

TABLE 7.3: Educational attainment of population 25 years and older by Delta status and state (2008–2012)

	DELTA COUNTIES		NON-DELTA COUNTIES		STATE	
	Less than baccalaureate (%)	Baccalaureate and higher (%)	Less than baccalaureate (%)	Baccalaureate and higher (%)	Less than baccalaureate (%)	Baccalaureate and higher (%)
UPPER DELTA						
Illinois	69.26	30.74	68.57	31.43	68.92	31.08
Kentucky	84.17	15.83	78.35	21.65	79.02	20.98
Missouri	85.39	14.61	72.76	27.24	74.23	25.77
LOWER DELTA						
Arkansas	80.76	19.24	79.65	20.35	80.20	19.80
Louisiana	78.86	21.14	76.74	23.26	78.58	21.42
Mississippi	79.54	20.46	81.57	18.43	80.03	19.97
Tennessee	76.67	23.33	76.46	23.54	76.51	23.49

Source: US Census Bureau, American Community Survey, 2008–2012, estimates (margins of error not shown here). Table created by the University of Missouri Office of Social and Economic Data Analysis.

TABLE 7.4: Estimated median family income ($) by Delta status and state (2008–2012)

	DELTA COUNTIES	NON-DELTA COUNTIES	STATE
UPPER DELTA			
Illinois	69,778	70,991	70,144
Kentucky	53,631	54,475	53,833
Missouri	57,833	61,748	59,395
LOWER DELTA			
Arkansas	49,925	52,070	50,453
Louisiana	55,648	60,856	56,047
Mississippi	48,329	48,417	48,300
Tennessee	54,852	55,453	54,737

Source: US Census Bureau, American Community Survey, 2008–2012 Estimates, with additional calculations provided by the University of Missouri Office of Social and Economic Data Analysis (margins of error not shown here). Table created by the University of Missouri Office of Social and Economic Data Analysis.

TABLE 7.5: Households in poverty with children under five years of age by Delta status and state (2008–2012)

	DELTA COUNTIES		NON-DELTA COUNTIES		STATE
	Number	Percentage	Number	Percentage	Number
UPPER DELTA					
Illinois	14,003	31.81	119,245	20.68	590,639
Kentucky	6,927	29.03	50,616	28.51	57,543
Missouri	10,212	33.38	57,509	23.25	67,721
LOWER DELTA					
Arkansas	21,528	32.98	29,061	27.56	41,445
Louisiana	46,868	30.49	14,098	23.48	60,966
Mississippi	25,014	34.34	22,567	32.66	47,581
Tennessee	22,692	31.04	55,622	25.58	78,314

Source: US Census Bureau, American Community Survey, 2008–2012, estimates (margins of error not shown here). Table created by the University of Missouri Office of Social and Economic Data Analysis.

TABLE 7.6: Percentage of live births with low birth weight (< 2,500 grams) by Delta status and state (2008–2011)

	DELTA COUNTIES	NON-DELTA COUNTIES	STATE
UPPER DELTA			
Illinois	8.34	8.42	8.42
Kentucky	9.01	9.12	9.11
Missouri	8.78	7.96	8.05
LOWER DELTA			
Arkansas	10.15	7.96	9.04
Louisiana	11.18	10.85	11.03
Mississippi	12.82	11.22	12.07
Tennessee	10.61	8.79	9.28

Source: University of Wisconsin Population Health Institute, *County Health Rankings and Roadmaps*, 2014, based on 2008–2011 data from the National Center for Health Statistics, www.countyhealthrankings.org (margins of error not shown here). Table created by the University of Missouri Office of Social and Economic Data Analysis.

To summarize the contemporary characteristics of the Delta region, we find that it has generally lost population, with the historically mostly rural Delta counties losing relatively more population than their non-Delta counterparts; it is home to a population more likely to live in poverty and faces challenges to health, well-being, and opportunity. However, generalizations provide only the big picture. When viewed close-up, the variations of life opportunities and outcomes between and within Delta counties across states emerge like the image of a freshly processed photograph. The most marked differences are between the upper and lower Delta subregions.

Zooming in for Snapshots in the Core Delta

Few areas of the Delta region have been studied as much as that portion of Mississippi often referred to as the Core Delta. There are eleven counties that fall fully in the flood plain of the Mississippi and Yazoo Rivers.[28] The region is commonly known for intensive large-scale agricultural production, rural isolation, limited educational attainment, low median household incomes, and high rates of poverty, with pronounced racial inequality. In recent years, much attention has been directed toward the interplay between race, education, and poverty in the region.[29] Further concern has been focused on how these and other factors influence poor health outcomes.[30]

In addition to the macro trends influencing the area as previously described, there are local challenges that continue to plague development in the Core Delta region, especially in regard to leadership and the inflexibility of cultural and institutional arrangements. Writing about one of the counties in the region in the mid-twentieth century, Frank Alexander argued,

> There are two dominant features of the culture of the people . . . one is Negro-white relations, the other is the plantation system of farming. Almost every phase of the people's thoughts and behavior is influenced by these two complexes. Schools, churches, families, law enforcement, public welfare, earning a living are all under the domination of the plantation economy. Similarly all of these institutions and activities are carried on within the definitions of white supremacy and racial segregation.[31]

With the history of negative race relations, there are more recent research accounts of white and black leaders not working together effectively. Even in the context of African American gains in political offices, whites continue to hold economic power and development initiatives are often viewed as race-specific.[32] Race relations continue to negatively influence institutional arrangements. As Mark Harvey argued from data obtained through evaluation of a multiyear comprehensive community change initiative in the Core Delta,

> The fact that the civic sphere of the Delta, i.e. the sphere targeted for intervention by consensus initiatives, is split along a racial line that neatly tracks the line dividing institutionalized political power from economic power suggests that scholars of rural poverty, social capital, and community development must re-examine the relationships among civic capacity, institutional power structures, and community with specific attention to race.[33]

Interestingly, in response to the trajectory of underdevelopment and inequality, numerous community-based efforts arose over time to improve livelihoods and serve as models for the nation as a whole. Among these was the community health center model. Building on work from the civil rights movement and elaboration from a model observed in South Africa, activists, cooperative organizers, and physicians from across the country worked together to establish the nation's first rural community health center in the Core Delta town of Mound Bayou.[34] This legacy of community-based initiatives continues throughout the region today, with numerous examples of collective action to improve education, create employment opportunities, and improve public health.[35] Although these instances of collective action tend to be localized, severely under-resourced, and vulnerable to broader economic forces, they may help to inform broader programmatic and policy-level initiatives to improve livelihood security under unfavorable conditions.[36]

Discussion and Looking toward the Future

From a sociodemographic perspective and the exploration of both historical trends and contemporary data, it is apparent that the area we know commonly as the Mississippi Delta has somewhat different characteristics relative to its counterpart across the respective states. These include population loss, especially due to out-migration, coupled with poorer economic conditions and health outcomes. For the most part, these represent the challenges associated with a history of underdevelopment and limited livelihood opportunities. These are wide-angle snapshots.

Zooming in closer with a more fine-grained focus, we see multiple "Deltas" with variation between them. In developmental terms, they range across a continuum of decline and vulnerability at one extreme and improvement and hope at the other. Delta counties' divergent developmental trajectories, especially between the upper and lower areas of the region, have produced a pattern where some places are spiraling up toward better outcomes while others remain stagnant or are spiraling down. Although a great deal of scholarship has focused on separating the Delta as a unique region, similar to efforts in noting the distinguishing characteristics of the Appalachian and Black Belt regions, it is now time to focus more attention on the variations internal to the region. Doing so will help us to understand the region with greater depth and clarity while also informing development initiatives as to what accounts for those differences. What combinations of structure and agency influence the opportunities for some places to advance while others continue to face challenges? To what extent do institutional arrangements shape the development pathways open to individuals and their organizations?

This shift toward multiple foci represents the maturing of Delta studies as a cross-disciplinary and inter-professional field, with some scholars looking toward what defines the Delta region from a macro perspective and others focusing on the existing meso- and micro-level patterns. In looking toward the future, we recommend specific attention to comparisons of the upper Delta and lower Delta, rural to suburban and urban areas, and comparisons of different approaches to development with the intent of informing program and policy evaluation efforts. Furthermore, we strongly encourage more attention to de Haan's advice to connect quantitative and qualitative data and to pursue meta-studies across localized case studies to advance livelihoods theory.[37] This would contribute to our taking more detailed and useful snapshots for the Delta region.

Notes

The authors would like to thank Amy Greer and Lance Huntley for their assistance with the research presented in this chapter.

1. John J. Green and Debarashmi Mitra, "Intersections of Development, Poverty, Race, and Space in the Mississippi Delta in the Era of Globalization: Implications for Gender-Based Health Issues," in *Poverty and Health in America,* ed. Kevin Fitzpatrick (Westport, CT: Praeger Publishers, 2013).
2. Mikko Saikku, *This Delta, This Land: An Environmental History of the Yazoo-Mississippi Floodplain* (Athens: University of Georgia Press, 2005).
3. The Delta Regional Authority (DRA) also includes parts of Alabama in its operating area, but those counties are excluded here because they are not part of the more commonly understood "Delta region" given their distance from the Mississippi River.
4. Leo J. de Haan, "The Livelihood Approach: A Critical Exploration," *Erdkunde* 66, no. 4 (2012): 345–57; Ian Scoones, "Livelihoods Perspectives and Rural Development," *Journal of Peasant Studies* 36, no. 1 (2009): 171–96.
5. Cornelia Flora and Jan Flora, *Rural Communities: Legacy and Change,* 3rd ed., (Boulder, CO: Westview Press, 2007); Isabel Gutierrez-Montes, Mary Emery, and Edith Fernandez-Baca, "The Sustainable Livelihoods Approach and the Community Capitals Framework: The Importance of System-Level Approaches to Community Change Efforts," *Community Development* 40, no. 2 (2009): 106–13.
6. Green and Mitra, "Intersections of Development, Poverty, Race, and Space."
7. Mary Emery and Cornelia Flora, "Spiraling Up: Mapping Community Transformation with Community Capitals Framework," *Community Development* 37, no. 1 (2006): 19–35.
8. De Haan, "The Livelihood Approach."
9. Saikku, *This Delta, this Land.*
10. John Barry, *Rising Tide: The Great Flood of 1927 and How It Changed America* (New York: Simon & Schuster, 1998); *Karen* M. O'Neill, *Rivers by Design: State Power and the Origins of U.S. Flood Control* (Durham, NC: Duke University Press, 2006).
11. Max Pfeffer, *"Social Origins of Three Systems* of Farm Production in the United States," *Rural Sociology* 48, no. 4 (1983): 540–62.
12. George L. Beckford, *Persistent Poverty: Underdevelopment in Plantation Economies of the Third World* (New York: Oxford University Press, 1972), 210–11; Charles S. Aiken, *The Cotton Plantation South since the Civil War* (Baltimore, MD: The Johns Hopkins University Press, 1998).
13. Aiken, *The Cotton Plantation South,* 7.

14. John J. Green, "The Status of African Americans in the Rural United States," in *Rural America in a Globalizing World: Problems and Prospects for the 2010s,* ed. Conner Bailey, Leif Jensen, and Elizabeth Ransom (Morgantown: West Virginia Press, 2014).
15. John J. Green, Eleanor M. Green, and Anna M. Kleiner, "From the Past to the Present: Agricultural Development and Black Farmers in the American South," in *Cultivating Food Justice: Race, Class, and Sustainability,* ed. Alison H. Alkon and Julian Agyeman (Cambridge, MA: MIT Press, 2011).
16. Beckford, *Persistent Poverty.*
17. James C. Cobb, *The Most Southern Place on Earth: The Mississippi Delta and the Roots of Regional Identity* (New York: Oxford University Press, 1992).
18. Gilbert Fite, *Cotton Fields No More: Southern Agriculture, 1865–1980* (Lexington: University Press of Kentucky, 1984).
19. Stuart E. Tolnay, "The African American 'Great Migration' and Beyond," *Annual Review of Sociology* 29 (2003): 209–32.
20. 2012 Census of Agriculture, United States Department of Agriculture, National Agricultural Statistics Service (http://www.agcensus.usda.gov/Publications/2012/).
21. James C. Cobb, *Industrialization and Southern Society, 1877–1984* (Chicago, IL: Dorsey Press, 1988), 27–28.
22. Green and Mitra, "Intersections of Development, Poverty, Race, and Space."
23. Arthur G. Cosby and Diana M. Bowser, "The Health of the Delta Region: A Story of Increasing Disparities," *Journal of Health and Human Services Administration* 31, no. 1 (2008): 58–71.
24. Richelle Winkler, Kenneth M. Johnson, Cheng Cheng, Jim Beaudoin, Paul R. Voss, and Katherine J. Curtis, "Age-Specific Net Migration Estimates for US Counties, 1950–2010," Applied Population Laboratory, University of Wisconsin-Madison, 2013, http://www.netmigration.wisc.edu/, accessed July 12, 2014.
25. Ed Diener and Eunkook Suh, "Measuring Quality of Life: Economic, Social, and Subjective Indicators," *Social Indicators Research* 40, nos. 1–2 (1997): 189–216.
26. [7] Anthony P. Carnevale, Stephen J. Rose, and Ban Cheah, *The College Payoff: Education, Occupations, Lifetime Earnings* (Washington, DC: Georgetown University, Center on Education and the Workforce, 2011).
27. [8] Nancy E. Reichman, "Low Birth Weight and School Readiness," *Future of Children* 15, no. 1 (2005): 90–116.
28. These core counties in Mississippi are Bolivar, Coahoma, Humphreys, Issaquena, Leflore, Quitman, Sharkey, Sunflower, Tallahatchie, Tunica, and Washington.
29. Cynthia M. Duncan, *Worlds Apart: Why Poverty Persists in Rural America* (New Haven, CT: Yale University Press, 1999).
30. John Green, Kathleen Kerstetter, and Albert Nylander, "Socioeconomic Resources and Self-Rated Health: A Study in the Mississippi Delta," *Sociological Spectrum* 28, no. 2 (2008): 194–212.
31. Frank Alexander, "Summary of Coahoma County, Mississippi, Reconnaissance Report" (US Department of Agriculture, Bureau of Agricultural Economics, 1944), 17.
32. Alan W. Barton and Sarah J. Leonard, "Incorporating Social Justice in Tourism Planning: Racial Reconciliation and Sustainable Community Development in the Deep South," *Community Development* 41, no. 3 (2010): 298–322; Mark Harvey, "Consensus-Based Community Development, Concentrated Rural Poverty, and

Local Institutional Structures: The Obstacles of Race in the Lower Mississippi Delta," *Community Development* 44, no. 2 (2013): 257–73; Albert B. Nylander III, "Rural Community Leadership Structures in Two Delta Communities" (diss., Mississippi State University, 1998); Michael J. Stovall, "Stability of and Differences in Black and White Leadership Structures over Time in Two Rural Mississippi Delta Towns" (thesis, Delta State University, 2005).

33. Harvey, "Consensus-Based Community Development, Concentrated Rural Poverty, and Local Institutional Structures," 269–70.
34. Bonnie Lefkowitz, *Community Health Centers: A Movement and the People Who Made It Happen* (New Brunswick, NJ: Rutgers University Press, 2007).
35. Kathleen Kerstetter, Molly Phillips, and John Green, "Collective Action to Improve Rural Community Well-Being: Opportunities and Constraints in the Mississippi Delta," *Rural Society* (forthcoming).
36. *Ibid.* Also, for example, see John Green, Antoinette Jones, and John Pope, "Underemployment and Workforce Development in the Mississippi Delta: Community-Based Action Research for Program Planning to Increase Livelihood Security," *Southern Rural Sociology* 20, no. 1 (2004): 80–106.
37. De Haan, "The Livelihood Approach."

CHAPTER 8

What Is the Mississippi Delta?

A HISTORIAN'S PERSPECTIVE

JEANNIE M. WHAYNE

What is the Mississippi Delta? It stretches from Lake Itasca in Minnesota to New Orleans in the Gulf of Mexico. The focus here is on the portion characterized as the "lower Mississippi River Delta," which borders the Mississippi River from Cairo, Illinois, to the Gulf of Mexico, a region that includes some of the most fertile farmland in the world. Though an alluvial plain rather than a delta in the technical sense, the area is commonly called the Mississippi Delta. Although it is marked by a problematic environmental history, for most historians the Delta is defined by the social, economic, and political forces that shape its people and culture. One of three images is likely to come to mind when pondering its history, especially that of the lower Delta. One popular view conjures up notions of a vast expanse of cotton fields and black slaves in the antebellum era that evoke Scarlet O'Hara's Tara, replete with aristocratic masters and indulgent mistresses. In fact, much of the antebellum Delta, particularly that along Arkansas, hardly harbored the kinds of enterprises that filled the imaginations of white southerners after the Civil War. Arkansas was on the edge of the southern frontier, and its plantation society had yet to evolve beyond its first decades of existence. The postbellum image of landless African Americans working as sharecroppers in grinding poverty is more accurate, but close observation will reveal a more complicated story, one involving landless whites, particularly by the end of the nineteenth century, moving into the region and vying with African Americans for places on the expanding plantation system. This system eroded in the face of the rise of scientific agriculture in the post–World War II era, and an entity called the "neo-plantation" arose, one which abandoned the traditional cotton monoculture that had marked the two previous periods and embraced a combination of three alternatives: cotton, soybeans, and rice. Ironically, contemporary neo-plantation owners are the greatest adherents to the plantation myth. Some have even erected great mansions that conform to the aristocratic past of their imaginations. A fourth era seems to be in the making: the "portfolio" plantation. Though still in its infancy and thus difficult to analyze, the portfolio plantation began to take shape in the last decade of the twentieth century and accelerated in the first decade of the twenty-first as investors, made jittery by the decline in residential and commercial real estate, rushed into agricultural land offered by financial firms as a safer alternative.

The myth of the antebellum plantation survives because just as the southerners who created the myth in the late nineteenth century needed to believe in a time when life was better, so too do present-day Delta inhabitants, both rich and poor. The harsh economic conditions in the post–Civil War South promoted the need to come to terms with the sacrifice of hundreds of thousands of lives during the conflict by constructing a myth of the glorious past, but the overwhelming poverty in the region today encourages contemporary southerners to embrace the antebellum myth.[1] In fact, there were relatively few grand plantation houses even in Virginia and the Carolinas (typically referred to as the Old South) and few great planters on the scale common in the Caribbean and Brazil. The overwhelming majority of southerners were black slaves or poor whites.[2] As rare as a plantation aristocracy was in the Old South, it was rarer still along the Mississippi River. Located as it was on the western fringe of the South, it developed late in the antebellum period, as planters moved into the area with their slaves and began to cultivate cotton.[3] Aside from some who grew wealthy in Mississippi and Louisiana and built mansions as testimony to their achievements, most Mississippi Delta planters invested their profits in buying more land and purchasing additional slaves. Although some amassed considerable holdings in both land and slaves, they were content to inhabit log cabins or simple wood-frame structures. Thus one can visit grand plantation houses in the Natchez region, but few are found west of the Mississippi. Indeed, throughout the Delta the terms "big house" and "mansion" were commonly applied even to the most modest, even crude, dwellings of slave owners.

Before the planters arrived, the Mississippi Delta was swampy, disease-ridden, heavily forested, and relatively remote. Today it continues to be remote, although less so, but otherwise nothing else of the above description applies. The present Delta is drained of water, no longer a malarial zone, and almost entirely bereft of its former magnificent forests. This transformation took place unevenly. The areas least remote from the older South and most accessible by river were the first to be subject to the process of reclamation. Planters from Virginia and Kentucky, for example, moved into southeastern Arkansas, often after pausing in Mississippi for a couple of decades to open up plantations there. On the other hand, places like northeastern Arkansas, home to the sunk lands which were created by the New Madrid earthquakes of 1811–1812 and which made it more difficult to reclaim, remained untransformed until the beginning of the twentieth century. This difference in timing meant that very different sorts of people were involved in the making of the antebellum and postbellum Delta. The most accessible Delta was transformed by antebellum planters and black slaves, both with Old South plantation origins, and the rivers were the highways to the market. The less accessible Delta, which was reclaimed in the post–Civil War period, was developed as much by small town businessmen or lumbermen turned planters, many of them with Midwestern origins. Only a few were southerners with Old South plantation backgrounds. These planter-businessmen came to depend upon railroads in the late nineteenth and early twentieth centuries rather than rivers to market their crops, and, in fact, it was the railroads which first penetrated the region's forests, followed quickly by lumber companies. The planter-businessmen who either accompanied or trailed the railroad and lumber industries recognized the opportunities offered by the cut-over land, created drainage districts to reclaim it from its swampy conditions, and brought in black and white sharecroppers and tenants to found their plantation enterprises. New Deal programs funneled cash into the hands of landowners and both encouraged and enabled them to begin the process of mechanization. By the 1960s, the "neo-plantation" had fully emerged—that is, a more highly capitalized operation that had abandoned the tenancy/sharecropping system—and relied on a few wage laborers.[4]

Although the transformation of the various Delta regions occurred at different times, characterizations of its consequences are remarkably similar. The antebellum plantation, the postbellum plantation, and the twentieth-century neo-plantation are all burdened by the accusation that they were retrogressive economically. Frederick Law Olmstead and other travelers through the antebellum South accused planters, particularly those in the Mississippi Delta, of being "backward," and some advocates of an industrial New South in the post–Civil War era accused postbellum planters of retarding the region's economic development. Although the accusation of economic inefficiency may have some merit when considering how plantation agriculture inhibited alternative modes of development within the region, historians have established quite convincingly that the Old South plantations were quite profitable for certain capable planters.[5] The postbellum plantation and the twentieth century neo-plantation have received less attention from historians, but a similar consensus seems to be emerging that they too were profitable for those who were fortunate enough to own considerable land. One of the most successful practitioners of postbellum plantation agriculture, Lee Wilson of Mississippi County, Arkansas, built a 50,000-acre plantation from a 400-acre inheritance but also amassed a heavy debt and was on the edge of bankruptcy more than once, a fact that brings into sharp relief the economic environment facing planters in the postbellum period. Despite the fact that after the Civil War cotton prices dropped after a brief recovery and then remained erratic, postbellum planters concentrated on cotton and failed to diversify.[6] They adopted draconian means to squeeze profit out of their plantation operations, and great fortunes were made by planters who learned to use the mechanisms available to control labor.[7] The neo-plantations that came into full flower after the transition to wage labor was complete in the mid-1960s did grow not only cotton but also soybeans and rice. But that was the extent of their diversification. Their use of machines and chemicals had reduced their labor needs drastically, and they too reaped huge profits from their neo-plantation operations. The antebellum and postbellum plantations and the post–World War II neo-plantations were connected to an international marketplace which made some planters wealthy. In the antebellum and immediate postbellum periods, the cotton they produced played a role in the industrialization taking place outside the South. While the charge that they were unprofitable is not supported by the evidence, the fact that they inhibited economic development within the region, principally by absorbing most the available capital, capital which might have been used to industrialize, is now beyond question.[8]

The history of another party to the making of the Delta, the African Americans, whether they were the slaves of the antebellum period or the freedmen of the post–Civil War era, has also been marked by historiographical controversy. Until the 1950s, African Americans slaves were regarded as either happy dependents or dangerous savages in need of "civilizing." Both of these views are inherently racist. The work of Kenneth Stampp, John Blassingame, Herbert Gutman, Eugene Genovese, and Steven Hahn, among others, has transformed the study of slavery and illuminated the slaves' struggle to maintain their dignity and humanity in the fact of enormous odds.[9] Historians refer to the "agency" of slaves while political scientist James C. Scott calls it "weapons of the weak."[10] Black slaves fashioned lives for themselves in the context of a constant struggle to survive. Direct confrontation with the overwhelming power arrayed against them would have been futile, if not suicidal. So they adopted more subtle means to wring concessions from those who held power over them. Eugene Genovese, in *Roll Jordan Roll*, characterized the relationship between planters and slaves as paternalistic. But this was no ordinary paternalism. The best and most efficient planters took care of their slaves because it was in their own best interest to do so, and they expected a kind of loyalty in return. According to Genovese, slaves interpreted this paternalistic relationship differently from their masters, who believed they were behaving benevolently. From the slaves' point of view, the planter was *obliged* to provide for them because of the work they had performed. So slaves were not working because they were loyal but because the paternalistic compact whereby planters *owed them* for their work required them to provide something in return. Slaves used a variety of techniques to better their circumstances, but because of their relative powerlessness, their methods were typically subtle (malingering, pretending to misunderstand directions, etc.). Slaves driven to extremes might burn a smokehouse or steal from a corn crib (to hide evidence of a theft, to seek revenge for some wrong inflicted upon them, or to diversify their diets), but rather than being confrontational, these acts by nature were designed to conceal the identity of the perpetrator. Since confrontation could result in flogging or worse, more subtle measures were more often resorted to. Running away or direct violence was relatively rare. Despite the value of the approach used by Scott and Genovese, much of the historiography in this vein portrays slaves as "reactive" rather than active. It does emphasize African American "agency"—that is, their active efforts to ameliorate their situation—but given the dearth of firsthand accounts written by slaves, we can know only very little of what their day-to-day lives were like behind the closed doors of their own cabins. In other words, the best literature on slaves places them only in the context of their interactions with whites.

Discussing black sharecroppers, the direct descendants of black slaves, is even more problematic. Historians argue over when the paternalistic system existing prior to the Civil War withered away after emancipation, with some arguing that it died immediately and others suggesting that it survived in one form or other both on and off the plantations.[11] Historians typically portray black sharecroppers as victims—both of violence visited upon them by white-sheeted nightriders and of systematical cheating and oppression by evil and pernicious planters. Few historians have recognized black sharecroppers' ability to wield the "weapons of the weak" just as slaves had, and until recently the prevailing view depicted them as hapless victims. A few historians are now attempting to do for twentieth-century rural southern history what Genovese and others did for the study of slavery. While this is preferable to the passive victim historiography that dominates sharecropper studies, it still understands black sharecroppers largely in the context of their exploitation by white plantation owners, and this distorts the image of what their lives were like apart from these unequal interactions. There is no question that the white South truly circumscribed the opportunities open to blacks, that segregation and disfranchisement greatly inhibited their progress, and that economic exploitation and violence were real and pervasive, but African Americans lived their day-to-day lives with each other. It is that story that is so difficult to penetrate. They led a necessarily semi-secret life, perhaps as much as did the slave. Historians have struggled with some success to penetrate the slave experience, but have rarely succeeded with that of black sharecroppers.[12]

As difficult as it is to understand the black experience, it is almost as difficult to penetrate that of ordinary whites within the Mississippi Delta. Looking to the antebellum period, we find little reference to them outside those living in the few towns that existed in the Delta. We know that some small white farmers within the Delta interacted economically with planters. They used the planters' gin, they sold fruits and vegetables to the plantations, and they sometimes hired black slaves. This is what gave small farmers within the Mississippi Delta, and elsewhere in the South, a stake in the slavery question, and some historians have characterized the relationship between them and the planters with whom they dealt as paternalistic. The paternal planter dominated not only his slaves but also the small white farmers with whom he engaged in business transactions, and small white farmers paid him deference in return for services rendered.[13] To be sure, there were whites within the undeveloped parts of the Delta who lived the lives of hunters and trappers, and they almost certainly escaped the paternalistic reach of planters. Unfortunately, we know almost nothing about them. They left little record of themselves not merely because they tended to be illiterate, but because they lived apart from the market, squatted on land rather than purchasing it, and rarely filed any legal documents—birth, marriage, death certificates.

In turning to the post–Civil War period, ordinary whites increased in number within the Mississippi Delta and thus became more historically significant and knowable. Like African Americans, however, their story is told largely in the context of their place in the struggle with planters. Thus we find whites on the edge of the Delta creating the populist movement and the People's Party in an attempt to address the economic problems confronting them in the face of the failure of the political system to speak to those issues, a political system dominated by planters.[14] Given the tendency of whites in the post–Civil War era to continue economic—and sometimes familial—relations with planters, they were less likely to become actively involved in political movements against them. Another view of ordinary whites in the Delta portrays them as vicious or desperate nightriders who tried to keep blacks from voting or drove black officeholders from office. In the years immediately following the war, they were viewed as the planters' co-conspirators in the effort to keep blacks on the plantations. By the end of the nineteenth century, in the wake of the populist movement's attempt to forge an alliance between blacks and disaffected whites, the Democratic Party, feeling threatened by such an alliance, successfully passed measures which imposed disfranchisement and segregation upon the black population and some poor whites. They did so, in part, by fashioning an offensive racist rhetoric that contributed to the rise of racial tensions and record lynchings of African Americans. However, the wrath of white nightriders was often directed at planters. Not only did whites in the Arkansas Delta attempt to drive blacks from plantations so they could secure the tenancies themselves, they sometimes destroyed planter property to make their point. Battles over the direction of development in the environmentally vulnerable sunken lands of northeastern Arkansas pitted poor whites against planters and powerful lumbermen, illuminating the fact that the unification of white classes against blacks was tentative and, perhaps, illusory. Taken together, these kinds of activities point to the variety of circumstances that ordinary whites experienced and suggest the dangers of generalizing too much about the "white" experience. Just as planters and African Americans lived autonomous lives, so too did ordinary whites.[15]

Having said all this, having indicated that we continue to know too little about the lives of each of these groups, it is nevertheless true that we can best understand the Mississippi Delta by examining the role that each of the actors played in it. Thus we are forced to talk about them in relation to each other and concentrate less on the worlds they built among themselves. In the end, the answer to the question "What is the Delta?" must address how people interact and, in the context of the social, political, and economic forces arrayed against them (or for them), how they created a particular society. The Mississippi Delta bears an especially heavy "burden of history." The harsh truth is that the people of the Mississippi Delta created a system which is today marked by extremes of wealth and poverty that more nearly resemble a third world country than post-industrial America. This is true of plantation regions the world over. Politicians have established commissions to study the Delta, social scientists and others have created elaborate models and strategies to reform the region, philanthropic groups have expended large sums on projects to enhance it, and some planters and businessmen have voiced concerns and engaged in their own remedies. But the solution remains elusive. The fact is that large-scale plantation operations, wherever they are and in whatever era, create disparity of wealth and income. Even though past experience does not offer much hope of altering this condition, people will undoubtedly continue to discuss solutions. And there are some modest success stories in the Delta. Communities that enjoyed the benefit of enlightened black and white leadership in the post–Civil Rights era and have a sufficient population to support a workforce established small industrial enclaves.[16] These are extractive industries, to be sure, and extractive industries do not typically create wealth that remains within a region, but the trend is encouraging; and again, history suggests that when people recognize opportunities and pursue solutions, progress can be made. The new portfolio plantation, a phenomenon which is only now coming into focus, rests on a shaky but hopeful foundation. Whether the portfolio planters—faceless investors who are far removed from the agricultural process itself—will have any interest in promoting the interests of the agricultural communities in the Mississippi Delta is, as yet, an unanswered question.

Notes

1. Historians have also demonstrated that northerners in the late-nineteenth-century North embraced the romantic view of the antebellum South. Their eagerness to do so reflected a period of national reconciliation. See Nina Silber, *The Romance of Reunion: Northerners and the South, 1865–1900* (Chapel Hill: University of North Carolina Press, 1993).
2. Kenneth Stampp, *Peculiar Institution: Slavery in the Antebellum South* (New York: Knopf, 1956).
3. Donald McNeilly, *The Old South Frontier: Cotton Plantations and the Formation of Arkansas Society, 1819–1861* (Fayetteville: University of Arkansas Press, 2000); Orville W. Taylor, *Negro Slavery in Arkansas* (Fayetteville: University of Arkansas Press, 2000; Duke University Press, 1958).
4. Merle Prunty Jr. was the first to identify the neo-plantation, but the scholarship of Pete Daniel and Jack Kirby is closely associated with the term. Merle Prunty Jr., "The

Renaissance of the Southern Plantation," *Geographical Review* 45 (October 1955): 459–91; Pete Daniel, *Breaking the Land: The Transformation of Cotton, Tobacco, and Rice Cultures since 1880* (Urbana: University of Illinois Press, 1972); Jack Temple Kirby, *Rural Worlds Lost: The American South, 1920–1960* (Baton Rouge: Louisiana State University Press, 1987). See also Robert Leon Brandfon, *The Cotton Kingdom of the New South: A History of the Yazoo-Mississippi Delta from Reconstruction to the Twentieth Century* (Cambridge: Harvard University Press, 1967); Jeannie Whayne, *A New Plantation South: Land, Labor, and Federal Favor in Twentieth -Century Arkansas* (Charlottesville: University of Virginia Press, 1996); and Jeannie M. Whayne, *Delta Empire: Lee Wilson and the Transformation of Agriculture in the New South* (Baton Rouge: Louisiana State University Press, 2011).

5. U. B. Phillips, writing in the early twentieth century, was the first to question the profitability of cotton plantations under the slave regime. U. B. Phillips, *American Negro Slavery: A Survey of the Supply, Employment, and Control of Negro Labor as Determined by the Plantation Regime* (New York: D. Appleton, 1918). A host of historians in the second half of the twentieth century challenged this view. See especially Hugh G. J. Aitken, *Did Slavery Pay?* (Boston: Houghton Mifflin, 1971); and Gavin Wright, *The Political Economy of the Cotton South: Households, Markets, and Wealth in the Nineteenth Century* (New York: Norton, 1978). See also Gavin Wright's *Slavery and American Economic Development* (Baton Rouge: Louisiana State University Press, 2006); and Whayne, *Delta Empire.*
6. Sven Beckert provides a provocative analysis of the transformation of the global cotton market during the Civil War, a transformation that would negatively influence cotton prices for the long term. Before the Civil War, southerners dominated the market, and though they would reclaim their position of dominance, too many competitors drove the price of the commodity down drastically. See Sven Beckert, "Emancipation and Empire: Reconstructing the Worldwide Web of Cotton Production in the Age of the American Civil War," *American Historical Review* 109, no. 5 (December 2004): 1405–1538.
7. Pete Daniel, *Shadow of Slavery: Peonage in the South, 1901–1969* (Urbana: University of Illinois Press, 1972); and Harold Woodman, *New South, New Law: The Legal Foundations of Credit and Labor Relations in the Postbellum Agricultural South* (Baton Rouge: Louisiana State University Press, 1995).
8. James C. Cobb, *Industrialization and Southern Society, 1877–1984* (Lexington: University Press of Kentucky, 1984).
9. Kenneth Stampp launched a systematic attack on U. B. Phillips's racist interpretation of slaves as simple-minded or savage, but he tended to overplay the brutality of slavery to the point that the slave's personality was obscured. Historians responded immediately with a consensus culminating in the 1970s: John W. Blassingame, *The Slave Community: Plantation Life in the Antebellum South* (Oxford: Oxford University Press, 1972); Eugene Genovese, *Roll Jordan Roll: The World the Slaves Made* (New York: Vintage, 1972); Herbert Gutman, *The Black Family in Slavery and Freedom, 1750–1925* (New York: Vintage, 1976); and, more recently, Steven Hahn, *A Nation under Our Feet: Black Struggles in the Rural South from Slavery to the Great Migration* (Cambridge: Harvard University Press, 2003).
10. James C. Scott, *Weapons of the Weak: Everyday Forms of Peasant Resistance* (New Haven: Yale University Press, 1985).
11. Jay R. Mandle, *The Roots of Black Poverty: The Southern Plantation Economy after the Civil War* (Durham: Duke University Press, 1978); Elizabeth Fox-Genovese and Eugene Genovese, *Fruits of Merchant Capitalism: Slavery and Bourgeois Property in the*

Rise and Expansion of Capitalism (New York: Oxford University Press, 1983); and Whayne, *A New Plantation South*. Planter Lee Wilson created his own kind of paternalistic ethos out of his early New South experience combined with progressive era manifestations of *nobles oblige*. Whayne, *Delta Empire*.

12. Daniel, *Shadow of Slavery*; Woodman, *New South, New Law*; Nan Elizabeth Woodruff, *American Congo: The African American Freedom Struggle in the Mississippi Delta* (Cambridge, MA: Harvard University Press, 2003).
13. Genovese, *Roll Jordan Roll*. For a more subtle treatment of the connection between planters and yeomen farmers within the Black Belt of South Carolina, see Stephanie McCurry's award-winning *Masters of Small Worlds: Yeomen Households, Gender Relations, and the Political Culture of the Antebellum South Carolina Low Country* (Oxford: Oxford University Press, 1995).
14. There is a lengthy historiography on populism, but for excellent sources on southern populism see Steven Hahn, *The Roots of Southern Populism: Yeomen Farmers and the Transformation of the Georgia Upcountry, 1850–1890* (Oxford: Oxford University Press, 1983); and Edward L. Ayers, *The Promise of a New South: Life after Reconstruction* (New York: Oxford University Press, 1992).
15. William F. Holmes wrote a series of articles on whitecappers in Alabama,Georgia, and Mississippi: "Whitecapping: Agrarian Violence in Mississippi, 1902–1906," *Journal of Southern History* 35 (1969): 165–80; "Whitecapping in Mississippi: Agrarian Violence in the Populist Era," *Mid-America* 55 (1973): 134–48; "Moonshining and Collective Violence: Georgia, 1889–1895," *Journal of American History* 67 (1980): 588–611; "Whitecapping in Georgia: Carroll and Houston Counties, 1893," *Georgia Historical Quarterly* 64 (1980): 388–404; "Moonshiners and Whitecappers in Alabama, 1893," *Alabama Review* 34 (1981): 31–40. See also Whayne, *A New Plantation South*; and Whayne, *Delta Empire*.
16. Charles S. Aiken, *The Cotton Plantation South since the Civil War* (Baltimore: Johns Hopkins University Press, 1998).

CHAPTER 9

Politics in Black and White

THE MISSISSIPPI DELTA

SETH C. MCKEE

Since white settlement commenced in the 1820s, the northwestern floodplain in the state of Mississippi, encompassing the most fertile soil in the state and typically referred to by political scientists and, particularly, scholars of southern politics as the Mississippi Delta, was, is, and remains consumed by the issue of race. From its inception as a land filled with the promise of generating great wealth through the cultivation of cotton, the scale of black labor required to produce such fortunes meant that white Deltans were decidedly outnumbered by their African American help. In the time of slavery, it was in the Delta where the largest plantations were found and where the ratio of white residents to blacks most favored the latter. It is this simple numerical imbalance of the races that persists to this day, which primarily explains the nature of Mississippi Delta politics and why in the most racially oppressive state, those whites with the most to lose through black enfranchisement historically wielded the most power.

Adopting historian James C. Cobb's geographic definition of the Delta as consisting of sixteen counties located in northwest Mississippi, from the vantage of a political scientist it is easy to understand why Cobb deemed it the "most southern place on earth."[1] Writing in the mid-twentieth century, in his masterpiece *Southern Politics in State and Nation*, V. O. Key famously boiled down southern politics to an elegant proposition: "In its grand outlines the politics of the South revolves around the position of the Negro. . . . The hard core of the political South—and the backbone of southern political unity—is made up of those counties and sections of the southern states in which Negroes constitute a substantial proportion of the population. In these areas a real problem of politics, broadly considered, is the maintenance of control by a white minority."[2]

Key goes on to explain that most southern whites who were greatly outnumbered by African Americans lived in Black Belt counties, named for the dark, fertile soil in this region of Dixie. The entire Mississippi Delta is comprised of Black Belt counties, and because of the unrivaled numerical inferiority of the Black Belt whites residing there, their behavior epitomized Key's thesis of southern politics. It may be only a slight overstatement in the case of Mississippi and, specifically, the Delta, to agree with the late civil rights lawyer Frank R. Parker, who averred, "The history of southern politics is the history of racial suppression of the black vote."[3] Since the enactment of the Civil War amendments ending slavery (Thirteenth, 1865), granting citizenship to those born in the United States and equal protection of the laws (Fourteenth, 1868), and granting the right to vote regardless of one's race (Fifteenth, 1870), white Deltans were in the vanguard of denying political equality to their black neighbors.

Through violence, economic and social deprivation, and legislative enactment of, judicial approval of, and executive enforcement of a Jim Crow system of second-class citizenship for African Americans, white Deltans safeguarded white supremacy in the Magnolia State for over a century—from the end of Reconstruction in the 1870s to the 1980s. And to this day, despite marked gains in political participation and the election of African Americans, because of the extreme and systemic poverty of black Deltans, these Mississippians continue to endure substandard living conditions whose roots are firmly tied to a history of racial subjugation. The politics of Mississippi, and especially the politics of the Delta, is the politics of race. To be sure, much has changed since the days of the Democratic Solid South, but the racial polarization of black and white political preferences is a constant, and now that almost the entire population of African Americans is Democratic, whereas whites are overwhelmingly Republican, skin color remains the most reliable indicator for explaining the Delta's political culture.

In his chapter on Virginia, V. O. Key subtitled it "Political Museum Piece." The South Carolina chapter was subtitled "The Politics of Color," and the Mississippi one was subtitled "The Delta and the Hills."[4] Sixty-five years later all three subtitles are applicable to Mississippi. The much higher percentage of African Americans in the Delta still distinguishes it from the rest of the state, and the politics of color best describes Mississippi because it is truly a political museum piece. In *The Rise of Southern Republicans*, the most comprehensive account of how the Republican Party took control of southern congressional politics, Earl and Merle Black present a series of telling graphical depictions of the demographic foundations of southern states.[5] Plotting the southern states along a vertical axis for the percentage urban and the horizontal axis for the percentage black at four different points in time—1880, 1930, 1950, and 1990—Mississippi stands alone in every time period for its combination of the least urban and highest black population.[6] As most of the South has progressed economically and become more urban and racially diverse, by comparison the Mississippi Delta seems almost frozen in time.

Since the antebellum period the only notable economic revolution in the Delta was the switch from manual labor to mechanized agriculture. King

Cotton now competes with a handful of other crops (i.e., soybeans, rice, and corn), but when catfish farms constitute one of the largest employers it is implausible to say that the fundamental agricultural identity of the Delta has changed.[7] Likewise, except for a notable but miniscule population of Asian Americans (Chinese and Indian), blacks and whites continue to comprise the lion's share of Delta residents. The lack of economic change, racial diversity, and in-migration, coupled with population decline, has made the Delta an ideal setting for the issue of race to dominate its politics into perpetuity.

A look into the past and up to the present shows how race dominates Delta politics. Despite constitutional safeguards set in place to protect African American political rights after the Civil War, the only guarantee of protection was administered through military Reconstruction.[8] But the official end of Reconstruction, brokered through a compromise to resolve the disputed 1876 presidential election between the Democrat Samuel Tilden and the Republican Rutherford Hayes, removed the presence of northern soldiers and hence the federal protection of black civil rights, opening the door for native white Mississippians to restore white supremacy through the vehicle of the Democratic Party.

In 1890 Mississippi was the first southern state to call a constitutional convention for the purpose of constructing a statutory framework designed to disfranchise African Americans and poor whites. Democrats dominated the convention and, not surprisingly, those most supportive of passing restrictive suffrage laws hailed from Black Belt counties in the Delta. In his careful and exhaustive analysis of the enactment and subsequent effects of suffrage restrictions passed in the South after Reconstruction, J. Morgan Kousser estimated that before passage of the 1890 Mississippi Constitution, "Negro turnout [went] from about 30 percent in the 1888 presidential race to virtually nothing thereafter."[9]

From the late 1800s until the 1960s, the poll tax, literacy test, understanding clause, and white primary provided ample barriers to ensure the elimination of African Americans from the Mississippi electorate and, of course, from holding elective office.[10] Additionally, the massive reduction in white turnout removed any credible outside threat to the Democratic Party from a Populist or Republican opposition, cementing a one-party system whose guiding purpose was to safeguard the interests of Mississippi's elites—the white Delta planters. As late as 1964, the year before passage of the Voting Rights Act (VRA), only 6.7 percent of Mississippi's adult black population was registered to vote.[11] And since 1882, when Republican John R. Lynch last served in the US Congress, Mississippi did not elect another African American to the US House until 1986 when Democrat Mike Espy won a US House seat in the Delta-based majority-black Second Congressional District.[12]

After the US Supreme Court struck down the white primary in 1944, Mississippi's white ruling class shifted to the literacy test and understanding clause to do the heavy lifting of blocking black political participation. However, for most black Mississippians a culture of deference to whites made it unlikely that many would even attempt to fight for their constitutional right to vote.[13] A passage detailing the behavior of a Mississippi registrar in *Southern Politics in State and Nation* offers an insightful anecdote.

> Few Negroes apply for registration except in two or three of the most urbanized counties. One of the gentler techniques of dissuasion in the rural counties is illustrated by the practice of the registrar of a county with over 13,000 Negroes 21 and over, six of whom were registered in 1947. The registrar registers any qualified person, black or white, if he insists. When a Negro applies, however, she tells him that he will be registered if he insists, but she gives him a quiet, maternal talk to the effect that the time has not yet come for Negroes to register in the county. The people are not ready for it now and it would only cause trouble for the Negro to register. Things move slowly, she tells the applicant, but the day will come when Negroes can register just as white people. Almost always, she says, the applicant agrees with her and departs in peace.[14]

But thanks to the courage of college students who organized lunch counter sit-ins in 1960 and then racially integrated interstate bus rides ("Freedom Rides") throughout the Deep South in 1961, the white massive resistance that prevailed since the 1954 *Brown v. Board of Education* ruling was finally met with a full-blown civil rights movement.[15] By 1964 a coalition of blacks and whites (mainly college students from prestigious northern universities) orchestrated "Mississippi Freedom Summer," whose goal was to bring national attention to the Magnolia State's militant resistance to local blacks' attempts to register to vote. And although hardly any African Americans were added to the voter rolls, especially in the Delta, the violence that ensued made national headlines and turned northern white opinion in favor of major reforms to secure black civil rights. President Lyndon Johnson successfully led congressional enactment of the 1964 Civil Rights Act, which, to put plainly, forbid racial segregation in all manner of places of public accommodation. But it was not until the following summer, in 1965, that black political participation would be safeguarded by the Voting Rights Act (VRA).

The 1965 VRA was a revolutionary piece of legislation. As mentioned, in 1964 only 6.7 percent of black Mississippians were registered to vote. By 1967 black registration had surged to 59.8 percent, reflecting the hunger of African Americans for genuine political empowerment.[16] After seventy-five years of formal exclusion from the Mississippi electorate, African Americans would finally have the opportunity to vote without obstruction. But the influence of the black vote was almost instantly neutralized by a new strategy that Mississippi's white ruling class engineered in the 1966 session (and special sessions) of the state legislature. Now that African Americans could no longer be denied the franchise, their votes were systematically diluted so that in all but a few instances (like the tiny all black Delta town of Mound Bayou) they would be rendered ineffectual in the goal of choosing candidates of their choice.

In all, the completely white Mississippi legislature passed thirteen bills in 1966 whose purpose was to alter the state's electoral laws so that black votes would be overwhelmed by white votes throughout the state. In congressional elections the majority black Mississippi Delta was cut up into three districts that each ran to Mississippi's eastern border so that they all contained a majority of white voters. This arrangement was not rectified until the 1980s, when the state finally drew a majority black Delta district, which African American Mike Espy won in 1986. In state legislative elections, districts were redrawn to dilute the black vote by making most of them multi-member and constructed so that in those cases where majority-black counties threatened white voting majorities, these counties were combined with more populous majority-white counties. County and municipal elections employed the same minority vote dilution strategy, and in those cases where a majority black county was likely to elect an African American, many offices were made appointive.[17]

The minority vote dilution scheme was highly successful in limiting black political gains well into the 1980s. Nonetheless, because of the tireless efforts of local blacks and civil rights lawyers, a raft of US Supreme Court rulings eventually overturned all of Mississippi's legislated vote dilution measures.[18] As a consequence, because of the disproportionately black Delta population, in most local settings African Americans made substantial progress in electing candidates of their choice. And against this backdrop of massive political change, the Democratic Party was undergoing a transition from being the historic bastion of white supremacy to a racially integrated party.

At the 1964 Democratic National Convention in Atlantic City, New Jersey, the African American Mississippi Freedom Democratic Party (MFDP), led by the vocal leadership of civil rights activist Fannie Lou Hamer (she grew up picking cotton in the Delta's Sunflower County), challenged the credentials of the all-white Mississippi Democratic delegation. Although the challenge failed, by 1967, the "Loyalist" wing of the Mississippi Democratic Party managed to oust the Mississippi "Regulars" who comprised the all-white state and local faction of the party.[19] With the national Democratic Party's recognition of the integrated Loyalist faction, the Democratic Party in Mississippi was no longer the political redoubt for upholding white supremacy.

The racial integration of the Democratic Party in Mississippi caused a steady exodus of white conservatives. They were turned off by the national party's support of civil rights and increasingly attracted to the racial conservatism of the Republican Party, starting with presidential nominee Barry Goldwater in 1964.[20] But the white shift to the Grand Old Party (GOP) took decades to complete because so many white Mississippi Democrats refused to relinquish their affiliation with the party that historically defended white supremacy.[21] And because of the paramount role of race in Mississippi politics, it was the only southern state where a large faction of African Americans ran as independent candidates aligned with the MFDP. There was also a strategic motive for running as independents. Mississippi only had a runoff for candidates in party primaries (Democratic and Republican) who failed to garner a majority of votes. In the general election a simple plurality constituted election; and, therefore, if black voters united behind an independent black candidate there was a possibility of winning office if the white vote was split between white Democratic and white Republican nominees.

Of all the legislation passed in 1966 to dilute black votes, the one provision that failed was an effort to make party nomination contests open primaries with a runoff if the top vote-getter failed to secure a majority. The purpose was to deny an independent African American candidate from having the opportunity to win the general election with a plurality of the vote. The measure failed to pass because Governor Paul Johnson vetoed it; he thought such a reform would not withstand a legal challenge.[22] However, the costs of being an independent candidate were considerably raised by increasing the required number of signatures tenfold for securing a spot on the general election ballot. Likewise, the filing deadline for independent candidates was moved to an earlier date and independent candidates were forbidden from voting in party primaries.[23] Nonetheless, in those local settings (county and municipal) where blacks held clear voting majorities—particularly the Delta—many black independent candidates won elective office. But over time, especially as the Democratic Party integrated, the number of black independent candidacies steadily declined.

For a brief period in Mississippi politics and in the rest of the South, from the 1970s and into the early 1990s, the construction of black and white biracial coalitions under the Democratic Party label were large enough to beat back most GOP challenges in contests below the presidential level.[24] This was a positive development for black Mississippians because, even though they remained greatly outnumbered by white Democrats, on a range of critical legislation, a unified black vote often made the difference between the success or failure of proposed bills in the state legislature. As late as 1989, in the entire eleven-state South, Mississippi still had the smallest percentage of Republicans serving in its legislature.[25] But this state of affairs could not persist. The inexorable white march to the GOP eventually overwhelmed Democratic biracial coalitions that shrunk to the point of consisting almost entirely of African Americans. Now, Mississippi's state legislature is majority Republican in both the state house and state senate, and a decided majority of the minority Democratic opposition is comprised of African American legislators.

The rise of the Mississippi GOP has now reached a mature stage. Black and white Mississippians have self-segregated themselves almost completely in party politics. And because whites are the numerical majority in the state, Republicans essentially have a monopoly on statewide elections and any other contests where whites outnumber blacks.[26] In contrast, because whites are greatly outnumbered by African Americans in the Delta, here, black Democrats now control most local contests (county and municipal). Table 9.1 uses presidential elections to illuminate the partisan evolution of voting behavior in the Mississippi Delta's sixteen counties as juxtaposed with the rest of the state. Selected contests were picked before (1928, 1948, and 1964) and after (1968, 1980, and 2012) the re-enfranchisement of black Mississippians in order to point out two overriding themes: (1) white conservative voting dominated prior to black re-incorporation into the electorate, and (2) the black response to suffrage restoration in the Delta almost perfectly reflects a preference for a racially liberal Democratic Party. In other words, racial polarization in party voting is on full display in the presidential results shown in table 9.1.

Table 9.1 displays the presidential vote percentage for the candidate most preferred by white Mississippians. The first election shown is the 1928 contest between Republican Herbert Hoover and Democrat Al Smith. This contest took place in the Democratic Solid South era (the late 1800s to 1950s), and although Hoover managed to win five southern states, he was shut out of the Deep South where Democratic loyalty reflected protection of the white supremacist status quo.[27] Hence, in the Delta, with no African Americans enfranchised, Smith took a whopping 92 percent of the two-party vote. In Humphreys County, 1,021 votes were cast and all but 1 went to Smith. Compared to the rest of the state, where defense of the Jim Crow system was somewhat less intense, Smith garnered an impressive 80 percent, 12 percentage points under his Delta vote share.

In 1948 President Harry S. Truman openly courted black voters because they were seen as the key to winning several northern swing states where African Americans were able to vote. The Democratic reaction to Truman's liberal stance on civil rights was fierce. South Carolina governor J. Strom Thurmond and Mississippi governor Fielding Wright headed the ticket for the insurgent States' Rights Democratic Party. Also known as the Dixiecrats, this party's overriding purpose was to defend the South's segregationist political order.[28] Not surprisingly, the Dixiecrats dominated in Mississippi, where Thurmond captured 90 percent of the Delta's votes, 3 percentage points better than the rest of the state.

The 1964 presidential election was the last in which black Mississippians were excluded from the electorate. This election proved to be the turning point in southern voting patterns for the major parties. Because the Republican Barry Goldwater voted against the 1964 Civil Rights Act while serving as a US Senator from Arizona, whereas Democratic president Lyndon Johnson championed the legislation, white southerners sharply changed course in their partisan preferences. In fact, ever since 1964 a majority of the two-party presidential vote cast by white southerners has gone to Republican nominees.[29] Similarly, because of President Johnson's leadership on civil rights, African Americans

TABLE 9.1. The presidential vote in the Delta counties and the rest of Mississippi

Counties	Smith (D) 1928	Thurmond* (SR) 1948	Goldwater (R) 1964	Wallace (AI) 1968	Reagan (R) 1980	Romney (R) 2012
Bolivar	88	89	86	44	37	31
Coahoma	89	85	81	34	40	26
Humphreys	100	98	96	59	38	25
Issaquena	96	93	93	48	37	39
Leflore	95	93	94	49	44	28
Quitman	90	90	86	56	37	28
Sharkey	93	96	90	49	34	29
Sunflower	97	93	94	52	43	26
Tunica	96	95	91	34	30	20
Washington	86	82	74	41	46	29
Carroll	96	93	95	67	51	66
Holmes	94	96	97	41	33	16
Panola	92	89	91	52	41	46
Tallahatchie	99	87	92	60	39	39
Warren	84	86	82	51	58	49
Yazoo	95	96	96	61	47	43
Delta	92	90	88	48	43	35
Rest of state	80	87	87	67	52	59
Mississippi	82	87	87	63	51	56
Delta-rest	+12	+3	+1	-19	-9	-24
Size of the Delta vote	18%	17%	15%	18%	16%	12%
Total votes	151,568	192,190	409,146	654,509	870,370	1,273,695

*The votes for J. Strom Thurmond also include the 225 votes cast statewide for Progressive Henry Wallace. The votes for Thurmond and Wallace were combined under the category "other" in Richard M. Scammon, *America at the Polls: A Handbook of American Presidential Election Statistics, 1920–1964* (Pittsburgh: University of Pittsburgh Press, 1965).

Notes: The election data for 1928 and 1948 are from Scammon, *America at the Polls.* The election data for 1964, 1968, 1980, and 2012 were all compiled from Dave Leip's *Atlas of U.S. Presidential Elections* (http://uselectionatlas.org/, accessed April 1, 2015). The vote percentages for 1928, 1964, 1980, and 2012 are based on the two-party total (Democratic plus Republican), whereas the vote percentages in 1948 are based on the votes cast for J. Strom Thurmond, Henry Wallace, Harry S. Truman (D), and Thomas Dewey (R); the vote percentages for 1968 are based on the votes cast for George C. Wallace (AI), Hubert Humphrey (D), and Richard M. Nixon (R). The party abbreviations shown in parentheses for the presidential candidates are as follows: (D) = Democrat; (SR) = States' Rights; (R) = Republican; (AI) = American Independent. In Mississippi (as was also the case in Alabama, Louisiana, and South Carolina), States' Rights Democratic Party (also popularly referred to as the Dixiecrat Party) candidate Thurmond was placed on the ballot as the Democratic Party nominee and Truman was designated as the National

TABLE 9.1. Notes (continued)

Democratic Party nominee. In 1948, South Carolina governor Thurmond's Dixiecrat vice presidential running mate was Mississippi governor Fielding Wright. The "Size of the Delta vote" is calculated as the total presidential votes cast in the Delta counties divided by the total votes cast in Mississippi. The Delta counties are those identified by James C. Cobb, *The Most Southern Place on Earth: The Mississippi Delta and the Roots of Regional Identity* (New York: Oxford University Press, 1992), 335. The first ten counties listed in alphabetical order wholly reside in the Delta as defined by Cobb, while the last six counties listed in alphabetical order (Carroll through Yazoo) are partially situated in the Mississippi-Yazoo Delta. "Delta-rest" is the percentage of the vote cast in the Delta counties minus the percentage of the vote cast in the rest of Mississippi's counties.

permanently shifted to the Democratic Party. With blacks still excluded from the electorate, Goldwater took 88 percent of the Delta vote, a percentage point better than the rest of the state. This was the first time a Republican had won Mississippi's presidential votes since Reconstruction, when Republican Ulysses Grant prevailed in 1872.

Like 1948, the 1968 election was another three-way contest that included an openly racist third-party contender (Thurmond in 1948 and Wallace in 1968). Democrat Hubert Humphrey was racially liberal, whereas Republican Richard Nixon was the softer-edged racially conservative nominee who was squeezed out of much of the South by the raw racist appeals of Alabama governor George C. Wallace, the American Independent Party nominee. There is no question that white Deltans split their votes between Nixon and Wallace, but by 1968 most black Deltans were re-enfranchised and, given their numerical majority, Wallace only carried 48 percent of the vote. By comparison, in the rest of Mississippi where whites outnumbered blacks, Wallace won 67 percent of the three-way vote. With black voting rights reinstated, never again would the Mississippi Delta back a racially conservative presidential candidate.

In 1980 Republican Ronald Reagan easily defeated President Jimmy Carter, but his victory was very narrow in Mississippi because African Americans cast a bloc vote for the Democratic incumbent. Like in 1968, in the Delta, the preferred candidate of whites lost to the candidate preferred by blacks. But in the rest of the state a white majority delivered Reagan a slim win. Similar to 2008, racially polarized presidential voting once again peaked in the 2012 election. Given the presence of an embattled African American Democratic incumbent facing off against a white Republican, the loyalty of black Deltans to President Barack Obama was expected, as was the overwhelming white support for Mitt Romney. According to the 2012 national exit poll for Mississippi voters, 96 percent of blacks voted for President Obama and 89 percent of whites favored Republican Mitt Romney.[30]

Based on the individual-level survey data from the 2012 exit poll on Mississippi voters and the fact that no other state rivals the racially polarized voting witnessed in the Magnolia State, even county-level data provides a remarkably accurate portrait of racial patterns in presidential voting. In the

Delta's sixteen counties Mitt Romney only managed 35 percent of the two-party vote. Notice, however, that there is considerable variation in the percentage of the vote for Mitt Romney depending on the county. This variation is almost entirely explained by the racial composition of any given county. For instance, a simple bivariate correlation between the Republican percentage of the vote and the black percentage of each of the sixteen Delta counties yields an astonishingly high and statistically significant negative coefficient of -.983 ($p < .001$; two-tailed test). In other words, there is almost a perfectly negative relationship between the Republican share of the vote and the black percentage of a county's population. Two examples should make this eminently clear. First, Mitt Romney's worst performance in the Delta was in Holmes County, where he managed to win just 16 percent of the vote. According to 2013 US Census data, Holmes County was 82.9 percent African American.[31] Second, Mitt Romney's best performance in the Delta was in Carroll County, where he won 66 percent of the vote. Based on the same source, in 2013, Carroll County was 33.3 percent black.[32]

The remarkable changes in presidential voting patterns that have transpired in the Mississippi Delta directly reflect the consequences of black re-enfranchisement and how the two major parties have reversed positions on the issue of race since the 1964 election. Black votes not only count in contemporary Delta elections, they determine the outcome in most contests because a majority of blacks prefer candidates opposed by the white minority. But black political empowerment in the Delta cannot overcome the deeply ingrained and historically rooted social and economic deprivation that continues to plague African Americans. As explained by Frank R. Parker, "despite the election of a large number of black officials and although substantial progress has been made, black income and employment levels remain disproportionately lower than white income and employment levels . . . [and] these socioeconomic disparities are the result of economic factors that are beyond the power of black elected officials to influence or control."[33]

When cotton was king in the Mississippi Delta, white Delta planters were the kings of the South. Their outsized wealth and political influence steered the ship of state in Mississippi and, by extension, all of Dixie, thanks to their similarly situated Black Belt allies in neighboring southern states. As V. O. Key points out with admiring candor and yet marked disapproval of their actions, the small minority of Black Belt planters in the Delta and throughout the South caused a Civil War, violently put down a farmers' revolt in the 1890s, and held African Americans in political bondage from the turn of the twentieth century until the 1960s.[34] In the annals of American history such political cunning is perhaps unrivaled, and it enabled white Delta elites to control the power structure well into the 1980s. But the relentlessness of several generations of African Americans and their white allies finally democratized the Delta. Now, the majority race rules. Blacks have captured most of the elective offices in the region because of their voting majorities. But one still wonders if the tremendous political strides of black Deltans are tragically insufficient to overcome such enormous poverty. The racial divide between blacks and whites in Delta politics is as wide as ever. And unless the white "haves" reach out to the black "have-nots," political equality will never bridge the gulf of poverty that has haunted African Americans since the first slaves worked the land.

Notes

1. James C. Cobb, *The Most Southern Place on Earth: The Mississippi Delta and the Roots of Regional Identity* (New York: Oxford University Press, 1992), 335.
2. V. O. Key Jr., *Southern Politics in State and Nation* (New York: Alfred A. Knopf, 1949), 5.
3. Frank R. Parker, *Black Votes Count: Political Empowerment in Mississippi after 1965* (Chapel Hill: University of North Carolina Press, 1990), 194.
4. Key, *Southern Politics in State and Nation.* Historically, Democratic factionalism in Mississippi often split geographically between the Delta in the western part of the state and the rolling hills found throughout much of the eastern part of the state. The racism of Delta planters was considered more paternal and respectable than the unadulterated version practiced by the poorer whites of the hills.
5. Earl Black and Merle Black, *The Rise of Southern Republicans* (Cambridge, MA: Harvard University Press, 2002). Throughout this essay the South is defined as the eleven former Confederate states: Alabama, Arkansas, Florida, Georgia, Louisiana, Mississippi, North Carolina, South Carolina, Tennessee, Texas, and Virginia.
6. Black and Black, *The Rise of Southern Republicans,* 21.
7. In addition to the mushrooming of large-scale catfish farms, in some Delta counties along the Mississippi River (i.e., Tunica), casino/gaming is now the largest employer. See Sharon D. Wright Austin, *The Transformation of Plantation Politics: Black Politics, Concentrated Poverty, and Social Capital in the Mississippi Delta* (Albany: State University of New York Press, 2006).
8. Eric Foner, *Reconstruction: America's Unfinished Revolution, 1863–1877* (New York: HarperCollins Publishers, 1988).
9. J. Morgan Kousser, *The Shaping of Southern Politics: Suffrage Restriction and the Establishment of the One-Party South, 1880–1910* (New Haven: Yale University Press, 1974), 144.
10. The US Supreme Court struck down the white primary in the 1944 Texas case of *Smith v. Allwright,* 321 US 649 (1944). The white primary refers to the exclusion of African Americans from voting in the Democratic primary.
11. Parker, *Black Votes Count,* 31.
12. Ibid., 43.
13. According to V. O. Key, "In the South registration assumes special importance because of the peculiar regional suffrage qualifications. Registration authorities determine whether applicants meet literacy and understanding tests and thus have functioned as the principal governmental agency for Negro disfranchisement" (Key, *Southern Politics in State and Nation,* 560).
14. Key, *Southern Politics in State and Nation,* 566–67.
15. C. Vann Woodward, *The Strange Career of Jim Crow* (New York: Oxford University Press, 2001); Brown v. Board of Education of Topeka, 347 US 483 (1954).

16. Parker, *Black Votes Count,* 31.
17. All of these minority vote dilution schemes are detailed in Parker's *Black Votes Count.*
18. Parker, *Black Votes Count.*
19. Ibid.
20. Edward G. Carmines and James A. Stimson, *Issue Evolution: Race and the Transformation of American Politics* (Princeton: Princeton University Press, 1989); M. V. Hood III, Quentin Kidd, and Irwin L. Morris, *The Rational Southerner: Black Mobilization, Republican Growth, and the Partisan Transformation of the American South* (New York: Oxford University Press, 2012).
21. Black and Black, *The Rise of Southern Republicans.*
22. Parker, *Black Votes Count,* 62.
23. Ibid.
24. Alexander P. Lamis, *The Two-Party South* (Oxford, UK: Oxford University Press, 1988).
25. Parker, *Black Votes Count,* 147.
26. Seth C. McKee, "The Past, Present, and Future of Southern Politics," *Southern Cultures* 18 (2012): 95–117.
27. The Deep South is defined as Alabama, Georgia, Louisiana, Mississippi, and South Carolina. Compared to the other six southern states (referred to as the Peripheral or Rim South), the Deep South contains a much higher percentage of African Americans, and this mainly accounts for the greater racial conservatism of whites residing in this southern subregion (see Black and Black's *The Rise of Southern Republicans*; Hood, Kidd, and Morris's *The Rational Southerner*; Key's *Southern Politics in State and Nation.*
28. Kari Frederickson, *The Dixiecrat Revolt and the End of the Solid South, 1932–1968* (Chapel Hill: University of North Carolina Press, 2001).
29. Black and Black, *The Rise of Southern Republicans.*
30. Data accessed on August 17, 2014, from CNN's website: http://www.cnn.com/election/2012/results/state/MS/president.
31. Data accessed on August 17, 2014, from the following US Census Bureau website: http://quickfacts.census.gov/qfd/states/28/28051.html.
32. Data accessed on August 17, 2014, from the following US Census Bureau website: http://quickfacts.census.gov/qfd/states/28/28015.html. It is worth mentioning that in the Delta, Barry Goldwater did best in Carroll County in 1964 (97 percent of the vote) when blacks were disfranchised.
33. Parker, *Black Votes Count,* 165.
34. Key, *Southern Politics in State and Nation,* 9.

CHAPTER 10

The Folklorist's Delta

William Clements

A century or so ago A. H. Burnette from Rena Lara, Mississippi, heard "an aged cotton picker of the Mississippi River country" named Mrs. Nixon perform a ballad that she called "Mary and Willie," the words of which the London printer James Catnach had printed sometime early in the nineteenth century. The narrative folksong tells a romantic tale: Several years after Willie had taken leave of his beloved Mary to go to sea, a half-blind beggar turns up at her door and offers to tell her fortune. The good news is that Willie is still alive, but the bad news is that he is in such dire poverty that he cannot think of marrying the love of his life. Mary, though, effuses that her devotion persists despite his economic plight. The beggar then reveals himself to be Willie and advises Mary that she has passed the love test that he has set for her.[1] Ballad scholar G. Malcolm Laws catalogued this piece as "N 28" in his index of American ballads based upon British broadside originals. He noted that North American texts in addition to Mrs. Nixon's from the Mississippi River Delta had been reported from Missouri, Maine, Indiana, Massachusetts, and Tennessee. The song, in fact, relates a formulaic story, similar to that narrated in many broadside ballads, that has nothing particularly to do with the time or the place of Mrs. Nixon's performance of it.[2]

During the summer of 1976 folklorist Simon Bronner was doing fieldwork in northwestern Mississippi when he encountered a cycle of stories featuring a figure called "Bad Man Monroe," who could outfight, outdrink, and out-curse anyone in the vicinity. He was known for his violent nature, which expressed itself, among other ways, in overt defiance of the white establishment. Bronner tape-recorded one story about this character from bluesman Eugene Powell in Greenville, Mississippi. Like many of his contemporaries, Monroe made at least part of his living by working in the cotton fields—though he did so armed with a pistol. During one picking season, he came into conflict with Mr. Smith, the overseer, who also habitually carried a firearm. Though Monroe would turn in several hundred pounds of harvested cotton at day's end, he risked chastisement because he included too much trash with the actual bolls. Upon being warned that Mr. Smith was going to "get on" him, Monroe defiantly threw whole stalks of cotton into his sack for weighing. Mr. Smith was about to remonstrate when he realized the identity of the person who had filled the bag. Instead of scolding Monroe or docking his pay, the overseer meekly suggested, "Well, try and get it a little better."[3] Although the defiant and rebellious male had become a widely known stock figure in African American folklore by the mid-twentieth century, this story about Bad Man Monroe reflects specifics of the Delta experience: especially, the setting in one of the region's ubiquitous cotton fields at a time in the growing cycle that once was especially labor intensive and the traditional relationship between white overseer and black worker that Monroe upends.[4]

A widely reported urban legend that deals with one of the most popular themes in this modern genre of folk expression and that owes its dissemination as much to the press and the Internet as to oral tradition tells of a man's awaking in a hotel bathtub full of ice. He's feeling pretty bad, much worse than the hangover that his dissipation the previous night would account for. As he tries to rise from the tub, he notices a sharp pain in his lower left side and discovers a gash that has been hastily stitched together. The hotel's onsite physician in consultation with specialists determines that one of his kidneys has been removed, presumably to be marketed illegally. In fact, a rash of such occurrences has come to the attention of medical personnel in the area. The most usual setting for this story seems to be Las Vegas, but the location of the incident has also been reported as Atlantic City or, outside the United States, Hong Kong or Rio de Janeiro.[5] For people in eastern Arkansas and northern Mississippi, though, the kidney theft usually has occurred in Tunica, Mississippi, site of several casinos that cater to metropolitan Memphis, Tennessee, and other parts of the Mid-South. Yet while the setting of some versions of this urban legend is in the Delta's version of "sin city," the story has been told and retold internationally and apparently has little to do with the specifics of life in the region.

These three examples—a British broadside ballad at least two centuries old, a narrative from a cycle of stories about a notorious regional character, an urban legend known throughout the world but with a local setting—represent regional Delta folklore. Each, though, relates to the region in a very different way. Taken out of context, they might not even be considered "regional." But as folklorists know, some sense of place almost invariably figures into folklore and its performance. Appreciating how that sense of place operates must begin with the place itself. Often the place has been denominated a "region," but what exactly this means both in general and in the specific terms of the Delta has not always been unquestionably clear.

Recognizing the importance of place accompanied the emergence of the formal study of folklore in the late eighteenth and early nineteenth centuries. For example, Jakob and Wilhelm Grimm, viewed as forerunners of academic folklore studies, found motivation for collecting oral narratives largely in their desire to identify an imagined German-speaking community that transcended what they regarded as artificial political boundaries. Influenced by the nationalistic philosophy of Johann Gottfried von Herder, the Grimms sought a common German identity that manifested itself in the literature of a *Volk*, a vernacular verbal art that included orally circulating *Märchen* and *Sagen* and that reflected the unique *Volksgeist* of a German people who could make their distinctive contribution to *Humanität*, the highest and best that mankind could achieve, only by celebrating and developing their folk literature. Following the example set by the Grimms, European collectors and popularizers of what was perceived as place-oriented folklore attempted to create national identities through their attention to ballads, legends, folktales, and other orally expressed verbal art as well as customary traditions and material culture for the next hundred years and more.[6] When British antiquarian William J. Thoms introduced his neologism "folklore" in 1846, more than thirty years after the first edition of the Grimms' *Kinder—und Hausmärchen* was published, he invoked the connection between the local culture "of the olden time," which he hoped some British "James Grimm" would record, and a sense of being British.[7] During the first half of the twentieth century, American folklorists continued to stress the relationship between place and folklore, and while a few attempted to characterize a general "American" folklore, serious students of "unofficial expressive culture" realized that such a relationship was more realistically evident in terms of smaller, more homogeneous geographic entities. They consequently turned their attention to regions.

When Richard M. Dorson, who contributed significantly toward establishing folklore studies as a presence in American academe, developed his "hemispheric theory" of the nation's folklore—the idea that American folklore could best be appreciated within the context of distinctive American cultural movements—he focused particular attention on regionalism and illustrated its importance in his book *Buying the Wind*, which sampled the folklore of seven distinctive American regional populations: Maine "Down-Easters," Pennsylvania "Dutchmen," "Southern Mountaineers," Louisiana Cajuns, Illinois Egyptians, Southwest Mexicans, and Utah Mormons.[8] The people of the Mississippi River Delta are notably missing from Dorson's case studies, probably in part because at the time *Buying the Wind* appeared, the folklore of those living on the alluvial floodplain of the Mississippi River had not been sufficiently recorded for Dorson to sample. He himself had done some fieldwork in the Delta, collecting folk narratives from African American narrators in Pine Bluff, Arkansas, and Mound Bayou, Mississippi, but the Delta had not yet fully emerged into the folkloristic consciousness as a "region" upon which folklorists might concentrate research and preservation efforts.[9]

Dorson defines the regions whose folklore he presents inconsistently. Although he does not rigorously demonstrate distinctively shared cultural factors in association with particular geographical entities, possession of such factors is the criterion that marks off "Down East Maine," the German neighborhoods of southeastern Pennsylvania, the southern mountains, the Hispanic Southwest, and Cajun country in Louisiana. The political boundaries of the state of Utah demarcate the regional folklore of Utah Mormons. For Egypt in southern Illinois, a local sense of geographical identity seems to be the defining factor. Moreover, in some cases Dorson's "regional" folklore may not actually be regional at all. The material presented from Mormon Utah, for example, might more accurately be termed "religious" folklore since many of its themes and motifs as well as complete stories and songs comprise the folklore of members of the Church of Jesus Christ of the Latter-Day Saints throughout the world. Meanwhile, the lore of Maine Down-Easters seems just as much occupational as regional, and ethnicity is the defining feature of folklore that Dorson reports from Pennsylvania Germans, southwestern Latinos and Latinas, and the descendants of exiles from Acadia in Louisiana. As one critic of *Buying the Wind* has noted, that a group of people has connections with a particular local geography does not necessarily make their folklore "regional." Only where culture and that geography interact does a truly regional folklore emerge.[10]

Dorson's book suggests a recurrent problem with regionalism in American folklore studies. Although region has often ostensibly been an important delimiting element in the collection of American folklore, impressionism, vagueness, inconsistency, and local boosterism have often characterized the concept. In the most thorough historical survey of the concept of region in American folklore studies, William E. Lightfoot has noted the divergent uses of the idea: "Regions are defined willy-nilly on the bases of physical characteristics, states of mind, political maneuverings, uniformity of cultural traits, distinctiveness, collective personality, consciousness, and single-trait variables."[11] Lightfoot has suggested a methodology for studying American regional folklore that begins with identifying the region itself with more precision than has usually been the case. He notes that frequently regions are perceived in terms of a core and periphery. Pinpointing the core and working out from it, Lightfoot suggests, enables the folklorist to delimit the region's boundaries.[12] But how to effect that delimitation is not always clear-cut. As I have suggested elsewhere, when folklorists have bothered to define the region upon whose folklore they are focusing, they have used varying criteria as foundations for what they mean by "region."[13] Thinking of the Delta in terms of a couple of those criteria helps to clarify and synthesize how folklorists could conceive of Delta folklore as regional folklore.

The most frequently employed basis used by folklorists for demarcating regional boundaries is that which apparently informs most of Dorson's examples: the folklorist's perception of distinctive cultural traits evident in and responsive to a particular geographical environment. Akin to the "analytical categories" that folklorists have used to create their cross-cultural definitions of genres of folk expression, this approach to regionalism has the researcher noting the prevalence and co-occurrence of easily documented cultural forms, ones that he or she can record and, if so inclined, plot on a map to demarcate clearly regional boundaries.[14] Although Lightfoot refers to regions identified in this way as "ad hoc" and characterizes them as "hypothetical, contrived, artificial," mapping documented occurrences of cultural forms to demonstrate their spatial distribution and to define regions when maps of different forms were superimposed on one another was a specialty of many European folklorists during the twentieth century, especially those focusing their attention on material culture and folk speech.[15] Showing that a particular type of corner-timbering in log architecture co-occurred geographically with a type of ploughshare and other instances of material culture might establish the boundaries of a regional culture for the Swedish or Danish folklorist.[16] In the United States, this approach to defining a region, when it has gone beyond the impressionism that characterized most of Dorson's examples, has also usually involved material culture and, sometimes, language. The most influential example of such an approach to defining region has been Henry Glassie's early work in identifying patterns in material folk culture in the eastern United States.[17]

A folklorist hoping to establish the boundaries of a Delta region using this analytic methodology might begin by mapping examples of shotgun houses. These residential structures, characterized by floor plans in which rooms are stacked one after another in a straight line and probably preserving a West African architectural type, proliferate on both sides of the Mississippi River in urban and rural neighborhoods from New Orleans to Saint Louis. Mapping only shotgun houses (or any other single example), though, does not delineate a folk region. One should also consider the distribution of other cultural traits—say, the tradition of playing one-string musical instruments and the terminology (perhaps "diddly bow") used to name them. The folklorist might also consider the musical style often referred to as the "Delta blues" and attend to recurrent topics of folk narratives such as an unfortunate history of racial violence, the impact of natural disasters such as tornadoes and floods, and the ravages of the boll weevil and other pests on cotton crops. While no one specific cultural form, none of which may, in fact, be found exclusively in the Delta, will define the Delta as region, the co-occurrence of several forms concentrated in a more or less clearly marked geographical area allows the researcher to think in terms of regional identity from an analytic perspective. Using this approach, a folklorist could characterize the Delta region as a geographical entity where shotgun houses with diddly bows strung along their outside walls provide venues for a heavily chorded, intense blues style that favors the use of slides on open-tuned guitars by performers who tell stories about race-based injustice, the terrors of tornadoes and floods, and the insects and other vermin who can destroy their means of livelihood.[18] The geographical boundaries which distinguish such a region will be nebulous, of course, but a core and periphery may be identified where these and other features of a regional folk culture proliferate.

Though American folklorists have seldom engaged in precise mapping of cultural traits to hone their concept of region, what passes for an analytic perspective has usually generated their concept of region: a place noted by the recurrence of particular cultural forms which the researcher has discovered. However, people who live in those "regions" may not overtly express much sense of regional identity, or their sense of region may not correspond to the boundaries between culture forms that the scholar has noted. Instead, they may evince a sense of regional identity which corresponds with what folklorist Dan Ben-Amos has called an "ethnic construct" in his work with folklore genres.[19] Lightfoot's term is "ontic region," and some cultural geographers have used "vernacular region" to refer to geographical entities that are "perceived to exist by their inhabitants and other members of the population at large." Not the products of academic analysis, such regional concepts result from "the spatial perception of average people. Rather than being based on carefully chosen, quantifiable criteria, such regions are composites of the mental maps of the population." The "mental maps" upon which vernacular regions are founded may have only vague resemblance to the products of formal cartography, but they reflect how people perceive themselves in spatial terms and consequently shape sense of identity.[20] For example, the "Egypt" region of southern Illinois, which Dorson included in *Buying the Wind,* may not be distinguishable in terms of cultural forms from neighboring geographical entities, but local perception—associated with the city of Cairo at Illinois' southern tip near the confluence of the Ohio and Mississippi Rivers—has traditionally noted a regional distinctiveness. Aside from local legends reflecting specific places and historical circumstances in Egypt, most of the folklore reported there can also be encountered in similar combinations if one crosses the Mississippi into Missouri, the Ohio into Kentucky, or the Wabash into Indiana.

People in Mississippi identify the Delta with some precision: the triangle of land whose apex is Memphis, Tennessee, whose eastern edge lies where the hills begin a few miles west of Interstate Highway 55, whose western boundary is coterminous with the Mississippi River, and whose base more or less follows the path of the Yazoo River.[21] In journalist David Cohn's often cited characterization, the Mississippi Delta begins in the lobby of Memphis's Peabody Hotel and ends at Catfish Row in Vicksburg. Regardless of the fact that there are just as many shotgun houses and bluesmen who began to learn their music by playing a diddly bow on the Arkansas side of the river, this is the Delta from a vernacular perspective in Mississippi. Meanwhile, what an analytic perspective might identify as the Delta may be known by other terms, or it may be subdivided into vernacular regions such as the "Bootheel" in southeastern

Missouri or Buffalo Island in northeastern Arkansas. As Dorson noted, "This self-awareness of a regional identity, often symbolized by a nickname, may indicate the existence of a regional culture more surely than any markings on the physiographic or population map."[22] An individual's sense of regional identity on the vernacular level may shift with the circumstances when he or she is thinking about that identity. Someone in the Buffalo Island region of Arkansas may stress that identity when considering himself or herself vis-à-vis people who live in the Missouri Bootheel or on nearby Crowley's Ridge, but "Delta" might be the regional identity of choice if one is placing himself or herself within the whole of the state of Arkansas or of the South. Meanwhile, people from outside a region may have a vernacular sense of the region's boundaries. For example, Alan Lomax notes that people from the Mississippi hill country defined the Delta as "the great crescent of land south of Memphis that was once covered by the river in its flood and is now protected by the levee."[23]

Folklorists find benefit from both approaches to regionalism. Looking at regions through the eyes of the analyst does afford an opportunity to make connections between cultural forms and local geography, to understand the ways in which cultural constructs respond to natural and social forces to produce appropriately distinctive combinations. Those who have turned their attention primarily to the second syllable in Thoms's neologism "folk-lore" have perhaps been more likely to think of "region" in these analytic terms. But those whose emphasis has been on the bearers of that lore—the so-called folk—may find the concept of vernacular region more to their liking. If the folklorist as ethnographer is attempting to see "with a native eye," he or she will want to know how people actually think of themselves, and that includes a geographical identity which may in some cases be national but in the affairs of everyday living, even in an age of instant international communication, is likely to be local or regional. And how people think of themselves regionally does not always conform to the analyst's core and periphery.

We should also probably note that folklorists may sometimes think of a region as an "administrative unit." Conceptualizing an "Arkansas folklore" which corresponds to political boundaries which have little to do, in many cases, with quotidian existence, for example, has its value when one wants to familiarize the subject for the general public, publishers, granting agencies, or branches of government. For the Delta, the authorization of the Lower Mississippi Delta Development Commission by the US Congress in 1988 brought together a vast geographical realm that encompassed territory in seven states: not only Mississippi and Arkansas, but also Illinois, Missouri, Kentucky, Tennessee, and Louisiana. But beyond its political and economic expediency, this view of the Delta probably has little to do with folklore and folklife.

What constitutes the folklore of the Delta depends on how one defines the region. It also requires that one recognize that regional folklore relates to local topography, climate, flora and fauna, demography, and history in different ways. Examples of Delta regional folklore include material whose dependence on local particulars are tenuous and sketchy as well as materials which could have developed only in the Delta and require a knowledge of the region to be fully appreciated. Consider, for example, Mrs. Nixon's ballad "Mary and Willie." It relates to the Delta only through its performance situation. The singer was in the Delta when she performed the song. It tells a story that has no special connection to life in the Delta and, in fact—like the ballads of knights and ladies that drew regional folklore collector Vance Randolph to the Ozarks—tells a somewhat exotic story of a love that transcends not only the passage of time but the prospect of poverty. Just as Mrs. Nixon was *in* the Delta at the time she sung the ballad, we might think of this ballad—as well as other British ballads and European wonder tales encountered in the performance repertoires of Delta residents—as exemplifying folklore *in* the region. But also consider the cycle of Bad Man Monroe stories. Although they focus on a stock character, the one in which the protagonist faces down an overseer over his cotton-picking peccadillos could probably have developed only in the Delta. To appreciate it fully, one needs to understand the relationship dynamics between the local black population—only a few generations from slavery and much less than that from Jim Crow—and the white establishment, still in many cases burdened with the legacy of unquestioned white supremacy. Comprehending the story also requires some knowledge of an economic system in which cotton was king. Although he might try to subvert it, even a notorious figure such as Bad Man Monroe had to participate in that system. His defiance of the expected adherence to a set of economic customs—in this case, those that informed the exchange of a cotton picker's labor for a just recompense—defies assumptions that had characterized Delta life since the land had been cleared for cultivation by cotton entrepreneurs in the late nineteenth century. This tale, which dramatizes a successful challenge to those assumptions, represents folklore *of* the region.

Some Delta folklore seems to fall somewhere between that *in* and that *of* the region. Like British broadside ballads, it may evince motifs and formulas which recur elsewhere; but, like Bad Man Monroe stories, it may nevertheless depend upon, respond to, and mirror its local context. Such folklore results from a process whereby it "makes itself at home in a new geographical environment."[24] Regionalization may be quite superficial, localizing an urban legend in a specific regional venue, for instance, or attaching a widely known story about a social bandit who robs from the rich to give to the poor to a regional folk hero. Sometimes the process may involve much deeper changes, though, and involve a process for which folklorists have borrowed the biological term "oikotypification." For instance, an internationally known folk narrative motif has a local character effecting a bargain with an evil force in which he exchanges his spiritual essence for a boon of some sort.[25] While the outlines of the episode are not contingent on any particular locale, they assume regional significance in the Delta when the bargainers are a famous bluesman (usually Robert Johnson) and Satan and the terms of exchange are the bluesman's soul as barter for the ability to perform the blues with uncanny art and skill. Identifying the site of the bargain as the crossroads where US Highways 61 and 49 intersect in Clarksdale, Mississippi (perhaps the regional core in Lightfoot's terms), enhances the Delta connection. Regionalized folklore joins folklore in and of the region to constitute regional folklore.

Defining the Delta is no simple process for the folklorist. She or he must decide on the bases for demarcation, must determine whether to consider the region monolithically or in terms of its subregions, and must recognize that once regional boundaries have been marked, the focus for study—the folklore itself—will relate to the bounded area in varying ways. Looking at articles appearing in *Journal of American Folklore,* the principal outlet for publishing in folklore studies, over the past generation, Simon Bronner has suggested that regionalism has lost some of its centrality for American folklorists.[26] But other evidence—that of folk festivals, regional folklife initiatives, and even book-length publications—seems to indicate that regionalism still enjoys a firm following. Delta folklore received important attention from some of the twentieth century's most important documentarians of American folklore—Alan Lomax, William R. Ferris, Judy Peiser, and David Evans, to cite four especially significant figures.[27] As the Delta has emerged more and more into the general public consciousness, their efforts can provide a model for other students of regional folklore who should not only carry on their focus largely on traditional music but also explore other ways in which traditional genres of expressive culture (stories, songs, arts, crafts, and architecture, among others) in the Delta and of the Delta have joined regionalized folklore to reflect and to shape the lives of those connected to a specific sense of place.

Notes

1. Arthur Palmer Hudson, *Folksongs of Mississippi and Their Background* (Chapel Hill: University of North Carolina Press, 1936), 153–54.
2. G. Malcolm Laws Jr., *American Balladry from British Broadsides: A Guide for Students and Collectors of Traditional Song* (Philadelphia: American Folklore Society, 1957), 217–18.
3. Simon J. Bronner, "Bad Man Monroe Legends from the Delta Region of Mississippi," *Mid-South Folklore* 5, no. 2 (1977): 54.
4. This character type has been common in African American folklore from throughout the country, including the Mississippi River Delta. Alan Lomax provides other examples in *The Land Where the Blues Began* (New York: New Press, 1993), esp. 216–26, 280–81. For instances of expressive culture about this stock character beyond the Delta, see, for example, two books by Roger D. Abrahams: *Deep Down in the Jungle: Negro Narrative Folklore from the Streets of Philadelphia,* 1st rev. ed. (Chicago: Aldine, 1970), 61–85; and *Positively Black* (Englewood Cliffs, NJ: Prentice-Hall, 1970), passim.

5. Véronique Campion-Vincent, *Organ Theft Legends* (Jackson: University Press of Mississippi, 2005), 24–36.
6. A useful case study of the interplay between folklore studies and nationalism is William A. Wilson, *Folklore and Nationalism in Modern Finland* (Bloomington: Indiana University Press, 1976). See also Roger D. Abrahams, "Phantoms of Romantic Nationalism in Folkloristics," *Journal of American Folklore* 106, no. 419 (1993): 3–37.
7. William J. Thoms, "Folklore," in *The Study of Folklore*, ed. Alan Dundes (Englewood Cliffs, NJ: Prentice-Hall, 1965), 4–6. See also Jack Zipes, ed. and trans., The Complete Fairy Tales of the Brothers Grimm (New York: Bantam, 1987); Richard M. Dorson, The British Folklorists: A History (Chicago: University of Chicago Press, 1968), 75-86.
8. Richard M. Dorson, "A Theory for American Folklore," *Journal of American Folklore* 72, no. 285 (1959): 197–242; Richard M. Dorson, *Buying the Wind: Regional Folklore in the United States* (Chicago: University of Chicago Press, 1964). Dorson did not include the Ozarks among his regions because he believed that the region's folklore could be easily accessed through examining the publications of a single collector, Vance Randolph.
9. For Dorson's folktale collecting in the Delta, see Richard M. Dorson, *American Negro Folktales* (Greenwich, CT: Fawcett, 1967), 39–47. Recent brief surveys of Delta folklore as regional folklore include Richard Alan Burns, "A Folklife Survey of the Arkansas Delta," *Mid-America Folklore* 25, no. 1 (Spring 1997): 1–13; Richard Alan Burns, "Mississippi Delta," in *The Greenwood Encyclopedia of World Folklore and Folklife*, ed. William M. Clements (Westport, CT: Greenwood Press, 2006), 4:166–75; and William M. Clements, "Delta, Mississippi River," in *Encyclopedia of American Folklore*, ed. Simon J. Bronner (Armonk, NY: M. E. Sharpe, 2006), 1:288–91.
10. Suzi Jones, "Regionalization: A Rhetorical Strategy," *Journal of the Folklore Institute* 13, no. 1 (1976): 109.
11. William E. Lightfoot, "Regional Approach," in *Folklore: An Encyclopedia of Beliefs, Customs, Tales, Music, and Art*, ed. Thomas A. Green (Santa Barbara: ABC-CLIO, 1996), 2:705.
12. William E. Lightfoot, "Regional Folkloristics," in *Handbook of American Folklore*, ed. Richard M. Dorson (Bloomington: Indiana University Press, 1983), 188.
13. William M. Clements, "The Folklorist, the Folk, and the Region," *Missouri Folklore Society Journal* (1979): 44–53.
14. Dan Ben-Amos, "Analytic Categories and Ethnic Genres," in *Folklore Genres*, ed. Dan Ben-Amos (Austin: University of Texas Press, 1976), 215.
15. Lightfoot, "Regional Folkloristics," 186.
16. Robert Wildhaber, "Folk Atlas Mapping," in *Folklore and Folklife: An Introduction*, ed. Richard M. Dorson (Chicago: University of Chicago Press, 1972), 479–96.
17. Henry Glassie, *Pattern in the Material Folk Culture of the Eastern United States* (Philadelphia: University of Pennsylvania Press, 1968).
18. For shotgun houses, see John Michael Vlach, "The Shotgun House: An African Architectural Legacy," *Pioneer America* 8, no. 1 (1976), 47–56; and John Michael Vlach, *The Afro-American Tradition in Decorative Arts* (Cleveland, OH: Cleveland Museum of Art, 1978), 122–31. For diddly-bows, see David Evans, "Afro-American One-Stringed Instruments," *Western Folklore* 29, no. 4 (1970): 229–45. For the "Delta" blues, see William R. Ferris, *Blues from the Delta* (Garden City, NY:

Doubleday, 1978; New York: Da Capo, 1984); and Robert Palmer, *Deep Blues* (New York: Penguin, 1981). For legends and other narratives, see William R. Ferris, "Black Prose Narrative in the Mississippi Delta: An Overview," *Journal of American Folklore* 85, no. 336 (1972): 140–51; William R. Ferris, *Give My Poor Heart Ease: Voices of the Mississippi Blues* (Chapel Hill: University of North Carolina Press, 2009); and Lomax, *The Land Where the Blues Began.*

19. Ben-Amos, "Analytic Categories and Ethnic Genres," 215.
20. Lightfoot, "Regional Folkloristics," 186; Terry G. Jordan, "Perceptual Regions in Texas," *Geographical Review* 68, no. 3 (1978): 293; Wilbur Zelinsky, "North America's Vernacular Regions," *Annals of the Association of American Geographers* 70, no. 1 (1980): 1–16. The term "mental maps" comes from Peter Gould and Rodney White, *Mental Maps* (Harmondsworth, Middlesex: Penguin, 1974).
21. Ferris, *Blues from the Delta*, 3–9.
22. Dorson, *Buying the Wind*, 289.
23. Lomax, *The Land Where the Blues Began*, 34
24. Jones, "Regionalization," 110.
25. Stith Thompson identifies the most generalized form of this narrative pattern as motif M210, "Bargain with devil." More specifically, he cites M211, "Man sells soul to devil." And still more specifically, Thompson lists the boon that is offered in return for the soul: help in building a house or barn (M211.2), a trip in a flying boat (M211.6), and assistance in eluding capture (M211.7), for example. See Stith Thompson, *Motif-Index of Folk-Literature: A Classification of Narrative Elements in Folktales, Ballads, Myths, Fables, Mediaeval Romances, Exempla, Fabliaux, Jest-Books and Local Legends, vol. 5, L-Z,* rev. ed. (Bloomington: Indiana University Press, 1955–58), 39–40.
26. Simon J. Bronner, *Explaining Traditions: Folk Behavior in Modern Culture* (Lexington: University Press of Kentucky, 2011), 57–58.
27. Alan Lomax began his work in the Delta while assisting his father, John A. Lomax, in collecting traditional music for the Library of Congress Archive of Folk Song in the 1930s. He returned to the Delta on several occasions and described his experiences there and sampled his (largely musical) findings in *The Land Where the Blues Began.* The other three folklorists initiated field research in the Delta in the 1960s, and each has contributed not only written documentation, but also audio recordings and films. See especially two works by William R. Ferris, *Blues from the Delta* and *Give My Poor Heart Ease*; and one by David Evans, *Big Road Blues: Tradition and Creativity in the Folk Blues* (Berkeley: University of California Press, 1982). Judy Peiser is longtime director of the Center for Southern Folklore in Memphis, which documents and preserves Delta traditional culture.

CHAPTER 11

Mississippi Moan

THE BLUES MUSIC LEGACY OF THE DELTA

WILLIAM L. ELLIS

Music of the Lower Mississippi Delta is as broad and serpentine as its namesake river, a dynamic confluence of traditional and popular sounds that has informed much if not most American music for more than a century. Musicians and music cultures of this particular Delta participated significantly in the development and popularization of blues, jazz, rock and roll, certain gospel styles, funk, soul, hip-hop, and even Jamaican ska. The Delta, in fact, is a complex of musical identities rich in African, Caribbean, European, and Native American antecedents filtered to a great extent through the lens of the African American experience. And while the Delta is but one component within a larger cultural hearth, namely, the Deep South, and within African American culture overall, in the inception and diffusion of the above musical forms and more, it is arguably the single most significant area in the United States where such creativity took place.

Regional interaction and the passing down of musical traditions within families and communities played a central role in the music that developed in the Delta, but so did the migratory routes of rivers, rail, and roads, all means by which Delta musical culture spread, sometimes as quickly as it was being born. In many ways, circulation of the music through objectification (i.e., the 78 record, jukebox, and radio) was especially significant for musicians, not just in bringing Delta music to the world but in bringing the world to the Delta. Leflore County–born blues legend B. B. King, for example, was less informed as a Delta youth by local guitar players than he was by the 78 records he heard on his aunt's Victrola, namely, those by the first two black guitar stars on commercial record: Texas bluesman Blind Lemon Jefferson and New Orleans–born Lonnie Johnson. King related in his autobiography: "As a Delta boy, I'm here to testify that my two biggest idols—guys I flat-out tried to copy—came a long way from Mississippi. . . . I later learned about Delta bluesmen like Robert Johnson and Elmore James and Muddy Waters."[1]

On a macro level, Delta music can almost be thought of in a bifurcated sense, namely, the musical styles that developed in New Orleans and surrounding rural Louisiana near the mouth of the Mississippi—the river delta proper—as well as the styles that are found in the alluvial flood plain that begins some two hundred miles farther north, the region between the Mississippi and Yazoo Rivers commonly recognized as the Mississippi Delta. Connected though they intrinsically were by the river and, from the late nineteenth century onward, by the Illinois Central Railroad that spanned New Orleans to Chicago, two truly unique musical destinations resulted, based on, among other things, differing social, economic, cultural, and geographic circumstances.

Founded in 1718, New Orleans maintains a rightful claim as the "Cradle of Jazz," where several centuries of cultural and ethnic interaction gave way to early jazz, brass band, and second-line traditions, as well as later urban forms, including the Crescent City's brand of rhythm and blues, rock and roll, and funk. As James Collier wrote in his definitive history, *The Making of Jazz*, "from its inception [New Orleans] faced south, a member of the Franco-Spanish culture of the Caribbean rather than of the Anglo-Saxon world to the north."[2]

Delta, then, when used in conjunction with music, more properly designates the sounds that emanated several hundred miles north, i.e., the Yazoo-Mississippi Delta, or the Yazoo Basin—the flat, fertile terrain of northwestern Mississippi where King Cotton established a highly different set of social, economic, and cultural interactions for African Americans.

For starters, the region was settled much later than New Orleans, having been an impenetrable wilderness of swamp and oak, gum, and cypress forest, riddled with panthers, bears, disease, and weather extremes that made drainage and clearing problematic until the mid-nineteenth century, when huge investments of white capital on the backs of black slave labor established the Delta's lucrative cotton plantation system. The rich soil deposits, created by centuries of flooding, were ideal for the cultivation of cotton in such Delta counties as Yazoo, which made Mississippi the nation's top cotton-producing state by the 1850s.[3] Post-Reconstruction, Delta cotton production made further gains and by 1915 was responsible for half of the state's crop, tended as it was by a populace that was nearly all black and all tenant farmers.[4] The cruel irony, of course, was that while many Delta landowners got rich, black farmers now generations removed from slavery were still reconciled to a life of servitude, thanks to sharecropping agreements that made them beholden to the plantation store, perpetually in debt.[5] Other African Americans moved in and out of the Delta, taking on seasonal work at plantations but also at levee, lumber, rail, and turpentine camps, creating an itinerant workforce and culture in which the blues would eventually flourish. As Leland music figure Shelby "Poppa Jazz" Brown wryly noted, the blues existed in Mississippi because of "the way they used to plow the folks here."[6] It's no coincidence, in fact, that blues arose during the decades of post-Reconstruction, when black southerners were at their most disenfranchised, when paramilitary "Red Shirts" suppressed the Mississippi black vote—by 1892 there were only 8,615 registered black voters in the state despite its black majority.[7] This was when Mississippi barely trailed Georgia in the heinous practice of lynching, including 195 documented cases (of which 190 were black) in the 1890s alone.[8] Within Mississippi, lynchings were particularly concentrated where there were the most black men, in the workforce-heavy Delta and neighboring counties.[9]

Under such conditions, music was one area where black identity and self-assertion thrived. Spirituals were the key to forging black consciousness—to paraphrase Lawrence Levine—in the South through much of the nineteenth century, while a number of forms came to fruition during and after post-Reconstruction, including ragtime, jazz, blues, and gospel music, all of which shifted the black musical experience from one of collective solidarity to levels of individual recognition, achievement, and self-reliance.

Through much of the nineteenth century, fiddles and banjos were the common instruments of the hearth and stage in the United States, played by black and white musicians alike. It has long been common knowledge that the banjo has special African provenance in the plucked lute traditions of the *xalam*, *ngoni*, *guinbri*, and *akonting*.[10] But scholarship now suggests that even the European violin, in the hands of African American players, at least, maintained aspects of West African bowed lute traditions in terms of presentation and practice.[11]

Although black music making had long been noted in certain places, such as the Sea Islands and New Orleans, the Mississippi Delta yielded fewer observations until a number of early twentieth-century song-collecting sojourns into the region. In antebellum Mississippi, the rhythmic handclapping of patting juba was observed as were slaves playing fiddles in Natchez.[12] One of the earliest mentions came from a writer who observed in 1858 a slave in Grenada performing a "songbow," a musical bow of clear African origin.[13]

A number of African musical survivals have been noted in Mississippi and its Delta, including the prevalence of ad hoc instruments, typically assembled from utilitarian items or found objects yet meeting the complex timbral requirements of traditional African American and African music and often made with same burst of spontaneity as the music itself. These include the earth bows of western and central Africa, reinterpreted as the washtub bass, as well as the mouth bow, panpipes, one-string fiddle, kazoo, jug, washboard, and one-string zither.[14] Best known as a "diddley bow" but also referred to variously as a "jitterbug," "one-strand-on-the-wall," and even "rumpling-steel-skin," the one-string zither has been shown to be of Central African origins in terms of performance style, method of construction, and enculturative function.[15]

Many bluesmen cited their first instrument as a homemade variant of the one-string, which was commonly made with a broom wire, bricks, nails, two-by-four, and a medicine bottle or other object for sliding, sometimes using the side of a house as a resonator. Bukka White, Muddy Waters, Buddy Guy, and Elmore James were among the many bluesmen that made one-strings before graduating to the guitar. Clarksdale, Mississippi, musician James "Super Chikan" Johnson fashioned diddley bows (his term) by the age of four or five:

> We were just doing what we saw other kids do. Broom wire, baling wire. If we couldn't find any wire, we'd burn an old tire to get the wire out of the inside of the tire and stretch it out there and string it up. Put Prince Albert can, snuff bottles or whatever we could find on it. We used to have a band out in the yard, man. Get large buckets for drums and molasses buckets. We'd take a molasses bucket and put just a little water in the corner of it, and I had a cousin that would beat the hambone on the molasses bucket, and he blew a turpentine bottle at the same time. Or he'd beat the ham bone on his hip and blow the bottle. Somebody else beating on a big large can, and we'd all be plucking our diddley bows. It'd be four or five diddley bows around plus a bass—we used to have a foot tub or a number three tub with a broom handle and a cord tied onto the broom handle. We could pull it tight and pluck that cord and the old tub would make a sound—dum-dum-de-dum-dum. You could tighten up—de-doo-doodle-doo; you could lay off—dedoo-doo-doo. . . . We was making all kinds of noises, just banging away. . . . You hear that all across the field, man.[16]

As much if not more than other regions in the South, Delta blues confronted the ever-present state of violence and racial tension at the time, often through less-than-veiled lyrics about incarceration: "Viola Lee Blues" by the Memphis Jug Band and "Jail House Blues" by Robert Wilkins, for example, or "Parchman Farm Blues" and "When Can I Change My Clothes," both by Bukka White about his time at the notorious Parchman Farm penal facility. African American males could wind up at on a chain gang or work farm for any number of reasons, including the catchall violation of vagrancy, which often gave legal justification for the illegal displacement and conscription of working-age black men. The itinerant lifestyle of blues performers made them especially susceptible.[17] The theme of false detention found its way into the blues repertoire most notably as the blues ballad "Joe Turner," about the notorious bounty-hunting penal official Joe Turney (also the subject of August Wilson's play, *Joe Turner's Come and Gone*). Once when riding the rails on his way home, David "Honeyboy" Edwards was pulled off the train in Glendora and put onto a county farm for several months. As he recounted in his autobiography, "I did everything on that farm: worked in the field, chopped cotton, dug ditches barefooted with shovels. It was something like a penitentiary. They run us from daybreak till sunset, then locked us up every night after we was done working for the day in a great big old building, a prisoner cage."[18] Few offered a more harrowing description than Son House, himself a onetime Parchman inmate for murder, in a sixteen-bar blues song he first recorded for Paramount in 1930 as "Mississippi County Farm Blues." House gave another, more explicit version, however, for Alan Lomax a decade later. Mincing not a word, House sang a chilling declaration of all that was the Delta for black men. That he set the lyrics to the same folk tune as Blind Lemon Jefferson's 1927 hit "See That My Grave's Kept Clean (which also circulated as "Two White Horses in a Line" and "One Kind Favor") was even more telling. The implicit conclusion was imbedded in the arc of the melody: death was the inevitable outcome.

Son House, "County Farm Blues" (1941)

Now it's sad when you do anything that's wrong,
Now it's sad when you do anything that's wrong,
Now it's sad when you do anything that's wrong,
They'll sure put you down on the county farm.
* * *
Put you down under a man they call Captain Jack,
Put you under a man called Captain Jack,
Put you under a man they call Captain Jack,
He'll sure write his name up and down your back.
* * *
Put you down in a ditch with a great long spade,
Put you down in a ditch with a great long spade,
Put you down in a ditch with a great long spade,
Wish to God that you hadn't never been made.
* * *
On a Sunday the boys be looking sad,
On a Sunday the boys be looking sad,
On a Sunday the boys be looking sad,
Just wonderin' about how much time they had.[19]

Indeed, if the Mississippi Delta can be relegated to one overriding sound, it would be the blues, which remains both in terms of popular mythology and historical import the most significant cultural contribution of the region, a music that has traveled the world over and back, touching on everything it comes in contact with, from the earliest strains of jazz to the latest Malian guitar styles.

The origin of the blues remains vague. It likely developed sometime in the late nineteenth century somewhere in the Deep South and was created by young, restless, mostly male African Americans who trafficked the South's work camps, plantations, and entertainment districts, where music offered one of the few employment alternatives to tenant agriculture and the disheartening hold it had on black farmers.[20] What informed the blues is a bit more certain. The solo work song (field holler, moan, arhoolie, etc.), string-band music, black folk preaching, and ballad forms, especially the bad man or blues ballad, all helped give rise to the blues, a radically new genre for the time, notable for its asymmetrical stanzaic form and matter-of-fact, non-stereotypical portrayal of black life. The preferred instruments of early blues

FIGURE 11.1. Son House, Toronto, 1974. *Photograph courtesy of Dick Waterman.*

were the affordable and highly portable guitar and harmonica as well as the piano, which was popular not only in river city barrelhouses and bordellos but also at work camps where entertainment was a valued weekend commodity. Banjos, fiddles, mandolins, and the ad hoc instrumentation of the jug bands were also used, if with less frequency, in blues—Red Banks-born jug band leader Gus Cannon, for example, played banjo and jug. And while the Delta can't take credit for birthing the blues, the concentration of performers and recorded activity that took place in the Delta, not to mention the direct link it provided to northern electric blues and, subsequently, rock and roll, gives it arguably more blues provenance than most other areas of the Deep South.[21]

Blues and proto-blues strains were documented from 1890 on, including eyewitness accounts in river cities such as Florence, Alabama, St. Louis, Missouri, and Evansville, Indiana, indicating the music moved around via migrants, boatmen, and stevedores traveling and working on the river.[22] During a mound-site excavation in Coahoma County in 1901 and 1902, archaeologist Charles Peabody offered the first account of what we now recognize as blues from the Delta, mentioning such qualities as guitar accompaniment, spontaneity in performance, and themes of hard luck and love.[23] Another important early source came from sociologist Howard W. Odum, who collected plenty of blues verse in Mississippi's Lafayette County around 1905 through 1908; Odum pointed out that much of what he collected was found in variants in Georgia and Tennessee, testament as well to how fast the blues was traveling in its infancy.[24] The most famous early citation of the blues, however, came from W. C. Handy in 1903, during his tenure as a Knights of Pythias bandleader in Clarksdale. While waiting for a train at a Tutwiler depot, he encountered an African American man playing guitar with a knife in the slide manner and singing an AAA blues verse, "Goin' Where the Southern Cross' the Dog," about the Southern and Yazoo Delta railroads' destination of Moorhead. Handy famously called it "the weirdest music [he] had ever heard."[25] He was further inspired to pursue what would be a lucrative career composing and publishing blues after witnessing a ragtag blues trio (guitar, mandolin, and upright bass) in Cleveland, Mississippi, who played repetitive strains with "no clear beginning and certainly no ending at all" that Handy recognized as coming out of the levee camps and sugarcane farms and that earned the trio more money than his headlining band.[26] Handy took note and soon had great success with his own blues hits such as "The Memphis Blues," "St. Louis Blues," and "The Yellow Dog Blues" (which appropriated the line he first heard from that anonymous Tutwiler bluesman), thereby popularizing more than any other single composer/arranger a once traditional form via sheet music and the vaudeville stage prior to the era of blues recordings from 1920.

So what is Delta blues, a term so ingrained in the popular lexicon that it is sometimes conflated or confused with larger distinctions in blues? Broadly speaking, Delta blues is a subgenre of Deep South blues, the music of acoustic blues players—typically, guitarists, pianists, and harmonica players, as well as various ensembles such as jug bands—who lived, worked, and traveled in the first part of the twentieth century, a time period often demarcated by World War II, in a region encompassing no less than eight states, from western Georgia to northern Louisiana and up into Mississippi and its surrounding states (blues historians further differentiate Deep South blues from the contemporaneous counterparts, Texas blues and East Coast blues).

More specifically, Delta blues may refer at once to a regional music, a historic period, a school of like-minded practitioners, an overall style, and a repertoire. The region is perhaps easiest to define: that is, the alluvial floodplain that resides west to east from the Mississippi to the Yazoo River and south to north from Vicksburg to Memphis. Yet in terms of the music, the Arkansas Delta to the west of the Mississippi and parts of western Tennessee and even the Missouri Bootheel and southern tip of Illinois factor in as well (seminal St. Louis bluesman William Bunch, a decided influence on Robert Johnson and someone who recorded under various pseudonyms, most famously as Peetie Wheatstraw, likely hailed from the western Tennessee/eastern Arkansas axis).

Even adjacent areas of Mississippi and into Louisiana can fall under the aegis of a so-called Delta blues style, to the point that non-Delta players, through travel and contact, participate in and become known for the style. Tommy Johnson, for instance, lived much of his life in Crystal Springs near Jackson, though his time spent in the Drew area interacting with Charley Patton and Willie Brown has since earned him a reputation as a leading Delta blues performer. Likewise, the idiosyncratic, emotionally severe performances of Robert Pete Williams bear a generalized "Delta sound," although he came from the Baton Rouge area, the Lower Mississippi River Delta proper. Delta blues can also inadvertently get mixed in with its hill country cousin, the blues of counties to the east of the Delta: namely, Tate, Panola, Marshall, Lafayette, and the eastern portion of DeSoto. These counties were settled earlier than the Delta by black farmers living in relatively more isolated, subsistence-based conditions, and where older musical traditions had survived, such as fife and drum and the banjo alongside a hypnotic type of guitar blues found in such players as Mississippi Fred McDowell, Junior Kimbrough, and R. L. Burnside.[27] Indeed, some of the more intense traits of Delta blues share a strong affinity with the hill country style.

The fact is, blues, even within the Delta proper, exists as a more subtle and complex panorama of activity than can be explicated in this essay, namely, levels of traditions that build out from the individual to local, regional, and then more broad networks of contact and influence.[28]

The pre–World War II style of Delta blues, then, is as varied as its players, a blend of internal and external forces that makes it hard to delineate a single "Delta" sound. It can range in the prewar period from the fingerpicked,

Piedmont-like delicacy of Mississippi John Hurt to the mournful minor blues of Skip James, from the popular hokum double-entendre tunes of Armenter "Bo Carter" Chatmon to his sibling string band, the Mississippi Sheiks, whose 1930 hit "Sitting on Top of the World" quickly became a standard, informing performances by fellow Delta players such as Charley Patton ("Some Summer Day") and Robert Johnson ("Come On in My Kitchen"), and has since been covered by a litany of artists, black and white, including Howlin' Wolf, Ray Charles, Bob Wills, and Bill Monroe. Beyond the guitar, Delta blues was practiced by many masters of the harmonica (the "Mississippi saxophone" in blues parlance), such as Indianola-born Chicago star Jazz Gillum, and by a number of important piano stylists, such as Helena-linked, "'44' Blues" hitmaker Roosevelt Sykes. Delta blues can be a confusing label even within one performer. The most iconic of Delta bluesmen, for example, Robert Johnson, recorded in a number of blues styles ranging from riff-driven slide numbers to sophisticated urban arrangements and the kind of ragtime picking more associated with East Coast players. That he, like other blues players who made their living in part as entertainers, knew and played non-blues material such as cowboy songs, vaudeville hits, and even polka numbers further clouds the issue. But against the standardized twelve-bar blues lyric and harmonic form that emerged in a broader regional and historical sense, a certain Delta sound has been attached to early players including Robert Johnson, Tommy Johnson, Son House, Charley Patton, Willie Brown, Ishman Bracey, and Skip James, as well as Muddy Waters, John Lee Hooker, Elmore James, and others who later traveled out of the Delta to pioneer the northern electric blues scenes centered around Chicago and Detroit.

The Delta sound, at least in regard to prewar guitar-vocal styles, shares much with other blues guitar styles of the period, not to mention blues in general, though certain traits may appear more concentrated or intensified in a typical Delta performance compared to the Piedmont region of the Carolinas, Texas, or even other blues pockets of the Deep South.

Among these traits are the following:

(a) Forceful, passionate singing and the use of vocal extremes from deep growls to falsetto leaps, head voice, and exaggerated vibrato (Tommy Johnson, "Cool Drink of Water Blues"; Ishman Bracey, "Woman Woman Blues"; anything by Howlin' Wolf).
(b) Strong riffing and motivic repetition (Robert Johnson, "Preaching Blues").
(c) Stanzaic structures that trend toward a twelve- or eight-bar construction, albeit with internal phrasing that may clip or extend musical bars as befits a solo performer adhering to his or her own sense of timing and delivery.
(d) A highly percussive approach with the use of snap pizzicato on the strings, extraneous sounds that compliment and complicate the string's melodic lines, and incessant driving rhythms including syncopated

rhythmic patterning both as staccato melodic accents (Charley Patton, "Down the Dirt Road Blues") and walking bass motives (Tommy Johnson, "Big Road Blues"; Willie Brown, "Future Blues").

(e) Musical space through open, skeletal textures of contrastive bass and treble guitar lines against the vocal (Muddy Waters, "I Be's Troubled").

(f) The use of a slide device on the guitar strings.

(g) Reductive harmonic accompaniment to the extent that some performances are free of harmonic function, played either entirely in the tonic (Charley Patton, "Mississippi Boweavil Blues") or in a loose, heterophonic manner in which the guitar essentially mimics the vocal line (Robert Johnson, "Come On in My Kitchen"; Skip James, "Hard Time Killin' Floor Blues"). Some players, such as Robert Johnson, go so far as to deflect any implication of the V chord by over-accentuating the tonic note where the V chord change should be, which Johnson did to dramatic effect on such slide numbers as "Terraplane Blues," "Cross Road Blues," and "Stones in My Passway."[29]

(h) Profound philosophical conceits and existential wrangling in the lyrics (Robert Johnson, "Hell Hound on My Trail"; Bukka White, "Fixin' to Die Blues").

Coupled with the open-ended constructs and multiple levels of ambiguity found in the blues as a whole—from its allusive, coded, and often antithetical language to the neutral tones of the blues scale and to its unresolved sequences of dominant seventh harmonies—the Delta variety of prewar blues made artful reflection on the uncertainties and vagaries of African American life in the South, especially for those who (to paraphrase Robert Johnson) had rambling on their minds. It was, according to Robert Palmer in his landmark exploration, *Deep Blues*, a music bound by the great leitmotif of impermanence.[30]

We find further magnification of many of the above traits in the hill country blues, coupled as that style tends to be with a prominent droning quality, though the degree of open musical space tethered to equally stark emotional wresting in a typical Delta blues can set it apart from the average groove-oriented, country dance performances of hill country players such as Mississippi Fred McDowell, R. L. Burnside, and Junior Kimbrough.

A case in point is Son House's two-part masterpiece from 1930, "Dry Spell Blues."[31] Here, House, whose resume included both preacher and bluesman, displays many quintessential Delta blues qualities, including impassioned, emphatic vocals built on a pentatonic scale (B♭-D-F-G-A♭) that he manipulates to great effect via the neutral third-scale degree both in the voice and guitar; open, extreme textures on the guitar's bass and treble strings; use of a slide device on the strings; instrumental riffing as a structural component with only incidental harmony implied from the parallel thirds played on the slide; and the dark, disconsolate nature of the lyrics, which addressed a 1930 drought that devastated a large portion of the nation, including the Delta.[32]

FIGURE 11.2. Son House, "Dry Spell Blues, Pt. 1," Paramount 12990, 1930 (CD track time 0:05–0:37). House's guitar is in an open G tuning (D-G-D-G-B-D from low to high string), likely capoed to the pitch of B-flat, and partly played on the strings with a slide device. *Transcribed by the author from "Screamin' and Hollerin' the Blues": The Worlds of Charley Patton (Revenant 212, 2001); with kind permission from Revenant Records and the Hal Leonard Corporation.*

It's unclear when the first Delta blues performer was recorded. Blues, which had been steadily gaining popularity on the vaudeville circuit and through sheet music well into the 1910s, appeared as a structural device in various brass band and jazz recordings prior to the 1920 debut of "Crazy Blues" by Mamie Smith, the first recording of a blues song by an African American singer.[33]

But the first such hits in the heyday of the vaudeville blues craze were largely manufactured in the record label centers of the North, notably, New York and Chicago. Not until the mid-1920s, with the advent of electrical recording technology, did labels began taking field trips to the South to capture regional talent. Freddie Spruell and Sam Collins, both singer-guitarists who hailed from the Mississippi-Louisiana border, are but two candidates for the first blues recorded in what was essentially a Delta style: Spruell was already a Chicago resident when he made "Milk Cow Blues" in June 1926 for OKeh, while ten months later, Collins performed a slide guitar rendition of "Yellow Dog Blues" for the Gennett label, offering a glimpse into what W. C. Handy must have heard decades prior at that Tutwiler train depot. Also in 1927, William Harris recorded "I'm Leaving Town" in a Gennett session that bore the stamp of the Delta, notably in the guitar's strumming patterns and walking bass line, though lack of research leaves this intriguing figure only likely, if ostensibly, connected to the region.[34] The seminal moment, however, for Delta blues artistry happened when Victor label's talent scout Ralph Peer, who would soon audition Jimmie Rodgers and the Carter Family in the landmark "Bristol Sessions," first went to Memphis, where he recorded Tommy Johnson and Ishman Bracey, both accompanied by Charlie McCoy, on February 3 and 4, 1928. Johnson's first two performances, "Cool Drink of Water Blues" and "Big Road Blues," remain a two-sided classic essentially marking Delta blues' bow into popular music (Peer then did the same for West Tennessee blues a few weeks later with the recorded entrée of the Memphis Jug Band, the first act from the Volunteer State to have been recorded commercially).[35] Johnson and Bracey recorded for Peer through the advocacy of H. C. Speir, a white record-store owner and talent scout from Jackson, Mississippi. His ability in finding and promoting blues talent was indispensable in getting the music out of the Delta, and Speir would go on to introduce to labels and arrange commercial recording sessions for, among others, Charley Patton, Son House, Willie Brown, Skip James, and Robert Johnson. What we know of Delta blues from the period, in fact, is in large part a result of Speir's promotional instincts, so much so that his siring role of a Delta tradition on record makes him as crucial a figure to Delta blues as Peer was to country music and Sam Phillips was to rock and roll.[36]

Key recording dates followed, beginning with Patton's stunning first session on June 14, 1929, and his 1930 Paramount-label trip to Grafton, Wisconsin, with pianist Louise Johnson, Willie Brown, and Son House in his landmark debut a few years after learning the guitar. The Vocalion label released the influential 1929 two-sider, "Cottonfield Blues," parts 1 and 2, by Garfield Akers and Joe Calicott (from the Mississippi towns of Hernando and Nesbit, respectively), and put out in 1930 the hit version of "Bumble Bee," Memphis Minnie's signature number with Kansas Joe. Vocalion also put out the initial run of 78s by Robert Johnson, whose storied sessions took place in San Antonio in 1936 and Dallas a year later. He died in 1938 under still-puzzling circumstances (poisoning seems probable), only months before he would have performed at John Hammond's historic From Spirituals to Swing concert at Carnegie Hall, where several of Johnson's records played instead. Finally, the 1941/1942 Coahoma County field trips of Alan Lomax for the Library of Congress in conjunction with John Wesley Work III and Fisk University gave Son House a second opportunity to record and led to the discovery of McKinley Morganfield (aka Muddy Waters). It was here that House shared with Lomax one of his signature tunes, "Delta Blues," a song that never mentions the Delta other than in the title but makes clear from its pervasive feelings of anxiety and unrest that the Delta for African Americans was at once a physical place and a state of mind, a demoralizing condition that could only be assuaged by leaving.

And leave the Delta musicians did. During the last century, nearly eight million southern-born African Americans moved to all points north and west, with even more southern-born whites doing the same.[37] As one African American who fled the Delta put it, it was like "getting unstuck from a magnet."[38] Black Mississippians, in particular, made their way to Chicago, where by 1945 the city's South Side was a repopulated southern city of its own.[39] Big Bill Broonzy caught a train out of Arkansas to Chicago, telling Lomax, "It's always 'Boy' until you git too old, then they call you 'Uncle.' You never be called a man in the South, you know that!"[40] John Lee Hooker's exodus took him from Clarksdale to Memphis, Knoxville, Cincinnati, and, finally, Detroit by the age of thirty. "I left Mississippi," he said, "because everyone was going up north, and they was mixing, and I wanted to go to the big city. I wasn't getting anywhere in Mississippi. I was fourteen or fifteen years old, something like that. I ran away because they wouldn't let me go otherwise. I had to do that or no one would have ever heard of me. I'd be old in Mississippi and no one would have heard of me."[41]

Musicians of the greater Arkansas and Mississippi Delta found their way to Memphis, of course, but also to St. Louis (Pettie Wheatstraw, Charley Jordan, Roosevelt Sykes, Henry Townsend), New York City (where Brinkley, Arkansas, bandleader Louis Jordan became the most popular jump blues artist of the pre–rock and roll 1940s), Los Angeles (acclaimed jazz bandleader Gerald Wilson), and Detroit (where John Lee Hooker revolutionized boogie woogie on the electric guitar). Mostly, Chicago was that "land of California," to paraphrase Robert Johnson, attracting the likes of Casey Bill Weldon, Big Bill Broonzy, Robert Junior Lockwood, Memphis Minnie, Memphis Slim, Sunnyland Slim,

Robert Nighthawk, Koko Taylor, Earl Hooker, and Arthur "Big Boy" Crudup, not to mention the architects of northern small-combo electric blues—Muddy Waters, Howlin' Wolf, Willie Dixon, Jimmy Rogers, Elmore James, "Magic Sam" Maghett, Hubert Sumlin, Little Walter, James Cotton, and other figures—who made Chicago to bluesmen in the 1940s and 1950s what Paris was to expatriate American writers in the 1920s.[42]

Though one type of manual labor (farming) was replaced with another (factory work), and though racism knew no borders, the lure of a life removed from the South's Jim Crow laws, sharecropping, labor camps, and menace of lynching was its own reward. Albert King summed up the feeling for many in his 1976 single, "Cadillac Assembly Line," written by a fellow Delta musician, Clarksdale-born Sir Mack Rice, who as a Detroit transplant became one of his generation's greatest songwriters ("Mustang Sally," "Cheaper to Keep Her," "Respect Yourself"):

> Goin' to Detroit, Michigan, girl, I can't take you.
> Hey, I'm goin' to Detroit, Michigan, girl, you got to stay here behind.
> Goin' to get me a job on the Cadillac assembly line.
> ***
> I'm tired of whoopin' and hollerin' up and down the Mississippi road.
> Hey, I'm tired of whoopin' and hollerin' pickin' that nasty cotton.
> Gonna catch me a bus up North, I won't have to keep sayin' "Yes, sir, boss."[43]

For those who stayed, radio was crucial within the Delta for the opportunities it gave to blues musicians and in cultivating an audience for the music that, by the mid-1950s, spilled into the youthful mainstream as rock and roll. KFFA in Helena led the charge in 1941 with the popular weekday lunchtime show *King Biscuit Time*, which touted live music by Sonny Boy Williamson II, Robert Junior Lockwood, Pinetop Perkins, and other Delta-steeped players.[44] Other blues-spinning programs followed. These included Early Wright's Clarksdale show beginning in 1947 at WROX; the 1948 debut of Nat D. Williams and *Tan Town Jamboree* at WDIA, which attracted a young Peptikon-promoting B. B. King, Rufus Thomas, and other seminal names in Memphis rhythm and blues; Howlin' Wolf's spot on KWEM in West Memphis; and Bluff City deejay Dewey Phillips's *Red, Hot, and Blue* in 1949 with WHBQ—a show that unified Mid-South youth under one record five years later when Phillips played a white singer's reinvention of a Delta blues: a teenaged Elvis Presley inventing his rock persona by goofing on the Arthur Crudup number "That's All Right."

That record, made at Sun Records on July 5, 1954, is one of the pivotal moments in popular music history, a catalyst for Presley's own fame, not to mention the subgenre of early rock known usually as rockabilly, which melded, in its simplest terms, country music and blues.[45] Even before Presley, however, Sun's studio had been the cynosure draw for regional artistry, mostly of the blues variety. Among the many now-hailed performers, almost all from the greater Mississippi/Arkansas/West Tennessee Delta axis, who got their first or near-first professional shot through founder-producer Sam Phillips and his enterprise were bluesmen B. B. King, Howlin' Wolf, Ike Turner, Junior Parker, Bobby Bland, Rufus Thomas, Little Milton, James Cotton, Joe Hill Louis, Mose Vinson, and Doctor Ross, as well as rock and country performers Presley, Johnny Cash, Carl Perkins, Jerry Lee Lewis, Billy Lee Riley, Sonny Burgess, Conway Twitty, and Charlie Rich (Texas native Roy Orbison, who also got his start at Sun, was one of the non-regional exceptions). Countless others never made a mark, though the time and attention Phillips showed them in the studio was its own validation. "A lot of those people, their names never became anything close to a household word," Phillips had said. "Do you know how important that was at that time that I recorded them?"[46] Phillips, who started Sun Records in 1952, opened the studio in 1950 as Memphis Recording Service and licensed his first efforts to other labels, notably, Modern/RPM in Los Angeles (which signed away B. B. King) and Chess in Chicago (which got Howlin' Wolf). Chess was also the beneficiary of the 1951 number-one hit, "Rocket 88," recorded by Ike Turner and his Kings of Rhythm for Phillips, though it was released under the name of the song's singer, saxophone member Jackie Brenston. Fittingly about a car, the song was an important transitional record in capturing the sound of Delta blues on the move and frequently gets cited (indeterminable though it may be) as the first rock-and-roll record in part because of the distortion heard in band mate Willie Kizart's electric guitar.[47] Another notable off-and-on-again artist for Phillips, Memphis pianist Rosco Gordon, played the keyboard with an accented, clipped upbeat, something Phillips dubbed "Rosco's rhythm." It sent his single "Booted" to the top of the rhythm-and-blues charts in 1952 and by the end of the decade served as the model for two of ska's first hits in Jamaica, "Easy Snappin'" by Theophilus Beckford and "Boogie in My Bones" by Laurel Aitken, the latter sparking the fortunes of Island Records in the process.[48]

For many Delta musicians, Memphis was the first of several stops heading north. Founded in 1819, the city temporarily lost its charter after it was devastated by several waves of yellow fever that killed thousands and forced thousands more to flee in the 1870s. By the turn of the twentieth century, it had reemerged as a central figure in the Delta's cotton market and was equally invested through regional black patronage and repopulation, a dynamic by which Delta culture met up with urbanized reinvention.[49] In terms of the blues, that transformation could be found in the jug bands (Memphis Jug Band, Cannon's Jug Stompers) and guitar duos (Memphis Minnie and Kansas Joe McCoy, Frank Stokes and Dan Sane), both harbingers of later developments in blues and rock. The Memphis education system was an important element as well, and many of the city's soul and jazz musicians benefited from school music programs such as the one at Manassas High School, which fostered a generation of Bluff City jazz players including Charles Lloyd, Booker Little,

George Coleman, Harold Mabern, and Frank Strozier. The touchstone was Manassas teacher Jimmie Lunceford, whose school band became his ticket to New York, where his popular swing group became the house band at Harlem's famed Cotton Club following Cab Calloway's residency.

In terms of sacred song, gospel quartets, guitar evangelists, sermons, and congregational singing were all commonly recorded expressions of black Christian faith in the South, with the sanctified movement exerting a particular pull on Baptist minister Charles H. Mason, who rebranded Memphis's Church of God in Christ as a Pentecostal denomination after attending the Azusa Street Revival. One outcome of this move was heard in C.O.G.I.C.'s spirited musical performances, which welcomed secularized instruments such as guitars, mandolins, drums, and even jugs (even as secular forms such as the blues were frowned upon), thereby presaging later ecstatic expressions in rock, soul, and modern worship music.

Since the days of W. C. Handy, who launched his song-publishing empire in Memphis after relocating from Clarksdale, the city's vibrant entertainment district on Beale Street attracted musicians and music patrons from across the Mid-South. Beale Street is where Bessie Smith appeared for the world premiere of her 1929 film *St. Louis Blues* at the Old Daisy Theater with composer Handy in tow; where Memphis Minnie got signed to Columbia Records after being heard at a barbershop;[50] where Robert Johnson had his now iconic pinstripe-suit photo taken at the Hooks Brothers studio; and where B. B. King, Bobby Bland, Johnny Ace, and other "Beale Streeters" got their start playing clubs and amateur talent nights such as the one Rufus Thomas co-hosted at the Palace Theater.

The success of Sun, which emerged out of the same independent record label boom that began with Cincinnati's King Records and included other start-ups such as Specialty in Los Angeles, Atlantic in New York, and Peacock/Duke in Houston, had an impact on others willing to start a label in Memphis. Begun in 1957, Hi Records found early success with instrumental dance hits by Ace Cannon, Presley bassist Bill Black, and famed band leader Willie Mitchell, who would take the fortunes of Hi into the sphere of 1970s soul music, producing hit after hit by Al Green (born on the Arkansas side of the Delta in Forrest City), Ann Peebles, Otis Clay, and Syl Johnson. The short-lived Gold Wax label also released a number of classic soul sides, none more enduring than "Dark End of the Street" by James Carr. The biggest label was Stax, a decidedly southern alternative to the Hitsville machine of Motown; it found initial success through a distribution deal with Atlantic. Formed in 1957 as Satellite and rechristened Stax in 1961, "Soulsville U.S.A." as it was dubbed, attracted regional and national talent alike, releasing hits by Otis Redding, Booker T. and the MGs, Rufus Thomas and his daughter Carla Thomas, the Mar-Keys, William Bell, Eddie Floyd, the Staple Singers, Johnnie Taylor, Isaac Hayes, Albert King, Big Star, and many others. Atlantic acts Wilson Picket and

Sam and Dave also made some of their biggest records at Stax. The label's lasting influence has been multifold: a key player in the development of southern soul; a cultural catalyst for black pride ("Soul Man," "Respect Yourself," "I Am Somebody," the 1972 Wattstax concert); and a hit-making company lauded in 1974 as the fifth-largest black business in America.[51] At Stax, Hayes earned distinction as the first African American to win in a music category at the Academy Awards, a 1972 best original song win for "Theme from Shaft." Decades later, Memphis made another Oscar milestone when Three Six Mafia became the first African American rap act to win a musical Oscar, also for best original song for "It's Hard Out Here for a Pimp" in 2006.[52]

Today Memphis is touted as the "Home of the Blues" and the "Birthplace of Rock and Roll," the latter largely because of Presley and Sun Records, and the former from the logo used by Pace and Handy Publishing when they operated in the city. Perhaps a less marketable if more accurate catch phrase would be "a" home of the blues and "a" birthplace of rock and roll, though another way to think about Memphis as music capital is as the direct beneficiary of Delta culture.

Finally, when speaking of the Delta blues, a parallel history to the larger processes of recorded activity and migratory movement should be mentioned, namely, the transference of repertoire and style from one musician to another. No more crucial lineage exists in this sense than the one once centered round the Dockery cotton plantation between Cleveland and the Ruleville and Drew communities in the heart of the Delta, what David Evans delineated through intensive fieldwork as the "Drew tradition" in Delta blues.[53] There, Charley Patton acquired blues skills from the unrecorded Henry Sloan and subsequently mentored key figures including Tommy Johnson, Howlin' Wolf (who, ca. 1930, learned his first guitar tune, "Pony Blues," from Patton), Son House (who, in turn, greatly informed Robert Johnson and Muddy Waters), and Roebuck "Pop" Staples of the Staple Singers. Even a young John Lee Hooker got his first taste of Delta blues when stepfather and Patton musical cohort Will Moore taught his son the latter bluesman's piece, "Pea Vine Blues."[54] This history, passed literally from person to person, gave considerable shape to Delta blues in its initial phase. Then, as players moved north, they transformed the streets, clubs, and studios of Chicago and Detroit into hubs for early electric blues combos, which, in turn, ignited regional rock and roll in the 1950s (Chuck Berry and Bo Diddley at Chess, for example) and the 1960s (an entire swath of British Invasion acts, starting with the Rolling Stones, who recorded at Chess). This is a lineage straight from the Delta—less than a dozen men strong in its simplest terms—that sparked a musical revolution still in progress.

At this point in its long history, the blues has become largely a product of revivalism, even in the Delta. Granted, exponents remain, such as James "Son" Thomas's son, Pat Thomas, in Leland, and Clarksdale natives James

"Super Chikan" Johnson and Blind Mississippi Morris; and select juke joints, clubs, community get-togethers, and the like still support the music in something of a context, though compared to the Delta's heyday of activity, blues as regionally rooted and shared cultural experience has been dwindling for some time. The last living links to the Delta of Robert Johnson's time have passed away in recent years, all in their nineties: Henry Townsend and Robert Junior Lockwood in 2006 and David "Honeyboy" Edwards and Pinetop Perkins in 2011.[55] It can certainly be argued that the impoverished conditions that persist in the Delta, especially for its black population, give musicians reason enough to continue singing the blues, though such commentary by younger performers has been taken up to a certain degree in the arena of rap, related as the Delta variant is with its blues predecessors through contexts, messaging, rhetorical devices, and communal alliances.[56]

The upside is that the state of Mississippi and neighboring environs have capitalized on the area's historical musical legacy through a growing number of museums, research programs and institutions, and festivals. These include the University of Mississippi's Blues Archive in Oxford, the Delta Blues Museum in Clarksdale, the B. B. King Museum and Delta Interpretive Center in Indianola, the Highway 61 Blues Museum in Leland, the Greenwood Blues Heritage Museum opened by Robert Johnson historian Steve LaVere, the Howlin' Wolf Blues Museum in West Point on the eastern side of the state, the Delta Music Museum in Ferriday, Louisiana, the Delta Cultural Center in Helena, Arkansas, and Tunica's recently christened Gateway to the Blues Museum and Visitor Center. Helena's long-standing King Biscuit Blues Festival (aka the Arkansas Blues and Heritage Festival) also leads an array of annual live music events focused on the genre's regional connections, including Greenville's time-honored Mississippi Delta Blues and Heritage Festival, the Bentonia Blues Festival, the Highway 61 Blues Festival in Leland, the Mississippi Blues Fest in Greenwood, West Point's Howlin' Wolf Memorial Blues Festival, and both the Juke Joint Festival and the Sunflower River Blues and Gospel Festival (of which Robert Plant was a recent participant) in Clarksdale. In Memphis, a synergistic package of tourist attractions includes the Smithsonian Institution–affiliated Memphis Rock 'n' Soul Museum, Sun Studio, the Stax Museum of American Soul Music, and Elvis Presley's home, Graceland, the granddaddy of Mid-South music destinations. The Blues Foundation is also based in the city, where it holds its annual Blues Music Awards (formerly the W. C. Handy Blues Music Awards). Other annual music events include the Beale Street Music Festival and the Center for Southern Folklore's Memphis Music and Heritage Festival.

Then again, revivalism has allowed the opportunity whereby newer generations of Mississippi musicians are reconnecting with their own blues heritage. On his 2012 track "Praying Man," Meridian rapper Big K.R.I.T. makes explicit the continuum of the black experience in song by weaving a guest appearance on vocals and blues guitar by B. B. King around historic slave narratives delivered in contemporary hip-hop rhythms and rhymes. The Delta blues of time and tradition, it would seem, continues to resonate far beyond the delimiting confines of cotton plantations and a sharecropping economy that, against great odds, forged an art form now universal in appeal and resilient in its staying power.

Notes

1. B. B. King with David Ritz, *Blues All around Me: The Autobiography of B. B. King* (New York: Avon Books, 1996), 23–24.
2. James Lincoln Collier, *The Making of Jazz: A Comprehensive History* (New York: Dell, 1978), 58.
3. William K. Scarborough, "Heartland of the Cotton Kingdom," in *A History of Mississippi, vol. 1*, ed. Richard Aubrey McLemore (Hattiesburg: University & College Press of Mississippi, 1973), 322.
4. James C. Cobb, *The Most Southern Place on Earth: The Mississippi Delta and the Roots of Regional Identity* (New York: Oxford University Press, 1994), 99; and Peter M. Rutkoff and William B. Scott, *Fly Away: The Great African American Cultural Migrations* (Baltimore: John Hopkins University Press, 2010), 182.
5. In the first decade of the twentieth century, only 7 percent of black Mississippi farmers owned the land they worked. Julius E. Thompson, *Lynchings in Mississippi: A History, 1865–1965* (Jefferson, NC: McFarland, 2007), 43.
6. Shelby "Poppa Jazz" Brown, quoted in William Ferris, *Blues from the Delta* (Garden City, NY: Anchor Press/Doubleday, 1978), 41.
7. Thompson, *Lynchings*, 32.
8. National Association for the Advancement of Colored People, *30 Years of Lynching in the United States, 1889–1918* (New York: NAACP, 1919; reprint, n.p.: CreateSpace, 2010), 1.
9. Thompson, *Lynchings*, 35–37.
10. See especially Cecelia Conway, *African Banjo Echoes in the Appalachian: A Study of Folk Traditions* (Knoxville: University of Tennessee Press, 1995).
11. Michael T. Coolen, "Senegambian Influences on Afro-American Musical Culture," *Black Music Research Journal* 11, no. 1 (spring 1991): 15; and Michael T. Coolen, "The Fodet: A Senegambian Origin for the Blues?," *The Black Perspective in Music* 10, no. 1 (spring 1982): 74.
12. Dena J. Epstein, *Sinful Tunes and Spirituals: Black Folk Music to the Civil War* (Urbana: University of Illinois Press, 1977), 143, 136.
13. W. H. Venable, "Down South before the War; Record of a Ramble to New Orleans in 1858," Ohio Archaeological and Historical Society, *Publications* 2 (March 1889): 498; cited in Epstein, *Sinful Tunes*, 128.
14. For the washtub bass/earth bow connection, see Harold Courlander, *Negro Folk Music, U.S.A.* (New York: Columbia University Press, 1963), 206–7. For African connections to other instruments, see especially writings by David Evans, including "The Reinterpretation of African Musical Instruments in the United States," in *The African Diaspora: African Origins and New World Identities*, ed. Isidore Okpewho, Carole Boyce Davies, and Ali A. Mazrui (Bloomington: Indiana University Press, 1999), 379–90; liner notes to *Afro-American Folk Music from Tate and Panola Counties, Mississippi* (Rounder 18964-1515-2, 2000); "The Origins of Blues and Its

Relationship to African Music," in *Images of the African from Antiquity to the 20th Century*, ed. Daniel Droixhe and Klaus H. Kiefer (Frankfurt am Main: Verlag Peter Lang, 1987), 132–33; "African Elements in Twentieth-Century United States Black Folk Music," *Jazzforschung* 10 (1978): 85–110; "Africa and the Blues," *Living Blues* 10 (Fall 1972): 27–29; and "Afro-American One-Stringed Instruments," *Western Folklore* 29, no. 4 (October 1970): 229–45.

15. See Evans, "Afro-American One-Stringed Instruments," 229–45.
16. James "Super Chikan" Johnson, interview by author, August 17, 2007, Clarksdale, MS, digital recording.
17. Adam Gussow, *Seems like Murder Here: Southern Violence and the Blues Tradition* (Chicago: University of Chicago Press, 2002), 38.
18. David Honeyboy Edwards, as told to Janis Martinson and Michael Robert Frank, *The World Don't Owe Me Nothing: The Life and Times of Delta Bluesman Honeyboy Edwards* (Chicago: Chicago Review Press, 1997), 36.
19. Son House, "County Farm Blues," July 17, 1942, Robinsonville, MS, recorded by Alan Lomax for the Library of Congress. Lyrics are reprinted by permission of the Hal Leonard Corporation.
20. David Evans, *Big Road Blues: Tradition & Creativity in the Folk Blues* (Berkeley: University of California Press, 1982), 36. A dissenting voice in this generally accepted theory of the blues' origins is Elijah Wald, who makes an at-times compelling argument for blues less as black traditional song than a commercialized, largely urbanized creation from the outset, influencing popular and traditional players alike; see Elijah Wald, *Escaping the Delta: Robert Johnson and the Invention of the Blues* (New York: Amistad, 2004).
21. David Evans, "Goin' Up the Country: Blues in Texas and the Deep South," in *Nothing but the Blues: The Music and Musicians*, ed. Lawrence Cohn (New York: Abbeville Press, 1993), 41.
22. Observed by W. C. Handy. See W. C. Handy, *Blues: An Anthology* (New York: Albert & Charles Boni, 1926; reprint, Bedford, MA: Applewood Books, 2001), 206–7; and W. C. Handy, *Father of the Blues: An Autobiography* (New York: Macmillan, 1941; reprint, New York: Da Capo, 1969), 142–43.
23. Charles Peabody, "Notes on Negro Music," *Journal of American Folklore* 16, no. 62 (July–September 1903): 148–49.
24. Howard W. Odum, "Folk-Song and Folk-Poetry as Found in the Secular Songs of the Southern Negroes," *Journal of American Folklore* 24, no. 93 (July–September 1911): 258.
25. Handy, *Father of the Blues*, 74.
26. Ibid., 76–77.
27. See David Evans, "Hill Country Blues," *Living Blues* 38, no. 2 (April 2007): 76–81.
28. Evans, *Big Road Blues*, 262–63.
29. Of course, Robert Johnson could also play songs of great harmonic ingenuity, none better than "Love in Vain," in which he inserted inversions and secondary dominants for effective voice leading.
30. Robert Palmer, *Deep Blues* (New York: Viking Penguin, 1981), 275.
31. Lyrics are reprinted by permission of the Hal Leonard Corporation.
32. House's performance is such a singular achievement that blues scholar Luigi Monge has persuasively argued for its use as a blues prayer, a supplication to God by House for relief. See Luigi Monge, "Preachin' the Blues: A Textual Linguistics

Analysis of Son House's 'Dry Spell Blues,'" in *Ramblin' on My Mind: New Perspectives in the Blues*, ed. David Evans, 222–57 (Urbana: University of Illinois Press, 2008).

33. For a pre-recordings history of the blues, see the excellent research of Lynn Abbott and Doug Seroff, including "'They Cert'ly Sound Good to Me': Sheet Music, Southern Vaudeville, and the Commercial Ascendancy of the Blues," in *Ramblin' on My Mind*, ed. Evans, 49–104.

34. Also in 1927, Memphis-via-Hernando singer/guitarist Jim Jackson had a huge hit with the decidedly more urbane "Jim Jackson's Kansas City Blues" parts 1 and 2 (Vocalion 1144), an influential record on Delta and non-Delta performers alike from New Orleans blues/jazz pioneer Lonnie Johnson, who essentially recorded its two sides verbatim later that same year (including the many Memphis references), to Charley Patton on 1929's "Going to Move to Alabama" and the Mississippi Sheiks for their 1930 release, "Back to Mississippi."

35. Larry Nager, *Memphis Beat: The Lives and Times of America's Musical Crossroads* (New York: St. Martin's Press, 1998), 63.

36. Ted Gioia, *Delta Blues: The Life and Times of the Mississippi Masters Who Revolutionized American Music* (New York: W. W. Norton, 2008), 54.

37. James N. Gregory, *The Southern Diaspora: How the Great Migrations of Black and White Southerners Transformed America* (Chapel Hill: University of North Carolina Press, 2005), 330.

38. Isabel Wilkerson, *The Warmth of Other Suns: The Epic Story of America's Great Migration* (New York: Random House, 2010), 221.

39. Peter M. Rutkoff and William B. Scott, *Fly Away: The Great African American Cultural Migrations* (Baltimore: John Hopkins University Press, 2010), 171.

40. Big Bill Broonzy, quoted in Alan Lomax, *The Land Where the Blues Began* (New York: Pantheon, 1993), 436–37.

41. John Lee Hooker, quoted in Jeff Dunas, *State of the Blues* (New York: Aperture Foundation, 1998), 7.

42. Blues artist "Little Milton" Campbell (1934–2005), born in Inverness, Mississippi, had a lengthy career—including the Sun, Bobbin, Chess, Stax, and Malaco labels—that took him from Memphis to St. Louis to Chicago, then finally back to Mississippi.

43. Albert King, "Cadillac Assembly Line," Utopia 10544, 1976. Lyrics are reprinted by permission of the Hal Leonard Corporation.

44. *King Biscuit Time* still airs weekdays, among the longest-running radio programs in America. Even more remarkable, since 1951 it has featured the same deejay, the award-winning "Sunshine" Sonny Payne.

45. Charles Gillett, *The Sound of the City: The Rise of Rock and Roll* (New York: Pantheon Books, 1983; reprint, New York: Da Capo Press, 1996), 26–27. Gillett prefers "country rock" instead of "rockabilly," deferring to what he says participants called the style, though the term has carried other implications since the 1970s. Personally, I never heard "country rock" used by the many Memphis-based rockabilly musicians I knew and interviewed, most of whom hated "rockabilly" as well, preferring "rock and roll" as the default term for the music they played. Gillett offered what is still the best delineation of 1950s early rock and roll, grouping artists into five overarching styles: country rock (Elvis Presley and the Sun roster); northern band rock and roll (Bill Haley); New Orleans dance blues (Fats Domino, Little Richard); Chicago rhythm and blues (Chuck Berry, Bo Diddley); and vocal group rock and roll (i.e., doo-wop) (ibid., 23–35).

46. Bill Ellis, "Phillips on Wolf, B. B., Jerry Lee, Rufus . . . ," *The Commercial Appeal*, January 2000, F5.
47. Jim Dawson and Steve Propes, *What Was the First Rock 'n' Roll Record?* (Boston: Faber and Faber, 1992), 88–91. The guitar's distortion resulted from an amplifier that had become damaged during the band's drive from Clarksdale to Memphis.
48. William L. Ellis, "Sam Phillips and Early Reggae," *The Academic Minute*, WAMC Northeast Public Radio, Albany, NY, aired March 30, 2012.
49. Nager, *Memphis Beat*, 18–23.
50. Memphis Minnie (Lizzie Douglas) is generally considered to have been "born in Louisiana, raised in Algiers," as the line in her song "Nothing in Rambling" goes, though she more likely hailed from the Delta: US census records from 1900 indicate she was born two years prior in Mississippi and was living at the time with her family in Tunica County, directly south of Memphis.
51. Rob Bowman, *Soulsville U.S.A.: The Story of Stax Records* (New York: Schirmer Books, 1997), 317.
52. See www.oscars.org, accessed April 1, 2015.
53. Evans, *Big Road Blues*, 174–264.
54. Charles Shaar Murray, *Boogie Man: The Adventures of John Lee Hooker in the American Twentieth Century* (New York: St. Martin's Griffin, 2000), 35.
55. The same cannot be said necessarily of the filial and communal musicianship that continues in hill country territory, where the offspring of Junior Kimbrough, Otha Turner, and R. L. Burnside continue to make music for a younger generation, much of it in the blues vein.
56. See Ali Colleen Neff, *Let the World Listen Right: The Mississippi Delta Hip-Hop Story* (Jackson: University Press of Mississippi, 2009).

CHAPTER 12

Blues Musicians in the Mississippi Delta

A PHOTOGRAPHIC ESSAY

William Ferris

Beginning in the 1960s, I photographed blues musicians in the Mississippi Delta as part of my research on black folklore. These images were part of that effort and capture musicians and their friends with whom I worked. They range from celebrated artists like B. B. King and James "Son Ford" Thomas to unknown performers in Parchman Penitentiary.

The Mississippi Delta is considered the heartland of black culture and the birthplace of the blues, a music that has nurtured generations. Its sounds are as familiar to Delta worlds as the changing of seasons. These photographs capture the power of the blues and the worlds that inspired its lyrics. The music that began in the Mississippi Delta has touched hearts and inspired musicians throughout the world.

James Thomas, Leland, Mississippi.

James Thomas, Leland, Mississippi.

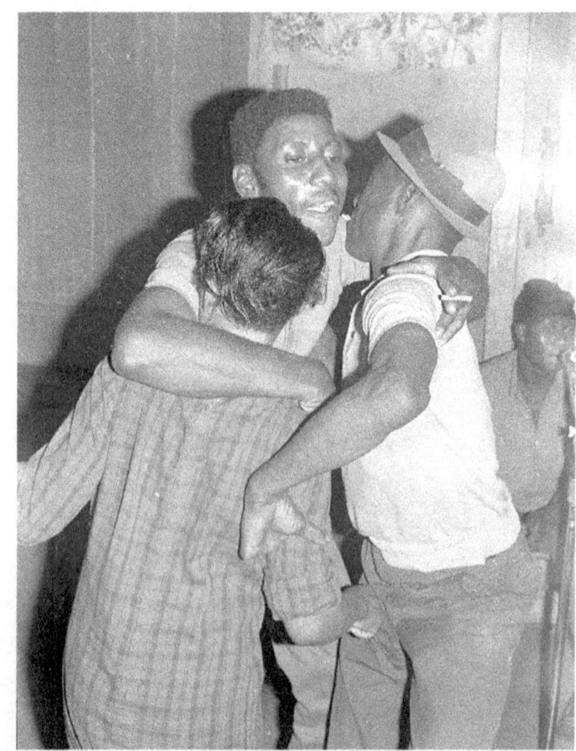

Dancers in Shelby "Poppa Jazz" Brown's home, Leland, Mississippi.

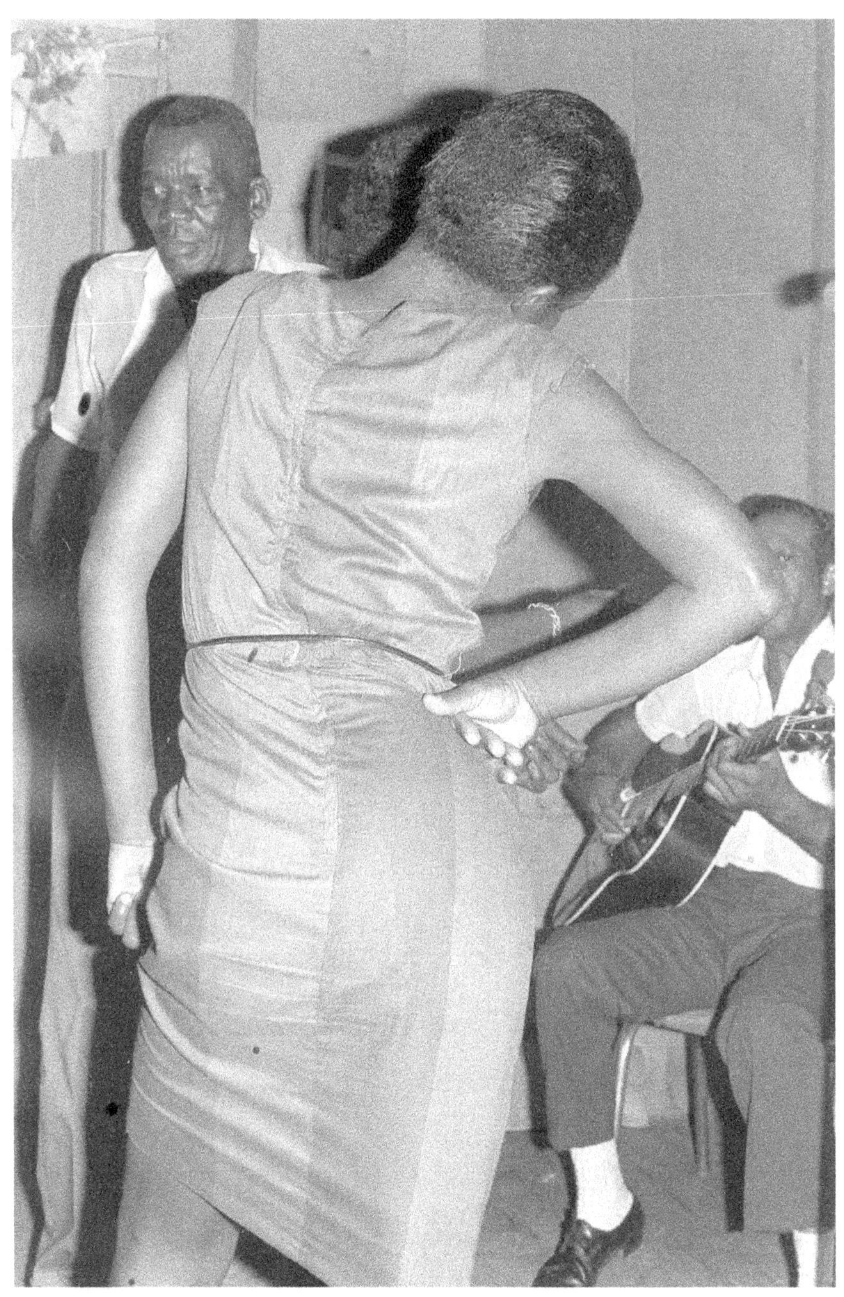

Dancers in Shelby "Poppa Jazz" Brown's home, Leland, Mississippi.

Dancers in Shelby "Poppa Jazz" Brown's home, Leland, Mississippi.

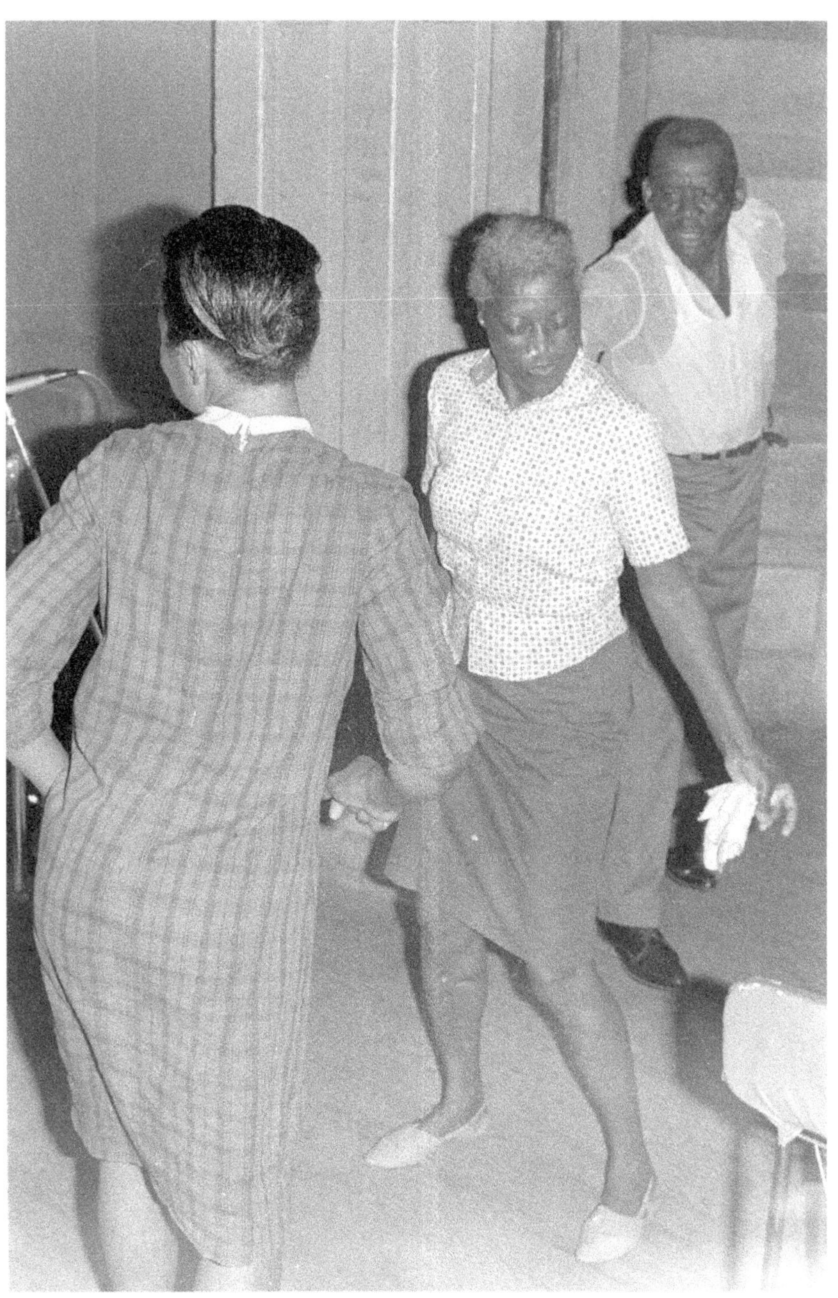

Dancers in Shelby "Poppa Jazz" Brown's home, Leland, Mississippi.

Gussie Tobe in Shelby "Poppa Jazz" Brown's home, Leland, Mississippi.

"Little Son" Jefferson in Shelby "Poppa Jazz" Brown's home, Leland, Mississippi.

Left to right, "Little Son" Jefferson, James "Son Ford" Thomas, and Gussie Tobe; *left front,* "Moonshine," in Shelby "Poppa Jazz" Brown's home, Leland, Mississippi.

Sonny Boy Watson, Stoneville, Mississippi.

James Thomas, Leland, Mississippi.

Lee Kizart's hat and guitar, Tutwiler, Mississippi.

Lee Kizart, Tutwiler, Mississippi.

Napoleon Strickland, Gravel Springs, Mississippi.

Othar Turner, Gravel Springs, Mississippi.

Othar Turner (snare drum), Napoleon Strickland (fife), and the Fife and Drum Band, Gravel Springs, Mississippi.

Louis Dotson, One-Strand-on-the-Wall, Lorman, Mississippi.

Louis Dotson, One-Strand-on-the-Wall, Lorman, Mississippi.

Parchman inmates, Camp B, Lambert, Mississippi.

Parchman prison guards, Parchman Penitentiary, Parchman, Mississippi.

Parchman inmates at B. B. King concert, Parchman Penitentiary.

B. B. King, Parchman Penitentiary.

B. B. King and Parchman inmates, Parchman Penitentiary.

Left to right, Jasper Love and Willie Love, Clarksdale, Mississippi.

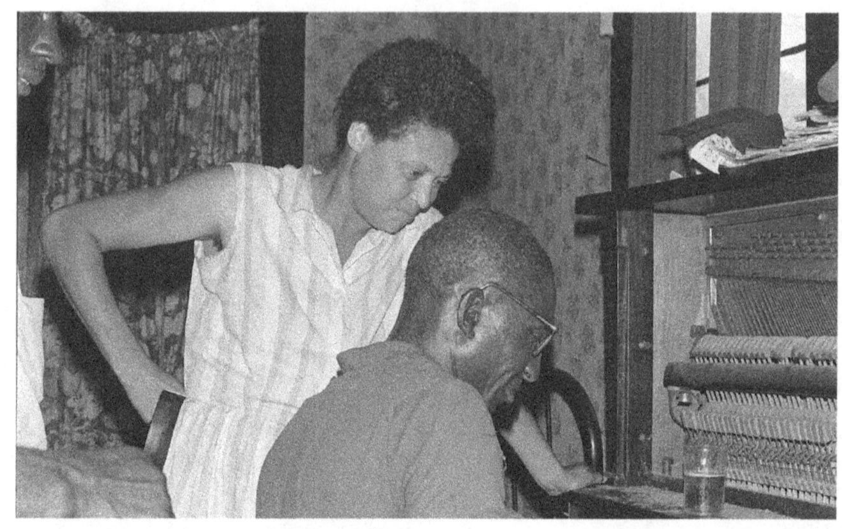

Maudie Shirley and Wallace "Pine Top" Johnson, Clarksdale, Mississippi.

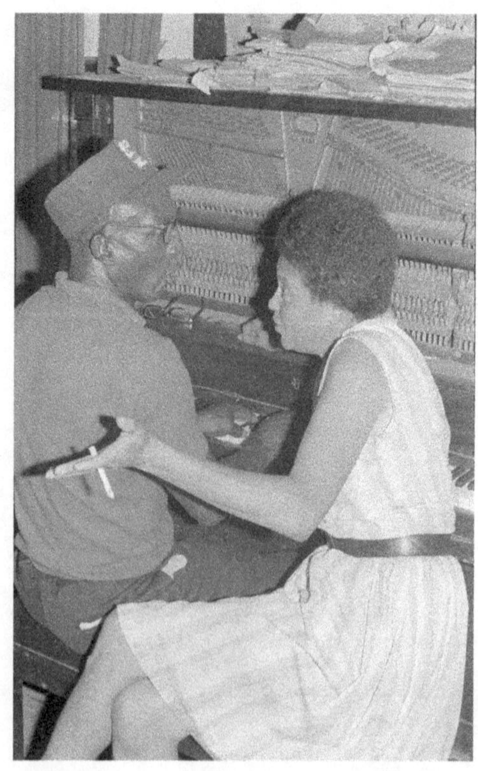

Wallace "Pine Top" Johnson and Maudie Shirley, Clarksdale, Mississippi.

Blues poster, Greenville, Mississippi.

Highway 61, south of Clarksdale, Mississippi.

WROX radio station, Clarksdale, Mississippi.

Rising Sun Café, Leland, Mississippi.

CHAPTER 13

Mississippi Blues Tourism

HISTORY, MARKETING STRATEGIES, AND TOURISM GOALS

Stephen A. King

What is the Delta? Arguably, for many blues fans, the words "the Delta" almost always precede "blues": the Delta blues of Charley Patton, Son House, and Robert Johnson. And the Delta, at least from the perspective of many listeners of the blues tradition, is certainly *not* part of the larger Lower Mississippi River Delta, a geographical area that encompasses a total of seven states, including Kentucky, Illinois, and Louisiana. No, the Delta can only be found in one special place, the Mississippi Delta, located in the northwestern region of Mississippi. Many of the state's most cherished blues musicians emerged from this largely unprivileged region, and, some would claim, the Delta is where the blues was "born." For some tourists, Mississippi, especially its Delta region, is hallowed ground, as the title of Steve Cheseborough's popular and indispensable travel book, *Blues Traveling: The Holy Sites of Delta Blues*, suggests.[1]

As I will discuss in the next section, although early efforts to transform Mississippi's blues culture and heritage into tourist commodities were largely met with bewilderment and scorn from local citizens, the state has now moved aggressively to promote deceased blues icons, including Robert Johnson and Muddy Waters, and living representations of a once thriving and innovative blues culture. Even a cursory glance at web-based and other mediated promotional materials will reveal the extent to which state, regional, and local tourism organizations are marketing Mississippi's blues history and culture to tourists. In general, many blues tourists are baby boomers: middle to upper class, well educated, white. International travelers, from Europe to Asia and beyond, also comprise the blues tourist demographic profile. Traveling primarily by car and occasionally by bus, tourists visit the gravesites of blues musicians such as Charley Patton, stand at the sacred "crossroads," at the intersection of US Highways 61 and 49 (where Robert Johnson allegedly sold his soul to the Devil in return for extraordinary musical talent), listen to local and national acts play at blues festivals, tour museums such as the B. B. King Museum and Delta Interpretive Center, and rent rooms at blues lodging establishments, including the Riverside Hotel and the Shack Up Inn. Mississippi even sports its own blues trail comprised of over 175 markers. The Mississippi Blues Trail's website promises tourists that the "sites run the gamut from city streets to cotton fields, train depots to cemeteries, to clubs and churches. We have a lot to share, and it's just down the Mississippi Blues Trail."[2]

Despite concentrated efforts to turn the blues into a tourism commodity in Mississippi as well as in Memphis, Chicago, and other "blues centers," relatively little scholarship has been published on the subject. In 1992, David S. Rotenstein published what appears to be the first scholarly examination of blues tourism in *Southern Folklore*.[3] Subsequently, scholars published work on individual aspects that comprise the larger blues tourism infrastructure, including festivals, museums, juke joints and blues clubs, and lodging opportunities.[4] In 2011, the University Press of Mississippi published my second book, *I'm Feeling the Blues Right Now: Blues Tourism and the Mississippi Delta*.[5] The goal of the book was to analyze and critique the rhetorical strategies employed to promote blues music as part of Mississippi's cultural heritage. This chapter serves to both summarize and update the findings from that book. As we will see, Mississippi's blues tourism industry employs mythic narratives, advances claims of authenticity, and engages in historical revisionism in order to satisfy a variety of economic and social goals.

History of Blues Tourism

Mississippi's blues tourism industry is a relatively new invention, dating back to the late 1970s. In fact, until recently, one would be hard pressed to call it an "industry" since blues tourism survived at the grassroots level for nearly a quarter of a century, supported by an informal network of African American and white entrepreneurs and volunteers with some assistance from local chambers of commerce and convention and visitors bureaus.

While it is impossible to historically reconstruct the specific location and time period that witnessed the arrival of the first blues tourists to Mississippi, it most likely occurred near the beginning of what is commonly referred to as the 1960s blues revival, a period that lasted from approximately 1959 to the early to mid 1970s. During this blues revival, tourists and non-tourists (e.g., folklorists, sociologists) visited what was presumed to be the genre's birthplace, the American South, in search of living remnants/blues musicians and surviving artifacts such as old recordings of a supposedly vanishing blues culture. At this time, Mississippi did not have a blues tourism infrastructure—the state and its local communities saw no reason to promote a music that emerged from the lowest stratum of the black community. Many whites had little to no interest in a cultural product that came from black people; and, conversely, for many religiously oriented African Americans, the blues was to be shunned—it was the Devil's music. There were no guided tours, no blues festivals, no blues museums, no blues markers, no signs directing blues tourists to points of interest. Nothing. Tourists were left to create their own tours as best they could.

By the late 1970s, however, a blues tourism infrastructure started to emerge from the creation of a blues festival. A civil rights group, Mississippi Action for Community Education (MACE), organized Mississippi's first major blues festival, the Mississippi Delta Blues and Heritage Festival (originally called the Delta Blues Festival). The 1978 festival attracted approximately two thousand blues enthusiasts and curiosity seekers to partake of local, regional, and national talent (e.g., Eugene Powell, James "Son" Thomas, John Lee Hooker) playing mostly acoustic blues in a field located in Freedom Village, a small hamlet roughly fifteen miles from the river city of Greenville.[6] MACE initially conceived of cultural events like the blues festival as a means to preserve the region's blues culture, to educate its local citizenry about the Delta's significant contribution to American music and the arts, and to increase "community consciousness, self-awareness, and pride."[7] As the festival's popularity increased and as its attendance soared (according to MACE, ten thousand people attended the 1980 festival), locals and organizers, while not abandoning the festival's original purpose, began to fully appreciate the festival's role in stimulating the local economy via tourism.[8]

In 1979, Sid Graves, the director of Clarksdale's Carnegie Public Library, decided to open Mississippi's first blues museum, the Delta Blues Museum. During its first years, the Clarksdale-based blues museum was housed in the library and struggled: an estimated 360 people visited the museum during its first year of operation.[9] In 1988, Clarksdale added a blues festival, the Sunflower River Blues and Gospel Festival (originally named Sunflower Riverbank Blues Festival), to its small, but growing, blues tourism offerings. Similar to the 1978 Mississippi Delta Blues and Heritage Festival, Clarksdale's 1988 blues festival attracted relatively few tourists and operated on a limited budget. That same year, two of the festival's organizers, Jim O'Neal (one of the founders of *Living Blues* magazine) and Patty Johnson, opened Stackhouse Mississippi Arts and Gifts/Delta Record Mart in downtown Clarksdale. The now defunct record store doubled as a recording studio and headquarters for O'Neal's Rooster Blues Records.

In 1990, blues attractions were limited—a handful of festivals, one museum, and a record store. Local response to these early blues promotional efforts ranged from curiosity to indifference to hostility. Blues tourism was a novelty, especially in an agrarian culture where its economic base was historically tied to cotton and other crops. Many local citizens did not understand how the blues could be turned into a viable source of economic development. Moreover, some whites were perplexed—and others enraged—to discover that some of the area's white citizens were involved in promoting and celebrating a music that, by its very nature, created opportunities for interracial gatherings (e.g., blues festivals). Although legal segregation was largely dismantled by the early 1970s, the Mississippi Delta still operated by strong (if unspoken) and rigid racial codes, manifested in de facto segregation practices and white privilege. At the same time, some African Americans were opposed to promoting the blues, if not on religious grounds then for painful historical reasons. The blues reminded some black listeners of a past (e.g., sharecropping and segregation) that should be forgotten. And many young African Americans simply could not relate to their parents' and grandparents' music. In contrast to rap music's often confrontational lyrics and postmodern sound, for example, the blues sounded old and passé.

Yet, for all its detractors, there *was* support for blues tourism. In 1981, then-governor of Mississippi William F. Winter declared September 19 as "Delta Blues Festival Day" to call attention to the growing success of the Mississippi Delta Blues and Heritage Festival.[10] Local newspapers, particularly the *Clarksdale Press Register* (and especially the work of feature writer and photographer Panny Mayfield), covered blues-related events, and the reporting was overall positive, sometimes celebratory. Nevertheless, it was largely the work and sacrifice of local blues promoters Sid Graves, Malcolm Walls, Worth Long, Patty Johnson, Jim O'Neal, John Ruskey, Skip Henderson, Andy McWilliams, and Howard Stovall, among others, and blues scholars, including William Ferris, and blues musicians (e.g., Wade Walton, Johnnie Billington) who kept the promise of a full-fledged blues tourism industry alive.

Another blues revival emerged in the early 1990s, creating the impetus for new interest in promoting the blues in Mississippi. In 1990, Robert Johnson's forty-one-track retrospective, *The Complete Recordings*, was released; the box garnered both critical and commercial success (selling well over a million copies) and ignited a new blues revival in the United States and Europe. With increased interest in Mississippi's blues culture, some local chambers of commerce and convention and visitors bureaus in the Delta actively started promoting the blues. In 1994, the Mississippi Delta Tourism Association formed with the expressed goal of promoting regional collaboration at a time when local efforts were marked by fragmentation and competition as various Delta communities were fighting over tourists and tourist dollars. This regional approach expanded a year later with the creation of "America's Blues Alley," an experiment involving a partnership between the Mississippi Division of Tourism and the Memphis Convention and Visitors Bureau. This interstate approach involved luring European tourists to fly via direct flights from Amsterdam to Memphis, where, in turn, they would (hopefully) visit Beale Street and travel south into Mississippi. By the end of the 1990s, state officials began to take blues tourism very seriously. In one example, the Mississippi Division of Tourism and the Mississippi Arts Commission funded a major study (eventually published in 2001) to assess the viability of creating a Mississippi Millennium Blues Trail and to evaluate current tourism practices.[11]

By the beginning of the twenty-first century, hard-fought local efforts to promote the state's blues heritage and culture finally morphed into the beginnings of an industry with support from state government, politicians, businesses, and other establishment institutions. By 2003 and 2004, Mississippi had its own officially sanctioned blues commission. The Mississippi Blues Commission, the state's most prominent blues organization, is responsible for coordinating the Mississippi Blues Trail, an ambitious project that involves conceptualizing, constructing, and ceremoniously unveiling blues markers throughout the state and beyond. Blues markers can be found throughout the various regions of Mississippi (Delta, hills, pines, coast, river, the capital), as well as in adjacent states, in major blues centers (e.g., Chicago), and as far west as southern California and as far north as Maine. By celebrating and remembering both prominent and more obscure blues artists (Robert Johnson, Jack Owens), record labels/companies (Paramount Records), radio stations (WROX), plantations (Dockery Farms), gravesites (Charley Patton), highways (Highway 61), and other blues symbols, the Blues Commission hopes the markers will "tell stories through words and images of bluesmen and women and how the places where they lived and the times in which they existed—and continue to exist—influenced their music." Potential tourists can order a blues trail map, download a blues trail app (iPhone), preview blues markers, and watch a brief (4:31 minutes), professionally shot and edited video featuring B. B. King, Little Milton, Charlie Musselwhite, Robert Plant, and Bonnie Raitt.[12]

While Clarksdale and Greenville were the only cities meaningfully involved in early efforts to promote the blues, other Mississippi towns are now heeding the call. The town of Indianola, for example, now fully celebrates its most famous citizen, B. B. King. Costing an estimate $15 million to build, the B. B. King Museum and Delta Interpretive Center opened its doors in September 2008 and attracted an estimated thirty thousand visitors during its first year.[13] Located in a poor African American neighborhood and built next to a cotton gin where King worked as a teenager, the 20,000-square-foot museum carefully blends the rustic and primitive (wood) with technology (film, interactive displays), tracing the ascendency of King's career from struggling musician to the world's "foremost ambassador" of the blues.[14] Published in the *Wall Street Journal* a month after the museum opened its doors, one review claimed the museum is the "most elaborate museum in the U.S. about a single living musician, but Mr. King's stature justifies the investment."[15]

In a more recent example, Tunica County, once one of the poorest counties in the United States, is now home to a $2 million project, "the Gateway to the Blues Visitors Center and Museum."[16] In January 2011, a 116-year-old, neglected train depot (which in recent years had been used to store grain and cotton gin supplies) was transported from Dundee, Mississippi, to Tunica.[17] Since trains and train stations are iconic symbols of blues history and because W. C. Handy and other noted blues alumni played at the depot, Penn Owen, chairman of the Tunica Tourism Commission, was convinced the project was a worthy one: "We knew that this forlorn and forgotten depot was the perfect icon for the new Blues attraction."[18] Subsequently, the 1,250-square-foot train depot was renovated and transformed into a new visitor's center. The center, which opened in November 2011, serves as the entrance to a proposed 3,800-square-foot blues museum; the museum will include artifacts from the Horseshoe Casino's former blues museum worth an estimated $350,000.[19] Officials hope the "New Gateway to the Blues" will more than double tourist traffic to the visitor's center, from approximately 40,000 to 100,000 visitors per year.[20] One writer proclaimed that Tunica's renovated visitor's center will "shine like a beacon along U.S. 61 in Tunica."[21]

Marketing Strategies and Tourism Goals

A systematic review and analysis of blues tourism promotional materials reveals the use of three broad and overlapping marketing strategies. First, Mississippi's official culture—groups and organizations (e.g., government, chambers of commerce, museums) that represent the status quo—relies on myth to sell its most well known cultural heritage. For example, tourists are led to believe that Mississippi "birthed" the blues, the music's "conception" and "birth" literally occurred in a cotton field. The Mississippi Blues Commission promotes this myth ("experience the blues where they were born").[22] So does the Mississippi Division of Tourism ("music lovers know that Mississippi is where the blues began").[23] More recently, the state is promoting itself as the "birthplace of America's music" based on the premise that the blues is the "foundation" for all American popular music. The Washington County Convention and Visitors Bureau, for example, asserts that the "music born in the Delta is revered worldwide and recognized as the roots of jazz, rock and roll, rhythm and blues, and hip hop."[24] While Mississippi is often identified as *the* birthplace, according to blues expert Paul Oliver, there is no evidence to prove the assertion that the music was actually "born/originated" in the state: "The precise origin of blues may never be determined. Broadly, it may be considered to be the South, or for some authors such as Robert Palmer, the Deep South, if the early development of the music is implied in the phrase."[25] Furthermore, Oliver argues that the "blues is not a music of state and county lines or river boundaries, but of a people. While the blues was taking shape in Mississippi other traditions were emerging elsewhere."[26]

This birthplace argument is best conceptualized as a mythic story because a myth is an important cultural narrative that exists in the rhetorical space between falsehood and historical truth. As communication scholar Robert C. Rowland argues,

> myths are the most important stories in a society. Therefore, if a story is in fact mythic there should be strong evidence that it is taken very seriously by the people who tell it. A story need not actually be factual to be considered mythic. However, there should be evidence that those who tell it, believe the story to be more than mere entertainment; that it is in some sense true.[27]

In other words, although mythic stories are not grounded in historical realities, these narratives are usually perceived, by their promoters and consumers, as possessing a sacred and literal truth.[28] Thus, while there is no scholarly evidence that proves the blues originated in Mississippi, many blues promoters and consumers believe the narrative to be true. This is not to say that Mississippi was not fertile ground for the development of the genre. Without a doubt, many of the blues' greatest players and singers were born and lived in Mississippi. In this sense, then, the birthplace argument has some historical validity. Blues expert Adam Gussow argues that Mississippi is the "birthplace of Charley Patton, B. B. King, Muddy Waters, Robert Johnson, John Lee Hooker, Howlin' Wolf, Sonny Boy Williamson, and countless other foundational blues performers whose names are familiar to blues fans around the world."[29] However, promotional materials make it plain which meaning predominates—Mississippi is the "mother" that gave birth to the music. Since most blues scholars generally agree the blues emerged in multiple locations throughout the South during the late nineteenth century, it would be accurate to assert that blues has many birthplaces.

Drawn to the supposed birthplace of the blues, some tourists seek authenticity, a second marketing strategy. My work has focused almost exclusively on one type of authenticity, constructive or rhetorical authenticity. Dismissing the notion that authenticity is somehow imbedded within a person or object (objective authenticity), constructivists argue that all authenticity is essentially a symbolic phenomenon. In this sense, authenticity is nothing more than a rhetorical "claim that is made by or for someone, thing, or performance and either accepted or rejected by relevant others."[30] By employing authenticity-related terms such as "original" and "real," promotional materials promise tourists that they will "discover" and "experience" authenticity in the "birthplace of the blues." "When visitors travel to Mississippi to discover our state's blues musical heritage," promises the Tourism Division of the Mississippi Development Authority, "they find the real, authentic experience here—especially in places like Clarksdale in the Mississippi Delta."[31] Noting the state's important role in the early development of the blues, the Mississippi Delta Tourism Association reassures tourists that "you can still feel that authentic vibe of Mississippi Delta blues history."[32] The tourism company Delta Music Experience features a package tour called Real Deal Delta Experience. Promotional material employs religious imagery to capture the attention of tourists who seek authenticity: "To feel The Blues in your heart, experience The Blues in your soul. The soul of The Blues lives in The Delta and Delta Music Experience takes you there!"[33]

Some promotional efforts combine both rhetorical strategies of myth and authenticity, as evident, for example, in a recent ad published in *Living Blues* magazine that blends stereotypical images of the blues as sad and blues musicians as downtrodden with claims about the state being the birthplace of

America's music and its realness, its blues authenticity. Tourists are encouraged to visit Mississippi, "to Find Your True South":

> Until you've sat in the dark and smoky Saturday night audience of a real juke joint can you hear the real blueness of the blues or feel just how tired, just how lonesome, just how brokenhearted a man or woman can be. Hear true blues in the true birthplace of America's music—the real, authentic Mississippi Delta. It's playing all the time at Po' Monkeys, Club Ebony, and all the tiny holes in the wall here in Mississippi, the only place in the world to Find Your True South.[34]

Given the emphasis on location and history, Mississippi blues tourism is as much about remembering and reconstituting the past as it is about informing tourists about what they will actually see and experience during their visit. This, in itself, creates an ethical dilemma: how can blues promoters and tourism officials speak of Mississippi's blues history without, at the same time, acknowledging, in an honest and meaningful way, the state's horrific treatment of its African American population? As Paul Oliver reminds us,

> though many Mississippi blues singers are now recognised as great folk artists, they were unprivileged members of the community in their home districts, subject to exploitation by "The Man," the white landowner who at "settlement" time evaluated the crop, and subject to the rigours of a hard life growing cotton in the fertile, black Delta lands. . . . The sheer crudity of segregation in Mississippi, the barbarity of the measures to enforce it, the rich and yet despairing landscape with its low, red clay hills and the monotony of the flat bottomlands combine to give the state a perverse fascination. It is occasionally beautiful, but mostly it was elemental, cruel even, stifling in its feudalism.[35]

At the same, do tourists—many of whom are essentially traveling in the pursuit of a pleasurable escape from the habit of daily life—really want to be reminded of how black Mississippians, including blues musicians, endured the slave-like working conditions of sharecropping, racial violence (e.g., beating, lynching), state-sponsored segregation, and political corruption and disenfranchisement? Besides, this admission of guilt would certainly complicate one of the goals of blues tourism—to change the general public's negative impressions of the state. To solve this problem, state, regional, and local tourism organizations engage in a third rhetorical strategy centered on the concept of collective or public memory. Public memory is "a body of beliefs and ideas about the past that help a public or society understand both its past, present, and by implication, its future."[36] Official culture re-presents the past in order to serve institutional and ideological goals. As rhetorical scholars Greg Dickinson, Carole Blair, and Brian L. Ott put it, "groups tell their pasts to themselves and others as ways of understanding, valorizing, justifying, excusing, or subverting conditions or beliefs of their current moment."[37] In essence, blues tourism narratives generally engage in historical revisionism, a process that involves selective remembering and (deliberately) forgetting the past. Not surprisingly, blues narratives often avoid specific references to the state's officially sanctioned mistreatment of its African American population or even the struggles endured by its now-celebrated blues musicians.

It should be acknowledged, however, that the state is making more of a concerted effort to use tourism as a mechanism to face its shameful past, including the development of a civil rights trail called the Mississippi Freedom Trail (the first marker, memorializing the 1955 lynching of Emmett Till, was unveiled on May 18, 2012). Still, the state's lack of a civil rights museum is a glaring omission, although efforts are underway to construct such a museum, which is tentatively scheduled to open in 2017.[38]

These marketing strategies appear to serve at least four goals. Blues tourism serves an economic imperative, given persistently high levels of unemployment and poverty, a declining tax base, massive and long-term outmigration patterns, and other economic woes.[39] Although statistics are not available specifically for employment and revenue generated from blues tourism activities, travel and tourism accounted for 2.8 percent of the state's GDP, was responsible for 84,225 direct jobs in fiscal year 2014 (a decrease of 0.1 percent from the previous year), and ranked fourth in direct private-sector employment. Approximately 22 million visitors (a 1.9 percent increase from the previous year) spent an estimated $6.09 billion dollars in 2014, leading the Mississippi Development Authority Tourism Division to claim travel and tourism as "one of Mississippi's largest export industries and a major contributor to its financial affairs and quality of life."[40] State officials, including former governor Haley Barbour and state senator David Jordan, are convinced that the blues can play a significant role in the state's overall effort to accelerate tourism as a revenue producer.[41]

As mentioned earlier, blues tourism also serves the goal of reconstituting and reshaping negative impressions of Mississippi as a desperately backward and lawless state suffering under the weight of a largely uneducated and ignorant population, abject poverty, racial segregation, and violence. Unfortunately, for state leaders and local citizens, Hollywood movies such as *A Time to Kill* and *Ghosts of Mississippi* seem to confirm rather than dispel these negative and, many would argue, unfair stereotypes. In short, Mississippi is fighting an image problem and blues tourism acts as a type of public relations campaign. Obviously, Mississippi's leaders and its local citizens would prefer the state be known as the birthplace of the America's music as opposed to the land of the Ku Klux Klan. Actor Morgan Freeman (who is co-owner of a blues club in Clarksdale) and B. B. King are prominent boosters for the state, and both praise Mississippi's efforts to promote its blues heritage and encourage tourists to visit the state. Freeman wrote in a 2004 issue of *Living Blues* magazine, "If you haven't been to Mississippi, you've missed one of the joys of life. . . . We're not what we've been painted to be."[42]

Preservation is a third goal associated with blues tourism. This should come as no surprise since cultural or heritage tourism is ultimately an industry dedicated to preserving fragments of the past. From museums to festivals, blues artifacts and objects, songs, places, and people are preserved for tourist consumption. For many in the business of selling the blues, preservation is an imperative in the blues world that warns of the eventual collapse and death of the genre; "Save the Blues!" is a familiar slogan. Apparently, the owners of the Shack Up Inn, a tourist haunt located on the Hopson Plantation just outside of Clarksdale's city limits, believed old sharecropper shacks were worth saving. A popular and highly successful alternative lodging spot, the Shack Up Inn is a collection of renovated sharecropper shacks designed to appeal to tourists interested in "comfort" and "authenticity."[43] Most shacks are identified by a name, and some shacks have "blues" names (e.g., Crossroads Shack, Electric Blue Shack). Largely due to limited employment opportunities and poverty, many blues musicians lived in sharecropper shacks: wooden structures that were bereft of basic necessities such as running water and electricity and offered residents minimal protection from the weather and other potential external threats. The appeal of the Shack Up Inn motivated other blues entrepreneurs to open up a similar lodging venture called Tallahatchie Flats near Greenwood. The company's website describes the wooden dwellings as "preserved examples of the small rural homes that once dotted the Delta countryside where so many great blues artists were born and raised and wandered."[44] No item, it appears, seems out of reach for blues promoters to preserve and commodify for profit.

Finally, blues tourism can function to promote racial reconciliation. Reconciliation typically involves various communication behaviors: mutual dialogue (speaking and listening), acts of contrition and truth-telling, forgiveness, and some form of restorative justice (e.g., reparation).[45] In other words, dialogue and debate replace violence, and acknowledgment of past crimes replaces denial and guilt. As opposed to formal, institutional reconciliation efforts (e.g., South Africa's Truth and Reconciliation Commission), reconciliation-like efforts within the context of blues tourism are certainly less structured, often implied, and unnamed. Reconciliation often takes the form of acknowledging and honoring African American musicians who often experienced the injustice visited upon many blacks who lived in the South during the early to mid twentieth century. Blues markers and the accompanying unveiling ceremonies, blues museums, awards ceremonies, and other memorializing opportunities serve, in one form or another, to remember, honor, and pay tribute to both musicians (and others associated with the blues) and the music's organic ancestral home, the African American community. Blues festivals, events that typically bring tourists and locals, whites and African

Americans, together to share a common musical interest, act as a community-building function in a state where, unfortunately, whites and African Americans still often live segregated social lives. Fundraisers for ailing blues musicians and grants to assist musicians in distress may demonstrate another aspect of reconciliation. For example, in March 2010, the Mississippi legislature passed House Bill No. 1160, allowing the Mississippi Blues Commission to raise private capital to assist blues musicians who are "living in extreme poverty, in need of food, shelter, medical care and other assistance." Since the vast majority of Mississippi blues musicians are African American, most beneficiaries will likely be minority recipients. The Blues Musicians Benevolent Fund distributes grants (up to $1,000 per twelve-month period) to any "established blues musician" who can document a need for financial assistance.[46] Indeed, some local African American blues musicians currently live in poverty or barely above the poverty level.

Conclusion

Since its inception in the late 1970s, Mississippi blues tourism has essentially transformed itself from a bottom-up, grassroots effort led by African American and white entrepreneurs and enthusiasts to a top-down, institutionalized, state-sponsored industry. This is not to say that local efforts are now less important or have been sublimated under the dominion of the state. Indeed, locally organized blues events (e.g., festivals) and entrepreneurs (e.g., Roger Stolle and his Cat Head Delta Blues and Folk Art store) are, in many ways, the backbone of the industry. It is to say that the state now plays an important role in promoting the blues and that the message is no longer within the sole purview of local organizers and promoters.

In many ways, the growth of blues tourism has been impressive. In 1990, tourists could only expect to find a handful of blues festivals in the state. Today, according to the *Living Blues* 2012 festival guide, that number has ballooned to sixty-five.[47] The Delta Blues Museum is now just one of at least twelve blues museums in the state. Tourists now have their choice of blues lodging spots ranging from the unorthodox (Shack Up Inn) to the more traditional (Blues Hound Flat). Lack of signage once made it difficult for tourists to find points of interest. Now, Mississippi has a blues trail. In addition, the Mississippi Department of Transportation now posts brown signage highlighting blues-related attractions. With the Mississippi Blues Commission leading the charge, state, regional, and local tourism organizations pitch the state's blues history and culture. While once a seasonal industry from April to October, events of interest to blues tourists now span the calendar year.

With all this activity, one would assume that blues tourism is reaping profits. The proliferation of blues-related businesses, increased attendance at blues museums, and expansion of the number of blues festivals indicates, on one level, that the industry is achieving (some) its economic development goals.

Indeed, blues tourism has led to economic revitalization in some communities, particularly in Clarksdale.[48] Yet, as I argued in *I'm Feeling the Blues Right Now*, without reliable and comprehensive data, it is difficult to determine the bottom line associated with blues tourism and challenging to determine the number of blues tourists who visit the state annually. Moreover, official tourism data from 2004 to 2014, which tracks important statistical indicators (e.g., number of out-of-state visitors, state/city-county tourism tax, and direct jobs), looks like a mountain range (with peaks and valleys) rather than an upward-ascending slope.[49] It also should be pointed out that while some blues festivals and museums see an annual profit, others just break even, struggle to stay solvent, or capitulate to market-driven forces. And tourism, as the nation has witnessed during and after the great recession of 2008, is a volatile business. Since the automobile is the best method to visit the state's decentralized tourism sites, tourism is susceptible to oil and gas price spikes as well as economic downturns and natural disasters. Blues tourism is also vulnerable to waxing and waning interest in the blues. Since the 1960s, there have been sporadic blues revivals (early to mid 1980s, early 1990s, and the "Year of the Blues," 2003); these revivals typically witness an upswing and then declining interest in the music, as well as, one could argue, interest in visiting the state in search of blues sites and related events.

It is also equally difficult to assess the impact of the "birthplace" campaign on the general public's perception of the state. Some anecdotal evidence suggests that blues tourism has played a role in altering perceptions of the state.[50] But a comprehensive study on this issue is needed before rendering any reasonable claims about the state's success in changing public opinion. And what about racial reconciliation? There is no doubt that events that sponsor diversity (e.g., blues festivals) and ceremonies that honor African American blues artists serve positive social ends. At the very least, blues tourism facilitates community building; and the work of intercultural communication researcher James Steven Sauceda confirms that aesthetic events, such as festivals, can work to challenge prejudice and negative stereotypes, often leading to a newfound acceptance and understanding of cultural differences.[51] Yet, at the same time, reconciliation efforts are undercut by multiple factors: the reality of significant economic disparity experienced by white and black Mississippians, de facto segregation practices (particularly in housing), tourism narratives that obfuscate past injustices in favor of a sanitized understanding of history, and the perception, among some, that tourism serves white economic interests at the expense of black labor.[52]

Knowing the potential source of revenue, blues promoters will continue to define "the Delta" in mythological terms—a timeless, holy place where the blues was born. Knowing the motivation of some tourists to consume authenticity, efforts to appeal to that elusive concept will continue unabated—travel to Mississippi to experience "real deal" blues musicians playing in a rustic, authentic setting. No wonder there has been little interest among local and state blues promoters to "Disney-fy" the state's blues sites. Considering the motivations of some tourists to seek leisure and enjoyment (plus the goal of reshaping the public's perception of Mississippi), I suspect tourism narratives that minimize or fail to acknowledge past racial policies will also continue to pervade marketing and promotional strategies. In the end, it would be unfair to dismiss Mississippi's newfound effort to promote its blues culture and heritage as inherently misguided and exploitative. Blues tourism makes a number of positive contributions, including creating an opportunity for community and state pride, racial reconciliation and economic development, and it provides extra income for struggling local musicians. At the same time, scholars and other observers should continue to examine and critique blues tourism practices, making promoters and boosters accountable for the rhetorical messages that are constructed to serve official and ideological goals.

Notes

I would like to thank P. Renee Foster for her key suggestions and expert editing eye that improved the overall quality of this chapter.

1. Steve Cheseborough, *Blues Traveling: The Holy Sites of Delta Blues,* 3rd ed. (Jackson: University Press of Mississippi, 2008).
2. "List of Blues Trail Markers," Mississippi Blues Trail, accessed June 15, 2012, http:// http://www.msbluestrail.org/blues_marker_list.
3. David S. Rotenstein, "The Helena Blues: Cultural Tourism and African-American Folk Music," *Southern Folklore* 49 (1992): 133–46.
4. See, for example, David Grazian, *Blue Chicago: The Search for Authenticity in Urban Blues Clubs* (Chicago: University of Chicago Press, 2003); Stephen A. King, "Memory, Mythmaking, and Museums: Constructive Authenticity and the Primitive Blues Subject," *Southern Communication Journal* 71 (2006): 235–50; Paige McGinley, "Highway 61 Revisited," *TDR: The Drama Review* 51 (2007): 80–97; Jennifer Ryan, "Beale Street Blues? Tourism, Musical Labor, and the Fetishization of Poverty in Blues Discourse," *Ethnomusicology* 55 (2011): 473–503; Jeff Todd Titon, "The New Blues Tourism," *Arkansas Review: A Journal of Delta Studies* 29 (1998): 5–10.
5. Stephen A. King, *I'm Feeling the Blues Right Now: Blues Tourism and the Mississippi Delta* (Jackson: University Press of Mississippi, 2011).
6. Gregg B. Bangs, "Blues Festival Celebrates 'Dying Art,'" *Clarion-Ledger/Jackson Daily News,* October 22, 1978, 3A, 6A.
7. "The Mississippi Action for Community Education Delta Arts Project—Background," circa 1981, MACE, 119 South Theobald Street, Greenville, Mississippi, 38701.
8. "The Mississippi Action."
9. Danny McKenzie, "Delta Blues Museum Spreads Tuneful Tale across the World," *Clarion-Ledger,* March 27, 1992, 1B.
10. William F. Winter, Mississippi Executive Department, Jackson, "A Proclamation by the Governor," Blues Archives and Special Collections, University of Mississippi, Oxford, Mississippi.
11. Randall Travel Marketing, "Mississippi Millennium Blues Trail: Strategic Marketing Plan for Travel and Tourism," Mooresville, North Carolina, 2001.

12. "List of Blues Trail Markers."
13. "King Museum Still Drawing Crowds," *Commercial Appeal* (DeSoto ed.), September 13, 2009, DSA1.
14. B. B. King Museum and Delta Interpretive Center, accessed May 29, 2012, http://www.bbkingmuseum.org/.
15. Craig Havighurst, "B. B. King's Hometown Museum," *Wall Street Journal* (Eastern ed.), October 16, 2008, D7.
16. Yolanda Jones, "Old Depot to Become Blues Gateway," *Commercial Appeal* (DeSoto ed.), December 3, 2010, DSA1, DSA8.
17. Yolanda Jones, "End of the Line," *Commercial Appeal* (DeSoto ed.), January 20, 2011, DSA1-DSA2.
18. Tunica Travel, "Gateway to the Blues Visitor Center on the Move," accessed May 28, 2012, http://downtheroad.tunicatravel.com/2011/02/gateway-to-the-blues-visitor-center-on-the-move/.
19. Yolanda Jones, "Old Depot Gets New Life in Tunica," *Commercial Appeal* (DeSoto ed.), December 15, 2011, DSA1-DSA2.
20. Tunica Travel, "Gateway to the Blues Visitor Center."
21. Jones, "Old Depot," DSA1.
22. Mississippi Blues Trail, accessed June 8, 2012, http://www.msbluestrail.org/ index.aspx.
23. Mississippi Development Authority Tourism Division, accessed June 7, 2012, http://www.visitmississippi.org/search.aspx?q=blues+was+born+in+mississippi.
24. Washington County Convention and Visitors Bureau, "Greenville on the Mississippi," Greenville, Mississippi, Washington County Convention and Visitors Bureau, n.d.
25. Paul Oliver, *Barrelhouse Blues: Location Recordings and the Early Traditions of the Blues* (New York: Basic Civitas, 2009), 12.
26. Paul Oliver, *The Story of the Blues* (Boston: Northeastern University Press, 1969/1998), 39.
27. Robert C. Rowland, "On Mythic Criticism," *Communication Studies* 41 (1990): 105.
28. William G. Doty, *Myth: A Handbook* (Tuscaloosa: University of Alabama Press, 2004), 14
29. Adam Gussow, *Journeyman's Road: Modern Blues Lives from Faulkner's Mississippi to Post-9/11 New York* (Knoxville: University of Tennessee Press, 2007), 155.
30. Richard A. Peterson, "In Search of Authenticity," *Journal of Management Studies* 42 (2005): 1086.
31. Mississippi Development Authority Tourism Division, "True Blues," accessed June 8, 2012, http://visitmississippi.org/fa---clarksdale.aspx.
32. Mississippi Delta Tourism Association, "Music & the Blues," accessed June 8, 2012, http://www.visitthedelta.com/attractions/music_blues/default.aspx.
33. Delta Music Experience (tourism company based out of New Orleans), "A Musical Journey of Crossroads, Creativity & Culture," package tour, n.d.
34. "Find Your True Blues," advertisement, *Living Blues*, June 2012, 42.
35. Oliver, *The Story of the Blues*, 39.
36. John Bodnar, *Remaking America: Public Memory, Commemoration, and Patriotism in the Twentieth Century* (Princeton: Princeton University Press, 1992), 15.

37. Carole Blair, Greg Dickinson, and Brian L. Ott, "Introduction: Rhetoric/Memory/Place," *Places of Public Memory: The Rhetoric of Museums and Memorials*, ed. Greg Dickinson, Carole Blair, and Brian L. Ott (Tuscaloosa: University of Alabama Press, 2010), 6.
38. Margaret Ann Morgan, "Civil Rights Museum Planners Collecting Stories," *Jackson Free Press*, April 6, 2012, accessed May 28, 2012, http://www.jacksonfreepress.com.
39. Tom Charlier, "Delta Feels Massive Outward Migration," *Commercial Appeal*, May 27, 2012, A1, A5.
40. Mississippi Development Authority Tourism Division, "2014 Economic Contribution of Travel and Tourism in Mississippi, April 2015," p. 4, accessed May 5, 2015, http://www.visitmississippi.org/app/webroot/files/ECR%202014.pdf.
41. See, for example, Phil West, "Gov. Addresses Delta Summit," *Commercial Appeal* (Desoto ed.), June 17, 2010, DSA1.
42. Morgan Freeman, no title, *Living Blues*, March–June 2004, 9.
43. Shack Up Inn, accessed June 12, 2012, http://www.shackupinn.com/index.htm.
44. The Tallahatchie Flats Experience, accessed June 13, 2012, http://tallahatchieflats.com/welcome.html.
45. See, for example, John B. Hatch, "Beyond Apologia: Racial Reconciliation and Apologies for Slavery," *Western Journal of Communication* 70 (2006): 186–211.
46. "Mississippi Blues Commission Blues Musicians Benevolent Fund," Mississippi Blues Trail, accessed May 17, 2012, http://www.msbluestrail.org/pdfs/musicians_aid_form.pdf.
47. See "2012 Living Blues Festival Guide," *Living Blues*, April 2012, 82–85.
48. For an overview of renovation efforts in Clarksdale, see John C. Henshall, "Delta Blues at the Crossroads: Cultural Tourism and the Economic Revitalization of Downtown Clarksdale, Mississippi," *Thesis Eleven* 109 (2012), 29–43.
49. See, for example, Mississippi Development Authority Tourism Division, "Fiscal Year 2004 Economic Impact for Tourism in Mississippi, February 2005," accessed October 4, 2010, http://www.visitmississippi.org/press_news/docs/08_Final_Economic_Contribution_ Report.pdf; Mississippi Development Authority Tourism Division, "2014 Economic Contribution of Travel and Tourism in Mississippi, April 2015," p. 4, accessed May 5, 2015, http://www.visitmississippi.org/app/webroot/files/ECR%202014.pdf.
50. See, for example, Felix Ybarra, "'A Suburb of Mississippi: Talkin' Great Migration Blues with Mississippi Trail Boss, Alex Thomas," *Big City Rhythm and Blues*, October–November 2009, 11–12.
51. James Steven Sauceda, "Aesthetics as a Bridge to Multicultural Understanding," *Intercultural Communication: A Reader*, ed. Larry A. Samovar and Richard E. Porter, 8th ed. (Belmont, CA: Wadsworth, 1997), 417–26.
52. Alan W. Barton, *Attitudes about Heritage Tourism in the Mississippi Delta: A Policy Report from the 2005 Delta Rural Poll*, Policy Paper No. 05-02 (Delta State University: Center for Community and Economic Development, 2005), 22.

CHAPTER 14

Contemporary Women Musicians in the Delta

L. DYANN ARTHUR

In 2009 I realized what a valuable gift it would be to the American people (and the world) to increase the visibility of women musicians preserving and renewing traditional forms of American music. The musical past, as preserved in history books of the Delta, was filled with musicians representing a wealth of aural textures, eras, and genres. Yet cultural conventions produced, or recognized, women players in significantly smaller numbers than men. My current research considers the social significance of these percentages today. My argument, and my hope, is that introducing more role models for women, the type that men have had for generations, is likely the most culturally substantive answer to correcting the imbalance.

This essay provides a contemporary snapshot of women musicians active in the Delta today; how they are connected to their roots and how vernacular music is evolving, unmistakably from the point of view of women players. Future influences on Delta music and culture will certainly include the contributions of these artists.

Since it appeared the best way to change the landscape was to locate, document and support their efforts, a vast digital collection of oral history interviews and performances has been (and is being) amassed. I've named it "Americana Women."[1] It has become my quest, made with the unconditional support and assistance of my partner, my husband, Rick Arthur.

MusicBox Project was founded as a nonprofit organization in 2010 with the mission to document and preserve music history while furthering avenues of education, creation, and performance.[2] Research encompassing seven months and twenty-three thousand miles of travel conducting field recordings in nineteen states included a lengthy stay in the Delta.

The initial collection is currently housed at the Library of Congress in the American Folklife Center. It includes eighty-one oral history interviews and over a thousand songs performed by artists in a myriad of settings from individual performances and impromptu jams to live concerts and festivals. The broad collection captures the essence of the contemporary Americana roots

FIGURE 14.1. Rick Arthur, Kelli Jones (now Kelli Jones-Savoy), and L. Dyann Arthur (*left to right*) at Joel Savoy's recording studio, Savoy Faire, in Eunice, Louisiana. *Photo courtesy of MusicBox Project.*

music scene from A to Z (Appalachian to Zydeco). Rich variations in experience were brought into view as we recorded women from eighteen to ninety-three.

This essay is compiled largely from personal accounts and excerpts of the interviews conducted with the musicians we met in the Delta region. In all, recordings of 350 individual songs were logged and some twenty-five oral history interviews conducted during the course of our month-long fieldwork, which took us from Southwest Louisiana through Mississippi into Tennessee and Arkansas.

The following provides a glimpse of Delta life through the eyes and ears of local women today, expressing themselves with instruments in hand. They are as abundant in diversity and musical flavor as the soil underfoot. Let me now uncover some of the jewels we found lying scattered about this great landscape.

Southwest Louisiana: Bonjour Cajun Country and Bonsoir, Catin

Having left our home in the Pacific Northwest on April 5, 2010, we (L. Dyann and Rick Arthur) continued our fieldwork in the lower Delta regions on May 12. After filming in Beaumont, Texas, we headed east into Southwest Louisiana, to Lafayette Parish, where we spent several weeks getting to know the culture and the character of Cajun country.

Our first introduction to Cajun women musicians put us smack dab in the middle of a live gig performed by the band *Bonsoir, Catin* at Café Des Amie, a family friendly restaurant in Breaux Bridge. It appeared that everyone there knew each other. Their warmth extended easily to us. As the evening went on the space between the tables became compressed with dancers from five to eighty-five. The band was wedged onto a raised enclave inside large bay windows. With the drapes open they invited onlookers from the street like musical mannequins. The floor was pounding along with the hearts of musicians and dancers alike. The four women on stage smiled, knowing they had the adoration of their audience.

Kristi Guillory is a small woman of four foot ten, sporting a compact, power-packed demeanor. Her punchy accordion rang, and her strong, resonant voice sang out lyrics in French and English, slicing through the din. A remarkably talented musician with a BA in Francophone literature and an MA in folklore, she's also a wife, with a grade-school-aged daughter. When interviewed at her home in Lafayette, she proudly brought out pictures, sharing stories of her grandfather Jesse Duhon, whose *Dixie Ramblers* string band music during the 1930s and '40s was a catalyst for her taking up guitar at age eight.

At ten, during a company party "crawfish bowl," she discovered a local woman's accordion. Kristi recalled being the only kid at the party, and, wanting to play it, she picked it up, spending the next three hours alone in the living room (grown-ups outside carrying on) pecking out "Jambalaya" or whatever Cajun song was in her head. Soon she bought an old black Hebert accordion with her allowance money. She was drawn to Cajun music instead of rock and roll, the popular music of the day, because of that accordion and because she was "really good at it—right off the bat." And we see, she is still really good at it, as she plays a spunky original piece called "Blues A Catin." Guillory's uncle Paul Duhon, a bones player, easily coaxed her into the weekly Champions' gas station jam session, where she was introduced to some of the local greats: Dewey Balfa, Octa Clark, and Aldus Roger. The older people were attentive back in 1992, she remembers:

> With so few young people interested and the unusual fact that I was a young girl, I was showered with attention and an intense amount of nurturing. D. L. Menard would call "We're cooking supper at the house, bring the little girl." I remember—that was my name "the little girl." We'd go to hang out—they wanted me to learn everything. That community made it a no brainer.

Anya Burgess, fiddler with *Bonsoir, Catin* and *The Magnolia Sisters,* takes care of her young son Rubin during the breaks at Café Des Amie with her husband's help. She grew up in New England and began playing music at a very young age, allowing her to easily incorporate a number of styles into her playing, including old-time and country. She attended Bowdoin College, John C. Campbell Folk School, and Indiana University, where she studied folklore and completed a program in violin making and stringed instrument repair. By day she runs her own violin studio/shop on the banks of the Bayou Teche in Arnaudville, Louisiana. At night she plays on a fiddle that she herself has hand crafted, baby and husband at her side.

Christine Balfa, *Bonsoir, Catin's* legendary guitar player, Dewey Balfa's daughter, hails from a virtual dynasty of Cajun musicians. *The Balfa Brothers* band, including Dewey, Will, Rodney, and Burke, consisted of four out of nine children. They won national fame for their beautiful, traditional Cajun style, being one of the first Cajun bands to frequent the national folk festival circuit during the 1960s as proud ambassadors of the Cajun culture. Their distinctive musical talent was likely bequeathed them by their father, Charles, who played fiddle, accordion, and triangle.

Musical roots run deep in the family. Christine's mother, Hilda, is said to have once lamented, "Honey, if I had a nickel for every time I heard Dewey say, 'One more song!'" But it appeared she was as much a music lover as he. Fay Stafford wrote, "I have caught her off guard as she cooks and cleans, humming French songs to herself. And she confided to me that as a child she loved to hear her father play his accordion when he came in from the fields, saying 'I'd just dance and dance. I guess I was too crazy to learn how to play, me.'"[3]

Knowing Christine's family background, we were caught completely off guard to hear, repeated by *Bonsoir, Catin's* bassist Yvette Landry, of the now infamous incident when Christine was pregnant with Amelia, her daughter with then-husband Dirk Powell, another renowned musician. She had just stepped off the stage with her band *Balfa Toujours*; a man, congratulating her on her pregnancy, asked if she was having a boy or a girl. When she answered, "A girl," he replied, "Oh well, maybe the next one will be a musician."

Band members acknowledge that "it's pretty much given that Cajun music is male driven and dominated." Each has her own stories of being a woman in the face of the Cajun's boys' club, yet the all-female group seems more focused on musicianship and camaraderie.[4]

As Yvette Landry rocks comfortably in the chair on her back porch, the May sunshine slips lightly through the huge old shade tree in her backyard. This was her grandparent's house. There's a picture of her grandmother Viola Hebert Landry seated with her banjo and her brothers, posing as a family band. Yvette shares, "My grandma's oldest brother Wilton, who was head of the band, wrote a piece of music he wanted banjo on. But there were no boys left to play banjo. So he said to my grandma, 'You're in,' gave her a banjo, and said, 'You have to learn how to play that,' and so she did." Yvette discovered Viola also played guitar. And her aunt taught Yvette piano lessons as a very young girl—right there in that house.

Though Yvette's Hebert lineage traces back to the Acadians who settled in French Louisiana in the 1700s, it's the country music of her father's liking that seeps into her compositions. She breaks into tears while recounting how he inspired her to do a CD, and she was able to share with him songs she wrote for her recording *"Should Have Known,"* which she calls "homemade honky-tonk that'll knock you off your boots!" She does just that with her performance of her original composition "Saturday Night Special."

The Prairie Cajun Capital and Musical Magnolias

Out into the back country north of Lafayette, Rick and I travel to the small town of Eunice, Louisiana's "Prairie Cajun Capital." This marshy prairieland, spotted with rice ponds and fields and cane and crawfish farms, is where the Savoy family make their home, and they have maintained their music store since 1966.

Ann Savoy didn't begin life in a renowned Cajun family. She reminisces about being romanced by Marc Savoy, a musician and accordion maker with deep Cajun roots: "You must come to Louisiana where the moss is dripping from the trees. I would love to have a woman to serve, and that would serve me in return." Even though the landscape was not quite what Marc had depicted, they wed in 1976. Some thirty-six years later, Ann, too, has musical roots that are deeply Cajun. Ann surely embodies the beating heart of Cajun musical traditions that continue through three of her four children; Joel, Sarah, and Wilson. Youngest daughter Gabrielle has inherited Ann's finesse with the camera, pursuing professional photography.

We come up the long, tree-lined drive to their ancestral home, walking through the carefree garden onto the sun-soaked porch. The screen door squeaks lazily, offering us entry into a home visibly steeped in love of music and tradition. We video record "The Separation Waltz," written by none other the Cleoma Falcon—one of Ann's biggest female Cajun influences. *The Savoy Family Band* consists of Ann on guitar, singing in both French and English, Marc on his handmade Cajun accordion, elder son Joel on fiddle, and youngest son Wilson on piano and fiddle. They exhibit nuances only a family and lifetime of playing together brings.

After the music making, Ann shares hours of stories about her life. She's written the Botkin Book Award–winning *Cajun Music: A Reflection of a People,* chronicling the history of Cajun music through interviews, biographies, photographs, and song transcriptions.[5] Through the years she has made movies, recorded and produced records, performed, and taught. She fondly recollects:

> Jane Vidrine and I would sit with our babies at my kitchen all the time and we would sing. I was playin' the accordion and she was learnin' the fiddle. And the babies would play on the floor and that was the best way to pass the time. So we learned a lot of songs, dug up a lot of old songs, and

> started singing ballads and all kinds of cool ol' Cajun music. And, ah, we were playing at a boucherie, which is when you kill a pig in your yard and you make it into lots of different dishes like, you know, bacon and pork roasts and backbone stews and things. We were singing in the yard and Chris Strachwitz from Arhoolie Records, who is our dearest, oldest friend, happened to be there and he loved that harmony singing and he said, "Why don't you guys come make a record with Arhoolie?"

Ann mentions that the only woman she remembers seeing perform Cajun music in the dancehalls is Sheryl Cormier. She became interested in finding more women and documenting them. She recalls, "Alan Lomax was a really good friend of ours. He was at our house and I was talking to him about that situation." She tells us that he said, "Well yeah, you know, it's the unwritten rule in a traditional culture. There are no written rules, but the rule is, the law is, the woman takes care of the child, and when you break those rules, it's just like breaking a law." Ann continued, "It just was ugly; people didn't like it and they disapproved. So I found myself playing with my husband, Marc, which made it more acceptable because he was Cajun. He knew everybody around here; he was very well respected."

Today Ann, also very well respected, is one of four women (along with Jane Vidrine, Anya Burgess, and Lisa Trahan) who currently call themselves *The Magnolia Sisters*, one of Southwest Louisiana's finest Cajun bands. The original group formed in 1978. Ann Savoy, Mary Jane Broussard, and Jeanie McLerie (who now lives in New Mexico and continues a lifelong music career that also included the Harmony Sisters with Alice Gerrard and Irene Herrmann) were the only all-female group to play the CODOFIL (Council for the Development of French in Louisiana) Festival in 1980.[6]

We visit Mary Jane's cozy home in nearby Jennings, where she proudly introduces her son "T" Broussard, a Zydeco musician. She still plays traditional-style Creole accordion. T tells us he features his mom with *T Broussard and the Zydeco Steppers* regularly, saying his fans love and respect her. Recently, they worked together to play and produce *Creole Royalty—Mary Jane Broussard and Sweet La La*. Many local families like theirs have close ties, playing and teaching the old music, carrying on traditions even in the face of much musical change. The La La musical tradition can be traced to African American field workers who prayed and gave thanks by singing, clapping their hands, and stomping their feet in a syncopated style called juré, an important root of Zydeco. By 1900, the juré songs merged with Creole and Cajun influences into what came to be called La La. Musical house parties known as La La's were held in the prairie towns like as Opelousas, Eunice, and Mamou.[7]

Mary Jane tells us about her husband's cousin Queen Ida, her uncle on her father's side, accordionist Alphonse "Bois Sec" Ardoin, and, on her mother's side, her uncle, fiddler Carlton Frank. Though she's currently the only Creole woman playing accordion professionally locally, it's clear her roots extend deep into the past. We are charmed by her renditions of "Si Passe Davon Ta Porte (I Pass in Front of Your Door)" and "The Lake Charles Two Steps."

Mary Jane has now joined ranks with two of the current *Magnolia Sisters,* Jane Vidrine and Lisa Trahan, in the band *Sweet La La.* Here are two more women who exude understanding and reverence for the cultural heritage they bear.

Lisa Trahan describes a jam session in the early 1990s at Christine Balfa's place. She was still just playing the guitar at that time. Dewey Balfa said to her, "'Ma chère, you're really good—better than all those other guitar players—but ma chère you got to punch it.' I said, 'What do you mean?' And he actually grabbed my hand and started this (illustrated by moving her forearm briskly up and down), so I would get the rhythm right. And that was a huge lesson!"

At her home we are treated to the joyful sounds of Lisa on accordion, guitar, and vocals and in duets with her accordion-playing father, Harry Trahan. Later, with Bobby Michot on guitar and Kelli Jones on fiddle, their group, known as *L'Esprit Cajun,* serenades us with the "Opelousas Waltz" and the "Ossun One Step." Their music fills the kitchen with tasty Cajun sounds as richly inviting as any gumbo.

We meet up with Kelli later for an interview at Joel Savoy's (not a five minute walk from Ann and Marc's.) She is following in the footsteps of her father, Carl Jones, a well-respected, old-time musician. Kelli plays fiddle as comfortably as she plays guitar, sings, and composes. But dance is what brought her to school in Louisiana in the first place. She and Joel share musical talents in at least two other bands: *Double Date,* with Emma Leahy-Good and Linzay Young (just a note, both couples are now married), and *Jones and the Giants.*

It's an everyday occurrence in such close-knit traditional cultures to find family members playing music together. The connection from one generation to another is visible again as Jane Vidrine and her son Joe sit in their living room playing tune after tune, easily exchanging guitar, fiddle, and accordion. The duets between them are captivating, filmed in front of a cabinet filled with half a dozen exquisite handmade accordions.

The longer we stayed in Southwest Louisiana the more we heard Cajun music played the way roots music is meant to be played—as a happy release. We found it in places like the Whirlybird, a backcountry speakeasy in Opelousas. And then there were Vermillionville and Randol's Restaurant, where we recorded *Sheryl Cormier and Cajun Sounds.*

Though I hadn't located Sheryl Cormier through research, once made aware of her part in history, I couldn't wait to meet her. Her mother played drums (at weekly musical house parties, not in public) and her father played accordion when he wasn't working as a sharecropper. She recalled starting to play at age seven:

> One night we had a little fais dodo that had these musicians, and Mom made a gumbo. They were drinkin' wine, and what made me so brave

> with the accordion was I would serve the wine. I would go in the kitchen and someone would say can you fill this glass, and on and on. I would fill one and take a drink and bring it to them. I mean, I wasn't supposed to, by no means, do that. They went to the table to eat and I was brave. I'd had a couple of drinks and I picked up the accordion and I started to play. Daddy, he tells mamma, "I don't want 'em playin' with my accordion" and "Who's playin' with my accordion anyway?" because it kinda sounded like a song. She got up and came to the door and went "shhh" to me and said, "It's Sheryl." But I was brave because it was a no-no. He said, "You don't touch my accordion." They were expensive for them in those days, so I was forbidden to touch that accordion. But when he'd leave to go out into the field, I'd watch him for as far as I could see and make sure he was far enough before I would open the accordion case, and that's how I learned.

We first met up with Sheryl as she was leading a regularly scheduled jam session at Vermillionville, the Living History Museum and Folklife Park, serving to preserve the cultural resources of Lafayette Parish and the surrounding region. She and her husband, Russell, sat in a circle of about twenty happy souls playing fiddles, accordions, guitars, and triangles and singing and dancing.

This weekend was not an unusual one for her, playing four different gigs—even, and mind you this, after recent heart surgery. Russell, who began as her roadie, has now secured his position as their band's lead singer. Her son Russell Jr. plays drums. After hearing *Bonsoir, Catin* play Sheryl's *"Le Bottle,"* it was a treat to hear own her version firsthand, with so much clarity and skill. Another treat was the scrumptious home-cooked meal of fried chicken and macaroni and cheese, which they served with a side of southern hospitality.

Americana at Home in the New South

Johanna Divine's music can only be described as "Americana." Her influences range from bluegrass to Texas swing, with a big scoop of cowboy country plopped smack dab in the middle. Every bit as varied as her musical roots are her physical roots. She's lived from Alaska to Arizona, with the last couple of years in Lafayette. Sitting in her southern kitchen, decorated with a 1950s *Leave It to Beaver* flavor, she shares her current creative manifestations: the CD *Mile High Rodeo* and her co-writing and (musical) directing of the play *Le Rêve des Marionettes* (*Dream of the Marionettes*). Johanna appears an unstoppable musical force.

Johanna was born in Knoxville, Tennessee, in 1973. Her father played guitar, and three brothers became multi-instrumentalists alongside her. "My brothers play with a little more abandon than me in that they just feel confident in whatever comes out, and I think I always just needed to be a little more cautious in making sure it's right before putting it out there . . . for better or for worse." She considers how boys are encouraged to be out there and do their thing, "Oh, you're so great," where women are seen as the supporters. She thinks that's a little old school though, speaking affectionately of her niece Hanna, who, at eleven, is completely confident and open in her creativity, and being a girl doesn't inhibit her at all. Johanna's brother, a single father, has always just put an instrument in Hanna's hands and said, "Go! You can do this; of course you can do this. It's an instrument and it's meant to be played—by people."

Heading up the Mississippi River—Getting the Blues

It's up the road we must go. The insistent scheduling drives our motion into Mississippi along Highway 61, the "Blues Highway." We're filled with the increasing earthiness of the land in our nostrils as the green fields, square miles of them, roll by. Corn now, cotton now, and high cumulus clouds like fluffy Jiffy Pop pass overhead in rows. We slow down to pass a circa 1950 rake wheel tractor chugging along in the bright morning sun. We wind our way north along the great river with a stop in Natchez to visit an antebellum mansion. Walking the grounds, vestiges of the Old South seep into our souls. We're connected through music to the people who relied on it for their very connection to spirit and to their African homeland, forever a memory.

Our next interview is with Eden Brent, born quite nearly on the banks of the Mississippi Delta in Greenville. Her mother was a big band singer and songwriter in the 1950s. One of four children who all play instruments (one boy and three girls), Eden was playing her older brother's piano recital piece by ear at age four. Her parents begged his teacher, Molly Schwartz, to take her on, small as she might be, or they'd find somebody else who would. Eden tells us, "I wrote my first song when I was in the third grade partly because my older sister Jessica was writing tunes, so that inspired me to believe I could do it too."

She says, "There wasn't a lot to do [growing up]. We didn't get a lot of TV reception that far out in the county, so after dinner very often we would pass the guitar around the kitchen table and we would entertain ourselves." These days she entertains globally with her "whiskey-smoke" vocals and extraordinary piano work. Eden spent sixteen years learning from and partnering with the late blues pioneer Abie "Boogaloo" Ames, who ultimately dubbed her "Little Boogaloo." Eden's lifelong friend, journalist Julia Reed, says, "They were soul mate and road buddies. She was a young white woman of privilege and he was an aging black man in the Mississippi Delta, but theirs is a phenomenal story of mutual admiration and need." It's such a story that both PBS and South Africa documentarians have told it in film.[8]

Eden has performed from the Kennedy Center to Africa. She won the Blues Foundation's 2009 Pine Top Perkins Piano Award, the 2006 International Blues Challenge, and was a 2004 inductee on the Greenville Blues Walk. She even shared the bill with B. B. King at the 2005 presidential inauguration.

With a personality as full-blown as a juke joint holler, she relentlessly celebrates life and music. Her home, reflective of the tough blues lifestyle she leads, was cluttered with CDs, sheet music, unopened mail, and various instruments, including her impeccable Schimmel grand piano, which she takes to like a bird on the wing. To warm up her fingers she flies through "Boogaloo's Boogie," an instrumental at 125 beats per minute. Then she launches into one of her own poignant, double-entendre, inspired blues. Eden unabashedly laughs, "I wrote this song on a Saturday morning. I'd stayed up all night having a little vodka, or maybe a fifth or liter, and it was about eleven o'clock in the morning when I finally came up with this little number right here called 'My Man.'"

Memphis: More Than Blues on Beale Street

Eden tells us about her friend, the immensely talented, Memphis born and bred blues woman Di Anne Price, pianist and vocalist extraordinaire. We have good timing and catch *Di Anne Price and Her Boyfriends* (Tom Lenardo and Tim Goodwin) at Huey's Blues, Beer & Burgers, blocks off Beale and the Big Muddy. She tells us:

> I was born into this business. Daddy played ninety-nine instruments. Momma was a songwriter. My first gig was with my sister. I couldn't have been more than four or five. We played "Mississippi Mud"—one piano, four hands—on a riverboat. Daddy was born 1906, so early on people used to have dinner on the riverboat right here along the Mississippi. They would play at hotels, Elks Clubs, and "joints" in Mississippi, Arkansas, and Missouri. As I remember, 'cause I was a little child, I thought those women looked so beautiful. Back then ladies used to dress up, wear gloves and hats. I couldn't wait to grow up. Never knowin' that by the time I grew up you could go anywhere in your dungarees and T-shirt. It was a kinder time than I think it is now.

A lifelong regular on Beale Street, Di Anne has played the Blue Stallion, Hippodrome, and Club Handy and says, "I've always been able to work right here. I made my first recording in 1959. Its somewhere out in the world. . . . [W]e have seventeen projects just floating around out in the universe, you know how that happens, you don't have your own master, so that work is anywhere and everywhere." She proudly explains how she was honored as a Memphis Music Emissary by the US House of Representatives Congressional Record, by Honorable Marsha Blackburn of Tennessee, on March 26, 2009.

Di Anne tells us sincerely, "I'm always trying to help somebody." With her degree in sociology, with a concentration on the elderly and music, she's maintained a career for thirty years in music therapy, now helping Alzheimer's patients. "I'm always playing to the one person," she says, and it's evident from the way she connects honestly and intimately that she does just that.

We connected with another Memphis musician, Valerie June, at her folk's home an hour and a half outside the city. Valerie is a statuesque twenty-eight-year-old African American beauty with unforgettable Rastafarian hair charismatically rolling down her shoulders. She calls her music "Organic Moonshine

Roots Music," saying, "I remember being in my father's dump truck, riding down the street listening to Tracy Chapmen sing 'Fast Car' and singing at the top of my lungs. I was probably about three years old, just bellowing and loving it." Her star power talent is still rising; Valerie recounts meeting her now good friend Autumn Grieves:

> She was so nurturing to what I do. Not only her, but the Oscar-winning filmmaker Craig Brewer that did "Hustle and Flow" and "Black Snake Moan" on Memphis Music. I remember Craig saying to me, "You just have to be ready, the opportunity is gonna come, you don't know whether it's going to come tomorrow or a few years from now." Having people around to say things like that when you need 'em; it's huge.

Caught in the Devil's Bargain: The Crossroads, Clarksdale, Mississippi

Back in Greenville, Mississippi, Eden plays a rousing set at the "Walnut Street Opener" of the Highway 61 Blues Festival, an annual music event staged in Mississippi since 2002 and organized by the Leland Blues Project and the Leland Blues Museum. But Eden has got an out-of-town gig during the actual festival. She is the only woman player on the entire bill, a rather sorry state of affairs as we see it.

With so many interviews planned over the next three weeks at locations around the Delta, we seek out a convenient location. We find a funky little campground called Frog Hollow in Grenada. This affords us manageable travel distances to the Blues Archives at the University of Mississippi and to "The Crossroads"—Clarksdale—where we want to spend a great deal of time talking to musicians and breathing in the blues.

Clarksdale, Mississippi—where do I start? We meet some remarkably dedicated artists and approachable people there. The local blues community is intimate and interdependent. They depend on each other and need each other to survive. But work for musicians is at a premium, so the competition can be fierce. Rosalind Wilcox, aka "Mississippi Rose," explains: "There's not a lot of money in Clarksdale at all. There's these little bits of cheese that's out there and it gets real territorial."

The competition between the old black blues men and younger blues women harbors some harsh dynamics. Rosalind knows firsthand the weight of being a female performer:

> You don't know if you're being discriminated against because you're a woman or because you're a black woman or because you're a northern woman that moved to the south that's really got southern roots. There's a lot of angles on this stuff. I've never been pushy, and I think that's worked against me. When there's free shows, and that's some of the best shows in Mississippi, these free parties, the BBC shows up and *Rolling Stone*, and all these people are there. The men are very well tuned in. They know exactly

> who's there. And they push, and they make sure they're on that stage. Even if they get kicked off—they'll do thirty minutes—even if they're told they get ten. I've watched them over and over again. The women tend to hang back, and they'll say, "We gonna put you guys up in a minute." The minutes become hours and the hours feel like days, to quote a blues song. Seriously, by the end of the night the women go up. The BBC is packed up, *Rolling Stone Magazine* is gone, and you wanna cry.
>
> And on top of that, I will not name names, but they know who they are. There are guys that are still around who will either be drunk or pretend to be drunk and get up there on drums when you didn't ask for a drummer. They'll just drum really loud and drum you to death. Or they'll pick up a bass guitar and drown you out. They'll pick up another guitar or they will pick up your guitar and re-tune it when you just did. I've had all these things happen. But talk about wantin' to fight. When you've been practicin' and you've got your stuff together and some clown makes a fool out of you, it hurts.
>
> We played Memphis two years ago, one of the biggest festivals around, and one of the old culprits unplugged our stage. Once we packed up and left he plugged it back in and played. It's notorious. Things that go down in the blues circles will make you wanna just scream. It's just like animals; if you aren't aggressive enough you just get pushed back.

Rosalind is a thoughtful, intelligent, and talented visual artist, musician, and art teacher at the Coahoma Community College. She started young on piano and violin, drums and percussion. Born in the late 1950s in Chicago, she remembers her momma showing her how to make instruments: flutes, drums, and little one-stringed guitars. She picks one up that she's recently made from a cigar box, a lovely work of art. Everywhere around her eclectic home are her paintings and sculptures, textiles and handmade instruments.

She was a mother at seventeen. She started losing her eyesight at twenty-one due to a genetic disease, so she took up guitar, thinking she might not be able to paint anymore. But truly, she doesn't let her failing sight slow her down much. She composes original songs, several of which she performs for us. One unforgettably honest number is *"Ain't Nothin' Shy 'bout Clarksdale."*

Rosalind kindly shows us around Clarksdale our first day, putting us up in her guesthouse so we won't have to drive back to Grenada late in the evening (just to come back the following morning). We cruise down Blues Alley, see the local music store, Hopson Commissary, and the Riverside Hotel, where Bessie Smith died. We even make it to Rosalind's favorite Chinese restaurant for dinner.

Cathead Delta Blues and Folk Arts, Inc., is near Morgan Freeman's Ground Zero Blues Club and the Delta Blues Museum. The store is owned by Roger Stolle, a man who's made it his mission to support local blues music and the infrastructure necessary to keep it viable. Laura "La La" Craig is employed there days and plays with *Super Chikan's* Band and her own group, *La La's Electric Eighty-Eight,* at night.

La La Craig grew up in a family band and has been playing professionally since the age of seven in Oceanside, California. She says the country roots of those days have really stuck with her and laid the foundation for today. "You couldn't swing a dead cat at a Craig family reunion without hitting a songwriter and an excellent one at that." When she was twelve the family got a piano and life changed:

> It woke something up in me. Initially, I wanted to be a classical player and take lessons. I went to a teenage guy that had won some concerto competitions. He got me started and then sent me promptly to his teacher. There was supposed to be a long rigmarole to get in with this guy, but the first time he heard me he took me in. I practiced so much they would literally come in, in the middle of the night sometimes, and pull me out from under the covers, because they had taken me off the piano after nine hours. They would take my flashlight, because I'd be under there looking at the music, doing this (motions fingers playing a keyboard). I had literally lost my will to do anything else in the world but that.

After moving to Clarksdale ten years ago with her then husband, a bassist with well-known bluesman Super Chikan, La La Craig was performing at an open mic and Super Chikan said, "Maybe sometimes I could use a keyboard player on big gigs." He invited her to play at Ground Zero one night soon afterwards; she then joined his band fulltime:

> That was the beginning of my work with the best band ever and the end of my marriage, though we continued to play together in that band for two years. It's interesting to go on tour with someone you really are breaking up with and keep a good vibe on stage. I think for those two years I was the place setting: *"Fighting Cock"* number three. Like those faces in Facebook that don't have a picture? That was me in the keyboard position. Nowadays I talk to people that love what I do and think that I've become an integral part of the band and [they] say, 'I saw ya'all in 2004, but you weren't with the band.' And I say I've been with them since '03, and they say, 'No, I'd remember you.' I say, "No, I was at that festival or whatever." They remember me now, but then I was just blending in.

Before our next rendezvous La La proceeds to knock out a few numbers on her Kerzweil, set up at Cat Head. Her original "My Dog Loves Me" is a fast-paced blues ditty that she says "is about the two-legged variety." Not only has she thoughtfully suggested some other players to interview, but she mentions the open mic at Ground Zero Blues Club (GZBC) on Friday, a golden opportunity to showcase the local female talent—all in one place, at one time. What could be better?

We start to put the wheels in motion, heading down Blues Alley to GZBC. As with every public location where we're shooting a video, we'll need a location release signed. It's midday on June 2. Steamy Delta mugginess burdens the air. The club, like the little city itself, is virtually at a standstill. We drop off the form requesting approval to film Friday and lunch on fried green tomatoes, washed back with unsweetened tea.

We're back a few days later to interview bassist Torey Todora, who waits tables at the club in the afternoons. She grew up in Baton Rouge, Louisiana, sang in choirs, started playing bass with a punk rock band in New Orleans at nineteen, and taught herself guitar living in the woods in Austin, Texas. She ended up in Clarksdale totally by accident. About a year ago La La Craig recruited Torey to play bass in her band. Now Torey plays with five or six local bands. She sits, legs crossed, smoking a cigarette, and says, "Once I was introduced to playing music with other people, the feeling I get, it's better than anything. It's better than sex; it's better than drugs; you're not thinking about anything else. You're completely content and at peace with yourself when you're up there playing music with people."

The community of artists centered in downtown Clarksdale is by no means limited to musicians. We visit the Hambone Gallery, owned by Stan and Dixie Street, exhibiting art and music. Stan plays harmonica. Dixie plays drums. She's been with the *All Night Long Blues Band*, playing at places like Ground Zero, Red's Lounge/Juke Joint, and Tricia's Italian Restaurant & Pie Hole since 2009. Dixie is thoughtful and pleasant, a lanky redhead with a direct approach to subsisting as an artist while listening to the spirit that moves her. She told us, "In middle school I wanted to play drums and was told that girls don't play drums and I had to pick another instrument: oboe. I don't know why I would pick oboe; it's not a friendly instrument. But I guess I wanted to be different; maybe that's why I wanted to play drums."

During her seventeen-year marriage to slide guitar great Alex Gomez, she eventually decided to embody her childhood ambition, starting with just a snare drum. Though he was supportive, he was a difficult taskmaster:

> It was hard to be expressive because it had to be one way, and that's it. So it was tough. After the divorce I bought a whole [trap] set. And you know, when you're on the right track the heavens, the universe, goes yes, finally, do this. I took some lessons, but I have the personality of I wanna get it done, get to the point, so I taught myself to play and got part-time gigs. I was drawn to hill country blues; it's so primal. I can play this, sounds pretty simple. The gallery's closed. Stan's off doing a gig. I have time alone and I can make mistakes without someone hearing me and going "ugh." The more you play the better you're gonna get. It cannot *not* be achieved!

Hillbilly Music, Alive and Well

Our next foray takes us out of the flat Delta proper and into the hill country in the very northeastern corner of Mississippi, to the little town of Dennis. Here we've been invited to film at "Dennis Days" by the most hospitable and delightfully southern Lisa Lambert. In 1947, Bryan Sparks, Bufard Wells, and Nolan Wells put together a hillbilly band known as *the Pine Ridge Boys*. Having come back together after decades, they joined forces with song writer/musician Lisa and her husband, Scott Nunley, in 2007. Lisa's talent and energy have been instrumental in reshaping *Lisa Lambert and the Pine Ridge Boys*. Together, they've performed throughout the Mid-South and completed three albums.

Lisa and Scott host an annual gathering at their home after the outdoor concert, complete with music and a barbeque. Sitting comfortably, out of the sun on their wrap-around porch, the local gals share tales and tunes. There's Lisa and Annette Tucker on guitars, Peggy (Sparks) Adams on accordion, Laurie (Jones) Brosie on mandolin, Lauren Reister on upright bass, and Jan Pike on dulcimer. All the women chime in, sharing their love of the old-time, gospel, and hill country blues.

Treasure in the Arkansas Ozarks

Yet another sideline trek takes us out of Mississippi to the adjoining Arkansas hill country for an unforgettable experience with an unforgettable personality. I was introduced to Violet Hensley through Mary Katherine Aldin, an independent reissue producer and annotator who's been involved in the preservation of American traditional music since 1962. Violet has resided her entire life in the Ozark Mountains of Arkansas. The planning of a family reunion has gone on between us for months now. Violet's daughter Sandy Flagg helped round up as many of the kinfolk as possible. I assured her we'd be there with the camera and barbeque, if she would muster the troops.

Violet is in her element. She's the most lighthearted and engaging ninety-three-year-old imaginable, long ago dubbed "the Whittlin' Fiddler" not only because she whittled figurines, but also because she has whittled some seventy-three violins in her time. Astounding and inspiring us, she easily remembers not only when she made each one, but what it's made of, who bought it, for how much, and, usually, where the instrument is now. She can do the same with all of her children, grandchildren, and great-grandchildren. Today she is happily surrounded by twenty-two family members, most of whom play instruments as the day goes on.

I point at the camera as she picks up the fourth fiddle she ever made. Coyly, she smiles at the camera saying, "So, I tell that camera about this?" We all begin to laugh. "This is a fiddle I made in 1934. I cut five trees, wait a minute (points and counts), a maple for the back and back side of the neck, pine for the top, dogwood for the tail piece, walnut for the finger board, and persimmon for the pegs." I ask her about the autograph that reads "Ricky Skaggs 1996," to which she responds:

> Well, I've been demonstrating fiddle making and playing at Silver Dollar City since 1967, and Ricky was there in '96. I was the only one out of two

> hundred musicians that got to go back and meet him. Others had to set their instruments back there for him to sign, but he played on mine. He played me "Sally Johnson" like I played when I was a teenager. He handed it back to me, and they were near 'bout ready to go on stage, so I quickly did a little bit of "Rag Time Annie."

Though Violet Hensley's no movie star and about as down to earth as any human could hope to be, she's actually appeared on the *Beverly Hillbillies Show*, the *Art Linkletter Show*, and *Captain Kangaroo*, and Charles Kuralt stopped by her house once. She received the Arkansas Arts Council Living Treasure Award in 2004.

Born at Alamo, Arkansas, on October 21, 1916, she says matter-of-factly, "That wasn't a town. They had a schoolhouse and a gas station. I was born at home, about two miles from that schoolhouse." She had three sisters; one died shortly after being born. "No brothers. My mother died when I was eleven and we didn't have a stepmother." She recalls, "We had to be boys and girls too. We plowed the fields, cut timber, split rails, built fences, butchered the hogs, slopped the hogs, milked the cows, fed the chickens, and picked the cotton."

She started to play in 1929, listening to her father fiddle, getting some help from him, but mostly, she says, "I'd just try to hear it until it fit my ears. It has to get in my brain before I put it on the fiddle." When she was a girl they said "it was a sin for a woman to play the fiddle. They thought it was the Devil's instrument." Picking up a fiddle and shaking, it she adds, "That's why it's got a rattlesnake rattle in it; that scared the devil out." It doesn't appear anything could scare the devil or the life out of Violet Hensley though. And I want to be just like her when I'm ninety-three.

Back at Camp: Frog Hollow, Grenada, Mississippi

Back at Frog Hollow, the sticky Mississippi heat is taking its toll on the well-loved little pop-up Coleman tent camper we've been calling home, totting behind our 1999 Four Runner. Day after day the morning sky closes in, dumping gallons of rain, then gives way to hundred-degree sunshine that literally sucks up the puddles before our very eyes. The little swamp cooler is losing its battle. Everything we own is sodden. Bed sheets and towels won't dry; the carpet stays soggy and is beginning to smell like old socks. But being the dedicated documentarians we are, with another interview having presented itself on location, we hastily make arrangements to use the campground's community building, ashamed to have company in our own site.

Fortunately, Heather Crosse is an affable person. We set up our recording gear against a backdrop of shellacked jigsaw puzzles posing as wallpaper, completed by relaxing campers through the years. Heather was born in 1974 on the Red River in Louisiana and, though not from a musical family, started dance, piano, and singing early on. She auditioned for a musical play, got the lead, and "was hooked." She went through the Governors' Program for Gifted Children as a music major for three summers, taking college classes, while she was in sixth through eighth grades, at McNeese University near Lake Charles. She played alto saxophone in the junior high school band, gaining experience with jazz. Then, at eighteen, her guitar-playing boyfriend brought her to Clarksdale, where she instantly fell in love with the blues. Heather ended up in Hot Springs, Arkansas, for a while, where she was the bass player with St. Thomas Jenkins, an Arkansas bluesman who'd made his mark in Detroit in the 1970s. After moving to Clarksdale to live, she played with *Super Chikan & The Fighting Cocks* for four years and is now the leader of her own band, *Heavy Suga & the SweeTones*, on electric bass and lead vocals.

The Big Finale at Ground Zero Blues Club

The night before we leave Clarksdale is the Ground Zero Blues Club open mic (and impromptu ladies' night). Before the show we do a quick interview at Lambfish Gallery with Jacqueline "Jaxx" Nassar, a talented and confident eighteen-year-old who attended the Delta Blues Museum's Arts and Education Program. Jaxx is a multi-instrumentalist; having started the program on drums, she added bass and guitar. She recalls how, at eleven, one of the local bands tried to pay her eleven dollars for a five-hour gig; she smirks, "Because I was eleven and a white girl." But her mother intervened on her behalf. She learned the ropes at a young age, now commanding the respect she deserves.

The show at Ground Zero presents live-recording nuances we've learned to handle: Rick up front with the video camera and me by the soundboard, monitoring the direct line-in digital audio. To manage logging metadata for forty digital files, I'm back and forth verifying our data matches. Live situations present complications like unannounced performers who haven't signed releases and unannounced songs lacking composer and copyright information, all requiring follow-up. The venue is loud and the lighting poor, but we do our best to document all of the women who have come to play.

Tricia Walker traveled an hour from Cleveland, excited to add her contribution to MusicBox Project. Having just made contact with her through Greg Johnson, curator of the University of Mississippi's renowned Blues Archives, I suggested this open mic, which was to be our last recording in Clarksdale.

Tricia sat politely patient throughout the electric blues performances; and like Rick and I with acoustic instruments, also planning on sitting in, Tricia didn't feel the ambiance was conducive to her music. Knowing her references and considering she'd come so far, we couldn't just leave without including her. Instead we invited her to our rather shabby motel room after the showcase for an interview and some of her acoustic-style Delta blues.

At midnight we darted into the motel room, inviting a swarm of hungry mosquitoes to join our intimate concert. For the life of me, I will never forget the masterfully compelling performance of her song "*The Heart of Dixie*," the tender introduction, the warm, clear, and relaxed tone of her voice, the expressive dynamics of her precise guitar work. Nor will I ever forget the incredible professionalism she displayed as, unbeknownst to us, throughout the ballad a mosquito was feeding on her wrist! She ignored the assault until the instant the camera clicked off, immediately scratching and laughing wildly. This was why we had come, moments of unscripted life and music and artistic purpose. It was all so worth our efforts!

We left the heart of the Delta the following morning, short on sleep, a five-hour drive ahead of us, silently making our way up "Mississippi #1." The low-hanging fog swirling around, giving rise to melancholy feelings. Yes, we felt the blues. But when Tricia Walker's CD began "Ode to Billy Joe," we realized we were right where we were supposed to be. It had been the third of June when we'd crossed the Tallahatchie Bridge, and I'd thrown a flower off for my sister Deb, who shared my passion for music since early childhood. Sleepy, dusty Delta days, Carroll County, a copy of the original manuscript for that song given to us at the Blues Archives, a song I'd played most my life—now it all ran so much deeper. That is the magic of the Delta. Music and memories of time spent swirled in our heads like the fog as we headed toward other new adventures with more "Americana women." Our work would take us another four months, through the ever-changing, always-vibrant mosaic of American cultures. We would document the evolution of traditional music from the days of plantations and small farms filled with hardworking people who played music for comfort, joy, and community to the present moment in which women throughout the country echo and expand on that history.

Notes

1. "Americana Women: Roots Musicians—Women's Tales and Tunes" is the full title of the documentary (film) being produced by MusicBox Project. The current film work and music clips can be viewed on the MusicBox Project YouTube Channel: http://www.youtube.com/musicboxproject, accessed April 1, 2015. All of the quoted interviews are available in full, unedited, as both audio and video recordings. They are part of the "Linda Dyann and James R. Arthur, on behalf of MusicBox Project Collection," on file at the American Folklife Center, Library of Congress; the Southern Folklife Collection, University of North Carolina at Chapel Hill, NC; and the Blues Archives, University of Mississippi, Oxford, MS.
2. MusicBox Project is a 501(c)(3) tax-exempt public charity. The website URL is http://ww.musicboxproject.org, accessed April 1, 2015.
3. Ron Stanford and Fay Stanford, "The Balfa Bros.," in *J'Etais Au Bal: Music from French Louisiana*, 1972, http://www.hechicero.com/louisiana/lesartistes.html, accessed April 1, 2015.
4. Bonsoir, Catin, http://www.bonsoircatin.com/, accessed April 1, 2015.
5. Ann Allen Savoy, *Cajun Music: A Reflection of a People*, vol. 1 (Eunice, LA: Bluebird Press, 1984).
6. The Council for the Development of French in Louisiana (CODOFIL), created in 1968 by the Louisiana state legislature, was empowered to "do any and all things

necessary to accomplish the development, utilization, and preservation of the French language as found in Louisiana for the cultural, economic and touristic benefit of the state." See http://www.codofil.org/, accessed April 1, 2015.

7. Tom Dempsey, "Origins of Zydeco and Cajun Music," *Northwest Zedeco Cajun Music and Dance Association*, http://www.scn.org/zydeco/nwczHISTORY.htm, accessed April 1, 2015.
8. 1999 PBS documentary *Boogaloo & Eden: Sustaining the Sound,* Mississippi Public Broadcasting (MPB); 2002 South African production, *Forty Days in the Delta.*

CHAPTER 15

Heritage Tourism in the Delta

RUTH A. HAWKINS

The character of the Mississippi River alluvial plain, shaped by forces of nature and the imprint of diverse cultures, has led to a growing industry in the region in the form of natural and cultural heritage tourism. Efforts to preserve and promote aspects of the past throughout the seven states that comprise the Delta hold out hope for a brighter economic future for many rural communities whose glory days began to fade in the 1950s and 1960s.

While the Delta has some of the richest land in the country, it also has some of the poorest people. A Lower Mississippi Delta Development Commission report to Congress in 1990 likened the region to a third world country, and, over all, things have not changed significantly in the intervening years.[1] The economic distress is due in part to the failure of the Delta to diversify beyond its agricultural base. Thus, when mechanized agriculture reduced the need for field hands and tenant farmers beginning in the late 1940s, thousands were out of work with no place to go unless they could get to factories in the north.

Today, Delta farming operations have become large-scale and, in some cases, are managed by multi-national corporations. The need for less labor has led to some of the highest unemployment rates in the nation.[2] What remains, however, is the region's distinctive heritage, molded over time by struggles to tame the river and convert the land to agricultural production. The geography and natural environment of the region have influenced its history and culture in such a manner that a bond exists between the people and the land. There is a special resilience in the Delta, perhaps flowing from it residents' knowledge that what the river gives it also can take away.

The authenticity of the Delta way of life is of growing interest to tourists. Not only is tourism the world's largest service industry, but natural and cultural heritage tourism is one of the fastest-growing sectors.[3] Among domestic travelers during 2010, the most popular leisure activities were visiting friends and relatives, shopping, rural sightseeing, and visiting beaches.[4] It is significant that visits to rural communities were a high priority, while trips to large cities did not place in the top five. Additionally, the interest in visiting friends and relatives holds promise for the Delta. Many who left the region in search of jobs elsewhere, particularly African Americans, are now returning to maintain ties and attend family and school reunions. In terms of international visitors, the top-five leisure travel activities in 2010 were shopping, dining, city sightseeing, visiting historical places, and visiting amusement/theme parks.[5] While there are few major cities in the Delta, and none currently in the top twenty visited most often by international travelers, Memphis and New Orleans nonetheless draw millions of sightseers annually.[6] Visitorship to these cities is advantageous for the entire region, in that travelers to urban centers often make side trips to the more out-of-the-way historical places.

Many of the small towns and impoverished regions of the Delta—those that could not afford to tear down the old and build new—are finding that these authentic places are becoming major assets in the quest to experience "real" America. Shotgun houses that once belonged to sharecroppers, juke joints where many blues musicians began their careers, country stores with their pot-bellied stoves, drug stores on the square with old-fashioned soda fountains—all are becoming part of the heritage tourism mix for the Delta. It is still possible in some places to visit with elders of the town on their "whittling" benches in front of the local bank or barber shop. And in the right seasons, travelers might encounter hunters and fishermen at the community store or bait and tackle shop, stocking up on supplies and swapping tall tales with others who have returned from the deer woods or the many rivers, lakes, and swamps.

One problem the Delta faces is that many of the sites of interest are miles apart, with vast open spaces and fields in between, making it difficult to create the critical mass necessary to draw tourists in large numbers. However, with the growing interest in driving tours and long weekend trips, some of the transportation difficulties are beginning to be addressed.

As early as the 1930s, the US Congress recognized that the unique characteristics of the Mississippi River were of interest to tourists and took steps to create a route along both sides of the river to attract travelers and entice them to drive its entire length, stopping at various attractions along the way. In 1938 the Mississippi River Parkway Planning Commission was formed at the urging of Secretary of Interior Harold Ickes, with the governors of the ten states along the river appointing representatives to the commission.[7] This commission was instrumental not only in the early planning and development of the parkway, but in its construction, promotion, and marketing.

Though progress was interrupted by World War II, Congress approved funding for a feasibility study in 1949 that was completed in 1951. The study concluded that a parkway for the Mississippi River would benefit the nation as a whole. Rather than build an entirely new road, the report recommended that existing routes along the river in each state should be designated and linked as part of the Great River Road. The criteria noted, "The Great River Road should be located within designated segments to take advantage of scenic views and provide the traveler with the opportunity to enjoy the unique features of the Mississippi River and its recreational opportunities."[8] The criteria also specified that the Great River Road should offer a variety of experiences or themes and that it should provide convenient access to larger population centers.

Similar criteria guide ongoing development of the route today. Now known as the Mississippi River Parkway Commission, headquartered in Madison, Wisconsin, the mission of the commission remains to promote, preserve, and enhance resources of the Mississippi River Valley and the Great River Road. In the first few years of the twenty-first century much of the groundwork came to fruition when Great River Road routes through each of the ten states were separately designated and collectively promoted as the Great River Road National Scenic Byway.[9] The National Scenic Byway designation by the US Department of Transportation is reserved for special routes throughout the country that demonstrate significant natural, archeological, historic, cultural, recreational, or scenic qualities of interest to travelers along the byways.

From the beginning of these efforts, it has been clear to Great River Road planners that there are two distinctly different Mississippi Rivers in terms of character. The Upper Mississippi River, from Cairo, Illinois, to the headwaters at Lake Itasca in Minnesota, is characterized by locks and dams, continuous river views, and an abundance of recreational activities. The Lower Mississippi River, from Cairo, Illinois, to the Gulf of Mexico, is largely hidden behind tall levees and becomes the "Mighty Mississippi" at the point where it merges with the Ohio River. The result is a much different tourism experience for those traveling through the Delta region. While the lower river rarely shows itself, its handiwork is evident in the natural landforms and rich, alluvial soil that once nurtured only swamps, bayous, and bottomland forests and now also supports vast fields of cotton, rice, soybeans, and sugarcane. Though the Delta is defined in numerous ways, this essay focuses primarily on communities and attractions that have been influenced by proximity to the river. The ten broad themes that follow provide for a distinctly Delta heritage tourism experience.

River Heritage

Despite the fact that the Lower Mississippi is hidden behind levees, opportunities remain for travelers to get out on the river. At one time there were multiple steamboats providing overnight trips on the lower river, but when the *Delta Queen*, *Mississippi Queen*, and *American Queen* were retired from service in 2008, it appeared that an era had come to an end. As of 2012, however, the *American Queen* has been refurbished and is again plying the Mississippi as well as the Tennessee and Ohio Rivers. The grand sternwheeler, which offers seven- to ten-day cruises on the lower river, is a way to see the region in a manner that few get to experience. Adding to the mix, the European Viking cruise line announced in February 2015 that it will begin operations on the Mississippi River in late 2017. Other steamboats offer dinner cruises and short excursions from places such as Memphis, Tennessee; Tunica and Vicksburg, Mississippi; and New Orleans, Louisiana. Canoeing and kayaking are growing in popularity as ways to explore sections of the Lower Mississippi River, though heavy river traffic and strong currents require great skill. Outfitters and guides are available on both sides of the river to provide assistance.

Mark Twain once wrote, "The Mississippi will always have its own way; no engineering skill can persuade it to do otherwise."[10] Yet humans have tried, and these efforts to tame and control the river can be a source of fascination for visitors. For example, some pumping plants along its banks provide tours, while the Plaquemine Lock State Historic Site in Louisiana is specifically geared to helping visitors understand our complex relationship with the river. In addition to the lock, the area includes a lock house that serves as a museum and visitors center.

Numerous other museums throughout the seven-state area tell stories of the river and its changes over time. Among these are the New Madrid Historical Museum in Missouri and the Tunica River Park Museum and the Natchez Visitor Reception Center in Mississippi. One of the most interesting is the Mud Island River Park and Mississippi River Museum in Memphis, Tennessee. The park features a five-block-long exact scale model of the Lower Mississippi River, flowing 954 miles south from its confluence with the Ohio River at Cairo, Illinois, to the Gulf of Mexico. The adjacent museum is comprised of eighteen galleries showcasing ten thousand years of river history.

Natural Heritage

While landscapes created by the river have undergone tremendous change, it is possible in places to experience the Delta as it existed prior to human habitation. As part of the Mississippi River Alluvial Flood Plain, the region once was little more than forbidding swamps, along with dense bottomland timber forests. Meandering rivers constantly altered their courses through the region, creating lakes, bayous, and wetlands. Natural formations such as oxbow lakes (old meander belts cut off from the main channel of the Mississippi River) and swamps often are as interesting as the river itself. Lake Chicot State Park in Lake Village, Arkansas, has the distinction of having the largest oxbow lake in the nation, while the Louisiana swamps, with all their mystery and intrigue, have been enchanting visitors for years. The power of nature is visible in places such as Reelfoot Lake State Park in Tennessee, where the lake was created by the powerful earthquakes of 1811–12, which briefly sent the Mississippi River running backwards. Further evidence of the power of these quakes can be experienced on a drive through the Sunken Lands of Arkansas, where vast tracts of land sank fifty feet into the earth or disappeared into rivers and lakes.

Thanks to the US Fish and Wildlife Service, numerous National Wildlife Refuges exist throughout the region to preserve remaining wetlands and forests and make them available to visitors—whether they are hunters, fishermen, boaters, bird watchers, watchable wildlife enthusiasts, hikers, or nature lovers. There are state wildlife management areas as well, along with private lodges and camps for hunting and fishing. One of the most magnificent natural areas is the Atchafalaya National Heritage Area, composed of fourteen parishes in south central Louisiana. The area's centerpiece is the Atchafalaya Basin, a pristine region covering more than a half-million acres of bottomland hardwoods, swamplands, bayous, and backwater lakes. It supports nine endangered or threatened species, including the Bachman's warbler and the Louisiana black bear.

Native American Heritage

The entire Lower Mississippi Delta is rich in archeological sites, beginning with Paleo-Indian hunters and wild plant gatherers and evolving through the Archaic, Woodland, and Mississippian periods. These mound-building cultures developed complex societies that drew upon the land and water to create towns and social life, develop hunting, fishing, and agricultural patterns, and establish complex trade networks. Unfortunately, evidence of many of these civilizations has been destroyed by rivers, agricultural practices, and other forces of man and nature. Nevertheless, remnants of some of the larger community centers remain, along with museums to interpret these sophisticated prehistoric cultures.

The earliest site open to Delta visitors is Poverty Point in northeastern Louisiana, dating between 1650 and 700 BC and covering more than four hundred acres. Affiliated with the Smithsonian Institution, this National Historic Landmark is rare on this continent, with a complex system of concentric ridges and earthen mounds overlooking the Mississippi River flood plain. It is estimated that at least five million hours of labor were necessary to build the massive earthworks. Among other revelations, artifacts at this site provide evidence of an extensive trade network.

Just upriver from Poverty Point, Winterville Mounds in Mississippi is the site of a prehistoric ceremonial center built by a Native American civilization that existed from about AD 1100 to 1450. Interpretation at the museum indicates that the Winterville people lived away from the mound center on family farms in scattered settlement districts throughout the Yazoo-Mississippi River Delta basin, with only a few of the highest-ranking tribal officials living at the mound center.

In Memphis, Tennessee, the C. H. Nash Museum at the Chucalissa Archaeological Site was founded in 1956. Workers in 1938 encountered the Mississippian mound complex (AD 1000–1500) while developing the Jim Crow–era Shelby County Negro Park (now the T. O. Fuller State Park). Operated by the University of Memphis, the museum and site serve as a lab for training archaeologists and museologists, as well as being open to the general public. Educational programming and interpretive exhibits are related to both the Native American and African American heritage of the site's landscape.

Not far from Memphis, Parkin Archeological State Park in Arkansas is the site of a rectangular, planned village with plaza and surrounding mounds. There is evidence of a large Mississippian population (AD 1000–1550) that once included possibly upward of four hundred houses. Parkin is believed to be the Indian village of "Casqui" referred to by Hernando de Soto in the chronicles of his 1541 expedition. Another Native American village (AD 1100–1350) is open to visitors at Wickliffe Mounds in Kentucky. Along with exhibits showcasing the lifestyles of these early Wickliffe people, visitors have a spectacular view of the bluff area on top of a ceremonial mound, the largest mound on the site.

French Influences

For centuries various cultures have attempted to control the Mississippi River, and vestiges of their presence remain. The French influence is perhaps the most visible today on the Lower Mississippi River, ranging from the French town of St. Genevieve in Missouri to the Acadian and Creole cultures in Louisiana. Founded in 1735, St. Genevieve is the oldest permanent European settlement in Missouri and is said to have the largest collection of French Colonial buildings in the country, with three open to the public. Just across the river from St. Genevieve by ferry are other reminders of French Colonial life. Fort de Chartres State Historic Site in Prairie du Rocher, Illinois, includes remains of a fortress built by the French in the early 1700s. Visitors can see the partially rebuilt fort, along with artifacts discovered during archeological research. Nearby, Pierre Menard State Historic Site, home of Illinois' first lieutenant governor, is an example of French Creole architecture and offers a panoramic view of the Mississippi River.

The French influence remains alive in Southwest Louisiana, where today's natives of the region are typically either Cajun or Creole. Though the terms can be confusing, a Cajun is someone descended from the French-speaking Acadian exiles of Nova Scotia or other Canadian maritime provinces. A Creole historically was someone born in Louisiana during the French and Spanish periods, regardless of ethnicity. Thus, Creoles may be descendants of European settlers, enslaved Africans, or those of mixed heritage. Visitors can most readily experience Cajun and Creole cultures through their food, music (including numerous festivals), and architecture. The Vermilionville Living History and Folklife Park at Lafayette, Louisiana, is an excellent introduction to both Cajun and Creole cultures. There is also an Acadian Cultural Center at Lafayette that is part of the Jean Lafitte National Historical Park and Preserve. The Cane River Creole National Historical Park in Natchitoches, Louisiana, is part of the Cane River National Heritage Area and provides full immersion in the Creole culture.

Other places to experience the French influence on the Mississippi River Delta region are the Historic New Orleans Collection and the Louisiana State Museum, both in New Orleans. Founded by the French in 1718, New Orleans came under Spanish rule in 1762. Most of the iconic buildings in the famous

French Quarter actually were built during Spanish rule, thanks to two fires that destroyed most of the earlier French architecture. The Old Ursuline Convent, completed in 1752, is one of the few surviving structures and is the oldest building in the Mississippi River Valley. Louisiana was ceded back to France in 1801, but just two years later it became part of the United States. Even today, however, New Orleans can seem more like a visit to Europe.

Arkansas Post, birthplace of Arkansas and the first permanent European settlement in the Lower Mississippi River Valley, also was under French and Spanish rule at various times. In 1686 French explorer Henri de Tonti and his party built a small settlement and claimed the land for God and king. The French were traders with local Quapaws (Arkancas to the French) and named their encampment Pose de Arkansea—Arkansas Post. The site today is a national memorial and includes a visitors' center and museum.

Architectural Heritage

Plantation homes remain a visitor destination along the Lower Mississippi River. An entire stretch between New Orleans and Baton Rouge, known as the River Road, is lined on both sides with restored plantation homes open to the public. The earliest houses along this route were French Creole architecture, but only a few of these remain. Many were replaced in the years prior to the Civil War by large Greek Revival and Italianate structures belonging primarily to sugar barons who wished to show off their great wealth. Some of the most popular plantations along this route are Oak Alley, Laura, Drestrehan, Houmas House, San Francisco, and Evergreen.

While the emphasis at many plantation homes is on "the big house," numerous structures once made up the typical plantation complex, including extensive slave quarters. Few actual slave quarters have survived, but the Evergreen Plantation on the River Road in Edgard, Louisiana, is an exception. This plantation, a National Historic Landmark, has thirty-seven historic buildings listed in the National Register of Historic Places. It is considered the most intact plantation in the South and includes twenty-two of the original slave quarters. Emphasis at Evergreen is on the plantation's dependence on enslaved labor and, later, the labor of freed African Americans.

Other glimpses into the lives of African Americans along the River Road can be found at the River Road African American Museum in Donaldsonville, Louisiana, which focuses on the history and culture of African Americans in rural communities along the Mississippi River. African American museums can be found throughout the state, including the Northeast Louisiana Delta African American Heritage Museum at Baton Rouge and the Northeast Louisiana Delta Heritage Museum in Monroe, among others.

Another important stop before leaving Louisiana is the West Feliciana Historical Society Museum and Tourism Center, which provides information on seven plantations in the St. Francisville area that are open to tours.

Plantation homes also line the river in Mississippi, though these structures are testaments to wealth derived from "King Cotton," rather than sugar. Some of the oldest and most significant are in Natchez, Mississippi, where one of the most unusual is Longwood, the largest octagonal house in North America. The house was unfinished due to the onset of the Civil War and remains in its unfinished state today. Other popular Natchez houses include Melrose, Monmouth, Rosalie, Stanton, and Rosemont. In addition, some thirty antebellum mansions, mostly private residences, are open to visitors during the Natchez Spring and Fall Pilgrimages.

While large plantation homes at one time extended farther up river, today only the Lakeport Plantation in Lake Village, Arkansas, remains of all the plantations that once fronted the Mississippi River in Arkansas. Restored and opened by Arkansas State University in 2007, the house focuses on craftsmanship of the period, the lives of those who lived and worked on the plantation, and preservation techniques used in bringing the house back to life.

Civil War Heritage

William Faulkner wrote, "The past is never dead. It's not even past."[11] Nowhere is that more true than in the South, particularly as it relates to the rich Civil War heritage along the Lower Mississippi River. Control of the river was vital to both the North and the South during the Civil War. The South could not afford to be cut off from Confederate states west of the Mississippi River that provided many of its food supplies for the war effort, but with an insignificant navy, it was difficult for the South to protect long stretches along the Mississippi. New Orleans, Baton Rouge, and Natchez fell to the Union fleet early in the war, and capture of Island No. 10 (between New Madrid, Missouri, and Tiptonville, Tennessee) and Memphis left Vicksburg as the primary defensive position.

Vicksburg's high bluffs, deep ravines, and a hairpin turn in the river made it a difficult target for Union gunboats and Yankee infantry to attack. Early attempts to shell the city into submission proved unsuccessful, and other efforts to attack overland failed as well. Eventually, a Union blockade by river and land starved the Rebel army into surrender. Today the eighteen-hundred-acre National Military Park, established in 1899, commemorates this forty-seven-day siege and defense of Vicksburg.

The Battle of Helena in Arkansas took place the same day that Vicksburg fell, July 4, 1863, and resulted in another Union victory that affirmed control of eastern and northeastern Arkansas and made possible the capture of Little Rock later in the year. In 2012 a replica of the Union's Fort Curtis opened as part of a Civil War Helena project. The following year, Freedom Park opened to commemorate the African American experience in Helena, from fugitive slave to freedom and, for some, enlistment in the Union army and participation in the Battle of Helena. Interpretive markers throughout the city, ongoing exhibits at the Delta Cultural Center, and a Confederate cemetery also illustrate its historic importance. Other plans underway for Civil War Helena include restoration of the four artillery batteries and rehabilitation of the Civil War–era Estevan Hall as a visitors' center.

African American participation in the Civil War also is commemorated at Fort Pillow State Historic Park north of Memphis, Tennessee, on the Chickasaw Bluffs overlooking the Mississippi River. During that battle on April 12, 1864, Confederates successfully stormed the Union-controlled fort, killing 229 African American soldiers who allegedly had surrendered. News of the fight—labeled a massacre by northern soldiers—led to the battle cry "Remember Fort Pillow" for African American troops during the remainder of the war. Today the 1,642-acre park includes the battle site and a museum-interpretive center.

Further north on the river, Columbus-Belmont State Park near Columbus, Kentucky, interprets one of the strategic sites in the struggle to control the river. The heavily fortified area was called the "Gibraltar of the West." Visitors still can see the massive chain and anchor used by the South to block passage of Union gunboats on the river, as well as the earthen trenches deployed to protect more than nineteen thousand Confederate troops. The farmhouse that served as a Civil War hospital is now a museum.

At the south end of the river, Louisiana also has its share of Civil War sites for visitors, including the Port Hudson State Commemorative Area near Jackson. This National Historic Landmark features an elevated boardwalk above the breastworks in the Fort Desperate Area. Other facilities at the 909-acre site include a museum, three observation towers, and six miles of trails. Confederate Memorial Hall in New Orleans, Louisiana's oldest museum, has an outstanding collection of Civil War artifacts and personal effects of Confederate president Jefferson Davis and General Robert E. Lee.

Civil Rights Heritage

Much of the Delta's heritage for African Americans has been a painful struggle for rights and freedoms. To mark these efforts, several historic sites, museums, and initiatives exist throughout the region. They are dedicated to remembering the mistakes of the past and learning to apply their lessons to current issues. The most notable of these sites is the National Civil Rights Museum, the first in the nation, housed at the Lorraine Motel in Memphis, Tennessee, where Dr. Martin Luther King Jr. was assassinated on April 4, 1968. The museum traces the civil rights movement from 1619 to 2000 to assist the public in understanding the messages of this movement and its influence on human rights worldwide.

The sometimes-violent struggle for equality in education is remembered at the Little Rock Central High School National Historic Site. There nine African American students were admitted to the formerly all-white school in 1957 despite efforts by Arkansas governor Orval Faubus, using the Arkansas

National Guard, to prevent them from entering the school building. Today, exhibits at the site recognize the pivotal role played by the "Little Rock Nine" in desegregation of many public schools across the country in compliance with court-ordered desegregation.

Such civil rights violations have not been limited to African Americans in this country. After the bombing of Pearl Harbor, thousands of US citizens of Japanese ancestry were relocated to internment centers and incarcerated during the war solely because of their ethnic origins. Ten relocation centers were established, including Rohwer and Jerome in southeastern Arkansas. Though about eight thousand Japanese Americans were confined in each of these camps in the early 1940s, a cemetery with several memorial markers and monuments, along with a smokestack, is all that remains at Rohwer. Only a simple granite marker identifies the Jerome site. During 2012 interpretive panels were installed at Rohwer, including an audio tour narrated by George Takei, incarcerated at the Rohwer camp as a boy but better known as Mr. Sulu of *Star Trek* fame. To ensure that other Japanese American voices are heard, the World War II Japanese American Internment Museum has been established at McGehee, Arkansas, midway between the two former camps. The museum features an exhibit, "Against Their Will: The Japanese American Experience in World War II Arkansas," that draws from internees' memories of life in the camps.

Markers also are being placed throughout several Delta states to mark the Trail of Tears route taken by Native Americans forcibly removed from their homes. A 3,415-acre Trail of Tears State Park in Cape Girardeau, Missouri, stands as a memorial to the thousands who lost their lives on the journey to Oklahoma. The park is located at the site where nine of thirteen groups of Cherokee Indians crossed the river in the harsh winter of 1838–39.

Music Heritage

Perhaps the Delta is best identified through its music, which distinctly reflects the character of the diverse people who make up the region. Rooted in the "field hollers" of cotton workers and gospel music from southern churches, the musical forms of blues, jazz, country, and rock and roll grew out of the rich fertile landscape bordering the Lower Mississippi River. Cajun and Creole music add spice to the cultural mix.

There are numerous places to hear live music in the region, from Beale Street in Memphis to Preservation Hall in New Orleans. Beale Street, synonymous with blues greats such as B. B. King, has been a hub for Memphis music since before 1900 and includes numerous blues clubs and restaurants. Preservation Hall has an equal reputation for jazz and hosts nightly concerts with a rotating roster of bands, typically performing New Orleans–style jazz. Along with live performance venues, museums devoted to preserving and promoting the music heritage of the region exist throughout the Delta. In

Memphis, visitors can tour the legendary Sun Studios, become immersed in soul music at the Stax Museum of American Soul Music, or learn about early music pioneers at the Memphis Rock 'n' Soul Museum.

In between Memphis and New Orleans are other musical gems, including the Delta Blues Museum in Clarksdale, Mississippi, and the Delta Cultural Center in Helena, Arkansas. The Clarksdale museum has an extensive collection of blues memorabilia and provides educational programming as well, focusing on teaching young people to play musical instruments. The Delta Cultural Center in Helena traces the origins of blues music, along with stories of the river and the communities along its banks. The center is host to live broadcasts of the longest-running blues radio program, *King Biscuit Time*, which airs on Helena's KFFA radio. Further indicative of the important music heritage of the region, the Mississippi Blues Trail now has more than 150 markers commemorating a variety of notable sites, while a music trail in the Arkansas Delta, "Sounds from the Soil and Soul," provides a recorded driving tour and sound track featuring noted blues, jazz, gospel, and country performers from the region.

People

Many musicians, along with other notable figures, are commemorated through opportunities to visit their homes. Perhaps this interest has something to do with making famous personalities seem more like real people through insights into their daily lives, or perhaps it is a way of drawing inspiration from great artists. One only has to look at Graceland in Memphis to know that Elvis Presley's spirit has never left the building. Today it has more than six hundred thousand visitors annually and is one of the five most visited homes in the United States. It is the most famous home in the United States after the White House.[12] Presley's boyhood home in Tupelo, Mississippi, also is a popular destination, as is the B. B. King Museum in Indianola, Mississippi. Restoration of the Johnny Cash boyhood home in Dyess, Arkansas, has been completed, along with the restoration of other buildings in the historic Dyess Colony, where the Cash family was part of an agricultural resettlement project during the New Deal.

Musicians aren't the only ones who bring visitors. The South has its share of literary giants, and several of them have homes that are open to the public. William Faulkner's Rowan Oak in Oxford, Mississippi, is operated as part of the University of Mississippi. Alex Haley is buried in the front yard of his home that is open to visitors in Henning, Tennessee. Eudora Welty lived in her home in Jackson, Mississippi, for seventy-six years. Though Ernest Hemingway never lived in Piggott, Arkansas, he often visited the town while married to Pauline Pfeiffer, and his in-laws converted their barn into a studio to give him privacy for writing. The Pfeiffer home and barn-studio are now operated as a museum and educational center by Arkansas State University.

Agricultural Heritage

Fields of cotton, soybeans, rice, or sugarcane are the most common sights throughout the Delta, but it hasn't always been this way. The process of clearing swampland and converting it to agricultural production has required ingenuity, endurance, and a willingness to gamble with nature. Practically every small town museum throughout the Delta pays homage to rural life, while others document the process of converting the Delta into one of the most productive agricultural regions in the world. The Louisiana State University Rural Life Museum in Baton Rouge provides a comprehensive look at the material culture of nineteenth-century Louisiana through an exhibit barn with artifacts from rural life, plantation quarters and outbuildings with authentic furnishings, and Louisiana folk architecture interpreted through an extensive collection of buildings illustrating the styles of rural houses.

Other museums exist that demonstrate the history of cotton in the region. Two of these are the Louisiana State Cotton Museum at Lake Providence, Louisiana, and the Plantation Agricultural Museum at Scott, Arkansas. Both facilities feature exhibits and an array of outbuildings that trace the history of cotton from its introduction to the region in the early 1800s through mechanization of agriculture after World War II. The Frogmore Cotton Plantation and Gin in Louisiana (just across the river from Natchez, Mississippi) takes interpretation a step further and contrasts a working cotton plantation of the early 1800s with a modern cotton plantation and gin of today.

The newest museum devoted to this crop is the Cotton Museum at the Memphis Cotton Exchange, which opened in 2006 and is located in the historic Cotton Row district that once was a worldwide marketing center for the crop. A related museum that also opened in 2006, the Southern Tenant Farmers Museum in Tyronza, Arkansas, tells the stories of labor practices on cotton plantations during the early twentieth century. Some of these practices led to the 1934 formation of the Southern Tenant Farmers Union, the first integrated agricultural labor organization. The museum is housed in the building that served as informal union headquarters until offices moved to Memphis. Other crops have their associated museums as well. The Museum of the Grand Prairie in Stuttgart, Arkansas, interprets the history of rice in the region, while several museums in Louisiana spotlight the sugarcane industry.

Conclusion

Communities throughout the Delta often have more in common with each other than with other areas of their own states. This growing sense of regionalism has led to new efforts to develop heritage themes that promote tourism across state and county lines. For example, Mississippi, Arkansas, and Tennessee are developing a comprehensive blues heritage trail. Delta states have worked with Audubon in creating a birding trail along both sides of the river. The Delta region in Mississippi has been designated a National Heritage

Area, with common themes being developed. And in the Arkansas Delta, the Rural Heritage Development Initiative, in partnership with the National Trust for Historic Preservation, utilized a regional approach to heritage tourism and preservation-based economic development. As the interest in heritage tourism grows, it is likely that the experience will become more seamless for visitors, with Delta communities joining hands, rather than competing, to welcome travelers to the region.

Notes

1. David Wayne Brown, ed., *The Delta Initiatives: Realizing the Dream, Fulfilling the Potential* (Memphis: Lower Mississippi Delta Development Commission, 1990), 6.
2. Bureau of Labor Statistics, "Unemployment rates by county, June 2011–May 2012 averages," Local Area Unemployment Statistics, http://www.bls.gov/lau/maps/twmcort.gif, accessed April 1, 2015.
3. U.S. Travel Association, "U.S. Travel Answer Sheet," Power of Travel, 2012, http://www.ustravel.org/research/economic-research, accessed April 1, 2015.
4. Ibid.
5. Ibid.
6. Office of Travel and Tourism Industries, "Overseas Visitation Estimates for U.S. States, Cities, and Census Regions: 2011," U.S. Department of Commerce, May 2012, http://tinet.ita.doc.gov/outreachpages/download_data_table/2011_States_and_Cities.pdf, accessed April 1, 2015.
7. Karen Haas Smith, "The Great River Road Celebrates 60 Years," *Public Roads* 62, no. 3 (1998), http://www.fhwa.dot.gov/publications/publicroads/98novdec/great.cfm, accessed April 1, 2015.
8. Bureau of Public Roads, Department of Commerce, and National Park Service, Department of Interior, *Parkway for the Mississippi River* (Washington, DC: US Government Printing Office, 1951).
9. The routes through nine of the ten states are fully designated as National Scenic Byways, while a portion of the route through Missouri remains to receive official designation.
10. Mark Twain, *Mark Twain in Eruption* (New York: Harper & Brothers, 1940), 18.
11. William Faulkner, *Requiem for a Nun*, act 1, scene 3 (New York: Vintage Books, 1975), 80.
12. Nobert Bermosa, "The Most Visited Houses in America," http://nobert-bermosa.blogspot.com/2011/08/most-popular-houses-in-united-states-of.html, accessed May 4, 2015.

CHAPTER 16

What Is the Delta from a Foodways Perspective?

JENNIFER JENSEN WALLACH

The foodways of the Delta are the product of a complex and often dark history of cross-cultural interactions and engagements with the local environment. The food history of the region is a tale ridden with contradictions, characterized by both cultural fusion and racial antagonism, by abundance and want. The Delta, from the perspective of the region's foodways, is also a space where actual food histories can either be bolstered by or collide with a colorful food imaginary created and disseminated by scholars and other regional food enthusiasts.

Food historian Marcie Ferris Cohen offers the reminder that southern food history began thirteen thousand years ago when the first people arrived on the North American continent and began foraging and hunting for food. She wryly remarks, "Cornbread and fried chicken came *many* millennia later."[1] Embedded in her summation of what most people think of as typical southern foods is an admonition not to forget the region's foundational foodways and a reminder of the limitations of our knowledge of and collective memory about the eating practices in the area prior to European contact. Rayna Green points out that Native Americans living in the South "ate well and often from a huge and diverse larder" of meat, fish, nuts, vegetables, both those they gathered in the wild and those they learned to cultivate.[2] These foods helped feed later arrivals to the region from Europe, Africa, and elsewhere who combined their own culinary sensibilities with native techniques and ingredients to create the foodways of the Delta.

During the eighteenth century, residents of the European foothold of the Arkansas Post ate food that Morris S. Arnold describes as "surprisingly varied and widely available."[3] French and Spanish settlers adopted many Quapaw foods, including the corn porridge the French called *sagamité*. The substance, which became a dietary staple, was a flexible recipe prepared with whatever ingredients were available. It was often flavored with bear or buffalo fat and could contain scraps of meats and vegetables. They also enthusiastically traded with the Caddo for smoked buffalo tongues, which were considered a great delicacy.[4] Native Americans and European Americans alike took advantage of the culinary riches of the natural environment, dining on pheasant, duck, deer, wild greens, and fruits such as wild plums and mulberries.[5]

Native foods such as *sagamité* were prepared in various ways throughout the Delta region and gained both adherents and critics among the Europeans who encountered them. While living in Louisiana between 1718 and 1734, Antoine-Simon Le Page du Pratz came to admire the food; he said that to his "taste [it] surpassed the best dish in France."[6] Others were less enthusiastic about local food customs. In 1706, naval captain Jean-Baptiste Le Moyne de Bienville recorded that female settlers of Louisiana were particularly reluctant to abandon their traditional food customs, particularly their preference for wheat rather than corn as a staple grain. Referring to corn, which flourished in the region while wheat did not, he reported, "The women who are for the most part from Paris eat it reluctantly."[7]

If one item is emblematic of the diet of the Delta throughout time, it is indeed corn, a food that was originally cultivated in Mexico around 5000 BC, gradually worked its way northward, and became the backbone of the diet of native peoples throughout the hemisphere.[8] European colonists who encountered the grain were impressed with its ability to thrive in a variety of climates and its high yield and discouraged by their efforts to grow wheat, which failed to thrive in the Delta. French settlers in the port city of New Orleans were able to import some of their favored grain from Europe and eventually from French-controlled Illinois Country, further up the Mississippi River, but transport was costly and supplies too low to produce large quantities of the light, wheat bread many craved. Corn, and later rice, had to be accepted as substitutes when wheat was not available.[9]

The English colonists whose culinary encounters with native cuisines took place initially in the coastal regions of what is now Virginia and Massachusetts manifested a greater degree of culinary xenophobia than the colonial French and Spanish who dined on *sagamité* and buffalo tongue at the Arkansas Post or in New Orleans in the eighteenth century. They were accustomed to eating domesticated animals and thought of hunting for food alternately as a leisure activity practiced by English aristocracy or as New World savagery when practiced by Native Americans.[10] Furthermore, they were influenced by the pronouncements of botanist John Gerard in his influential 1597 *The Herball, or Generall Historie of Plantes* that when it came to corn, "the barbarous Indians, which we know no better, are constrained to make a virtue of necessity, and think it a good food: whereas we may easily judge, that it nourisheth but little."[11]

In contrast, historian Shannon Lee Dawdy portrays the eighteenth-century French who colonized parts of the Delta as being generally more omnivorous and open-minded about the exchange of European and native ideas about food. Most French Louisianans, she claims, "felt no shame in being corn eaters."[12] However, their openness to adopting new food items did not mean they necessarily embraced these foods in unchanged forms. Rather than avoiding Indian foods, the French settlers attempted to "civilize" them, by preparing them using French cooking techniques. For example, they transformed local meats into texturally familiar *pâté*. Dawdy argues, "Culinary practices comprised a material form of imperial hubris that reflected the larger ambition of Europeans to transform America into something civilized and consumable."[13]

Native peoples too adopted their food practices in response to exposure to new foods, for example, adding to their diet domesticated animals, newly introduced European fruits and vegetables, and alcohol.[14] Due to eventual European domination of the region, which increasingly lead to the death or displacement of native peoples, erasure as well as adaption occurred, leading to Rayna Green's somewhat flippant remark, "They all know, out there in Indian Country, that the loss of traditional diet and the cultural skills needed to maintain it has killed more Indians than Andy Jackson."[15]

Another important influence in the Delta came as a result of another history of exploitation and appropriation, which began with the arrival of people of African descent to the region. Enslaved Africans imported into Louisiana in the eighteenth century contributed what Charles Joyner has labeled an "African culinary grammar" to French-inspired cooking techniques, introducing, for example, a preference for spicier foods seasoned with hot pepper, as well as specific ingredients imported from Africa, such as okra, an essential ingredient in many versions of the iconic stew gumbo.[16] African American street vendors selling fritters known as *calas*, made with rice and sometimes with black-eyed peas, soon became a familiar sight on the streets of New Orleans. This popular street food has direct culinary antecedents in similar dishes made in parts of West Africa.[17]

The southern regions of the Lower Mississippi Delta developed Creole and Cajun cuisines that combined French, Spanish, Native American, and African cooking techniques to develop a distinctive style of home cooking characterized by robust Cajun sausages and gumbos as well as a sophisticated New Orleans–centered restaurant cuisine that drew on the abundance of the fish and shellfish of the Gulf to create classic dishes such as Oysters Rockefeller. Elsewhere in the Delta, the food was generally less varied and more simply prepared and comprised of the foundational foods of cornbread and also pork, a protein so commonly consumed from the early nineteenth century onward that "meat" could be used as a synonym.[18]

In places like Mississippi and Arkansas, where English culinary sensibilities were more predominant due to the increasingly Anglo origins of the white settlers, people of African descent also played a role in establishing the region's food customs. During the nineteenth century enslaved cooks were transported from the upper to the lower South as part of the internal slave trade; this

FIGURE 16.1. Louisiana residents at a crab boil in 1938. Due to their proximity to the Gulf of Mexico, residents of Louisiana have historically eaten more shellfish than people living elsewhere in the Delta. *Courtesy of the Library of Congress, Prints and Photographs Division, Farm Security Administration, Office of War Information Photograph Collection, Washington, DC, Russell Lee, photographer, #LC-USF33-011654-M1.*

involved one million forced migrants who brought with them the techniques and recipes they had developed as English and African ideas about food were combined in places like the Chesapeake. Fried chicken, destined to become one of the most totemic dishes in southern cooking, was a product that combined European recipes for breading poultry with West African techniques of frying fowl in hot oil. Africans also exhibited expertise at preparing vegetables such as turnip or collard greens seasoned with small pieces of meat, a technique common in West African styles of cooking, where meat was served in smaller portions than was typical in English fare.[19]

As Native Americans decreasingly posed a threat to white settlers, corn lost its taint as a "savage" food and became widely accepted as quintessentially "American" throughout the nation.[20] However, wheat was still preferred by many because it produced lighter and fluffier bread and superior baked goods. By the early nineteenth century, wheat flour was increasingly available in many regions of the country due to improved transportation networks and more efficient milling techniques. Delta residents, however, remained more isolated and had less access to inexpensive wheat. Furthermore, southern mills were largely designed to grind corn and not wheat, making corn the most commonly eaten grain during the first two centuries of European settlement in the region. Decreased flour prices and greater access finally enabled many Delta cooks to add flour biscuits to the Delta breadbasket in equal proportions by the twentieth century.[21]

African cooks demonstrated great expertise in preparing products made from cornmeal, a grain that some encountered for the first time in North America, where it was a staple of slave rations. Others brought techniques for preparing it from ancestral homes in the Gold Coast, where the New World crop had been introduced by European traders and had been rapidly incorporated into the local diet.[22] The simplest cornbreads were made with little more than meal, salt, water, and perhaps fat. These simple breads could be baked in the ashes of the fireplace. As settlers became established enough to move beyond bare subsistence and as transportation networks improved in the nineteenth and twentieth centuries, Delta residents living outside of urban centers gained greater access to a wider range of ingredients and could, if they wished, augment these recipes to create what historian Charles Reagan Wilson has labeled "second-generation" cornbread, which might contain wheat flour, eggs, milk, and leavening agents.

In addition to bread, cornmeal batter could be deep-fat fried to create hushpuppies. The whole grain could be transformed into hominy if lye was used to remove the hull from the kernel. The transformed grain could then be ground into grits, which were cooked into a thick porridge and became a common Delta breakfast. In areas of Louisiana inhabited by French-descended Cajuns, cornmeal mush might be fried to produce *couche couche.*[23] Simple breads could also be enlivened with fried pig skin or "cracklins," derived from pork, the other backbone Delta food.

Pigs were first introduced in what is now Florida by Spanish explorers in the mid-sixteenth century. Soon wild, domesticated, and semi-domesticated pig populations proliferated throughout the South. White nineteenth-century Delta farmers depended on the animal for their primary source of protein, as did enslaved residents of the region. From the intestines (known as chitterlings or chitlins) to the brains, every possible part of the animal was consumed. Scraps were made into sausages and hindquarters were smoked and transformed into ham, which could be preserved for long periods in the era before refrigeration.[24] While pork was a favored food throughout the class spectrum, what parts of the pig one ate took on class connotations. Prosperous whites were more likely to consume the choice parts of the animals, while poorer whites and enslaved Africans had to make do with the offal or the fat of the animal, which contained little protein.

Pork recipes that were developed in the antebellum Delta have been maintained and modified throughout the generations. Dorothy Brackin's 1969 cookbook *Arkansas Soul Food* contains numerous recipes for how to make the less desirable parts of the animal delectable. She offers instructions for making "Neck Bones and Rice," "Hog Jowl and Black-Eyed Peas," and "Brains and Eggs." She also offers the recipe "Chitlin Patties," a culinary adaption of less elaborate methods of cooking chitterlings that consist simply of cleaning them in vinegar and then slowly stewing them. Her preparation calls for the chitterlings to be ground or chopped, seasoned with thyme, and fried. For those adventuresome enough to hunt for wild animals or without the means to purchase or raise pigs, she also provides directions for how to prepare "Razorback Hog Roast" and "Wild Razorback Burgers."[25]

African American cookbook author Brackin's decision to label her recipes "soul food" is no accident; she wrote it during the era of the civil rights movement. "Soul" had become a word used to describe a distinctive black aesthetic in a variety of creative realms. "Soul food" was used to denote southern-inspired cooking in both northern and southern regions of the country, and the term was designed to invoke racial pride and point to the African origins of various southern recipes.[26]

Because the foodways of the Delta are a product of culinary fusion between peoples from various parts of the globe, food historians have been reluctant to ascribe sole culinary origins to any one group. Although the Native American origins of southern cooking remain largely forgotten, many argue that African American contributions have also been downplayed. Comparing the debate over food origins to the segregation-era music industry, a time when white artists often gained financially from recording songs written by African Americans who were denied airtime on white radio stations, culinary historian Jessica Harris continues to ask, "Who did the original and who did the cover?" In her estimation, "It's about acknowledging the unacknowledged."[27]

Although disentangling culinary influences from the hybrid cuisine of the Delta is a contentious and potentially impossible task, there is no question that the history of racism of the region is embedded in the history of the food culture of the South and the Delta, in particular. John T. Edge, director of the Southern Foodways Alliance, headquartered at the University of Mississippi, emphasizes what he sees as the potential of food as a means to "bridge the chasms of race and class that have long separated us."[28] Indeed, there are times that shared familiar food has certainly served this function. In his memoir *North toward Home,* transplanted white Mississippian Willie Morris recalls feeling the most at home he had ever felt in New York at a New Year's Eve meal of "bourbon, collard greens, black-eyed peas, ham-hocks, and cornbread" that he shared with the African American writers Ralph Ellison and Albert Murray.[29] Despite Morris's positive experience in the 1960s and Edge's optimism for future racial reconciliation around the dinner table, historically, the food habits of the Delta have highlighted a tremendous racial divide.

In the Delta, African American cooks and eaters have been subjected to competing stereotypes that sought to differentiate them from white residents of the region. They have been characterized both as innately talented cooks who are delighted to use their skills in the deferential service of whites and, alternately, as a group with a different and inferior sense of taste than white eaters. Writing in *Harpers* in 1887, Charles Gayarré expressed the common belief among many: "The negroes are born cooks [who] . . . from natural impulses and affinities, without any conscious analysis of principles, created an art of cooking for which he should deserve to be immortalized."[30] Historian Rebecca Sharpless exposes the careless racism of the belief that the skill of cooking was somehow an inborn trait, demonstrating that it was a skill that had to be taught.[31] However, the widespread acceptance that cooking skills were an essential racial trait was substantiated by the presence of African American cooks in white Delta homes whose labor was so vital and taken for granted that some white women did not know how to cook. *Mrs. Hill's New Cook Book,* published in 1872 by Annabelle Hill, expresses the anxiety many white women felt about facing the kitchen alone, something that seemed like a distinct possibility in the early post-emancipation period. Addressing her white readership, she urges them to develop cooking skills, mourning the fact that "'mother's cook' and 'trained servants' are remembered among the good spirits that ministered to the luxury and ease of by gone days."[32]

Depressed wages and the lack of other job opportunities made Hill's fear about the disappearance of the African American cook premature, and black cooks were seemingly ubiquitous in white Delta homes, even many of fairly modest means, throughout the first half of the twentieth century. Pioneering *New York Times* restaurant critic Craig Claiborne, who grew up in Laurel, Mississippi, fondly remembered the culinary skills of many African Americans who assisted his mother in running a boardinghouse. He recalled that "the talent and the palate of the American Negro" were a distinct advantage to the "old-fashioned southern kitchen."[33] Some southern white cookbook authors paid their African American cooks the dubious compliment of appropriating their recipes, often also using dialect and jargon to condescendingly attempt to capture their personas and publishing them without giving the cooks remuneration or proper credit for their culinary creations. This genre was popular throughout the South and the entire nation. One example from the Delta was Natalie V. Scott's 1929 *Mirations and Miracles of Mandy,* which begins with a quotation from the fictive Mandy, who declares, "My madam say she writin' mah cookin' down. Lawdy, put me in front of a cookin' stove, an' I don't need no prescription." In trying to document what she regards as the innate talents of the black cook who does not, like the presumably white readership of the cookbook, need written recipes, Scott confesses that she had "peeped here, watched there, borrowed and begged, and doubtless inadvertently stolen."[34]

This nostalgia for African American cooking skills was, however, counterbalanced by racist ideas about black people as comic eaters and as eaters with different taste sensibilities. In the late nineteenth and early twentieth centuries, American popular culture was rife with images of African Americans, who were generally portrayed with exaggerated or apelike features, dining with outsized glee on foods like watermelon, chicken, and opossum, common southern foods that often became associated with an allegedly unnatural black appetite.[35] Sheet music for "The Mississippi Barbecue," written by Dave Reed in 1904, contains an image of rural African Americans with tattered clothes and unnaturally large lips dancing in ecstasy to the backdrop of a large animal being turned on a spit. The singer moans, "Lordy I wish that it was time to eat. I'll have an awful fight controlling my appetite. It runs clean down to my feet."[36]

Not only were African Americans depicted as humorous and gleeful eaters in the white imagination, they were also portrayed as having an inferior sense of taste. Young Mississippian Richard Wright encountered this belief in the early twentieth century when his white employer served him stale bread and moldy molasses for lunch and then chastised him when he refused to gratefully accept the rotten food.[37] During the New Deal, an employee of the Works Progress Administration charged with recording regional foodways in

FIGURE 16.2. Drawing of African Americans dancing around a pile of watermelons, 1900. In the late nineteenth and early twentieth centuries, American popular culture was rife with images of Africa Americans who were portrayed with exaggerated features, dining with outsized glee on foods like watermelon, chicken, and opossum, common southern foods that became associated with an allegedly unnatural black appetite. *Courtesy of the Library of Congress, Prints and Photographs Division, Theatrical Poster Collection, Washington, DC, reproduction #LC-USZ62-24427.*

Mississippi unconsciously revealed the belief that African Americans had inferior taste sensibilities and would eat things that white eaters may disdain in a short description of "Mullet Salad," which proclaimed, "Mullet, not commonly prized as food, is commonly eaten by Negroes."[38] The competing ideas about black people as both talented cooks and people of different and substandard food sensibilities were combined in the kitchens of white households throughout the Delta, where black cooks inevitably infused saliva and sweat into the foods they prepared and tasted but who were given separate dishes and utensils to use while eating their own meals by employers who could not decide if their presence elevated or contaminated southern cuisine.[39]

Although black and white culinary encounters in the Delta were the predominant ones due to the nature of slavery and its legacy during segregation, the foodways of the region should not be described using a simple black and white binary. The Delta was home to people from other backgrounds whose culinary contributions outweighed their numerical minority. Marcie Cohen Ferris, whose Jewish ancestors arrived in the Delta in the 1920s, grew up in Blytheville, Arkansas, and recalled eating a diet that combined traditional Jewish foods like matzo balls and kugel with Delta foods like barbecued, pulled pork and fried catfish, eating "between these two worlds in a complicated culinary negotiation of regional, ethnic, and religious identity."[40] The influence of Italian immigrants on the cuisine of New Orleans is well known and exemplified in now-classical regional dishes such as the muffuletta sandwich, which combines cheese, cold cuts, and olive salad. A lesser-known community of Italian immigrants in Tontitown, Arkansas, helped popularize their food traditions in the state and pioneered cross-cultural pairings such as spaghetti served alongside fried chicken.[41] Beginning in the late nineteenth century, Chinese grocery store owners became a common sight in Delta towns. The immigrants planted vegetables such as bok choy in their gardens and also applied Chinese cooking techniques to local ingredients.[42] One of Helena's most popular restaurants, Habib's, was opened in 1888 by a Lebanese immigrant who became well known for his fruitcakes.[43]

One of the most emblematic but also mystifying manifestations of culinary fusion in the Delta is the "hot tamale," a cousin of Latin American versions of the portable food that combines ground corn and meat into a packet steamed in a corn husk or banana leaf. Beginning in the early twentieth century, tamale vendors, the majority of them African American, began appearing throughout the Delta selling a version of the treat made with course cornmeal instead of the more finely ground *masa harina* used in more traditional versions. Although Delta tamales are prepared with a number of proteins and are sometimes served with chili on top, the most basic version combines corn and pork, the two staples of African American diet since the era of slavery. Although it is unclear when the tradition of Delta tamales began, food historian Amy Evans speculates that black cooks may have learned the recipe from Mexican migrant laborers or perhaps from soldiers from Mississippi who fought in the Mexican American War.[44]

Throughout time Delta inhabitants have had access to a rich store of information about food preparation techniques that utilized the region's rich, abundant resources as well as increased access to foodstuffs imported from other regions of the country and the world. Delta gardeners grew a wide variety of crops, including green beans, butter beans, black-eyed peas, various greens, turnips, and cabbage. Sweet potatoes were grown in vast quantities and stored for winter, when gardens were sparse.[45] These staple items characterized the Delta diet from the antebellum era onward, and these homegrown items were served alongside a variety of imported and increasingly industrially produced foods. Prior to the Civil War, prosperous planters could afford to quench their thirst with imported wine and other liquor. After Emancipation, plantation owners continued to supply agricultural workers with rations of corn and pork, but rather than being raised locally, these staples were increasingly purchased from Midwestern producers.[46] In the twentieth century greater urbanization and more widespread access to automobile travel gave Delta residents more opportunities to purchase canned goods and other processed food

FIGURE 16.3. Child picking sweet potatoes in Mississippi, 1938. In the Delta sweet potatoes were grown in large quantities and stored for winter use, when gardens were sparse. *Courtesy of the Library of Congress, Prints and Photographs Division, Farm Security Administration, Office of War Information Photograph Collection, Washington, DC, Russell Lee, photographer, reproduction #LC-USF33-012008-M3.*

items.[47] Clarence Saunders made shopping for food supplies easier and more affordable when he opened the nation's first self-service grocery store, which he named Piggly Wiggly, in Memphis, Tennessee, in 1916. Instead of waiting for clerks to fill their orders, shoppers could now walk through the store and make their own selections.[48]

Locals, especially the most impoverished, whose ability to purchase or grow foods was the most limited, also found various ways to feed themselves utilizing wild resources at their disposal. Delta residents gathered pokeweed, a plant that can be toxic unless boiled in several changes of water. Connoisseurs claim that the shoots are similar to but superior to asparagus and that the leaves are reminiscent of spinach.[49] Hungry black and white southerners also hunted for small game such as squirrels, opossums, and raccoons. During the 1930s, employees of the Federal Writers' Project collected an Arkansas recipe for squirrel mulligan that called for a combination of the rodent with potatoes, sweet potatoes, okra, and whatever other vegetables were available.[50] They also documented the meeting of the Polk County Possum Club, where "governors and senators rub elbows with Ouachita Mountain backwoodsmen" while dining on the animal, which was traditionally served alongside baked sweet potatoes.[51] Nostalgia for these indigenous foods is still manifested in the Gillett Coon Supper, which has been taking place in Arkansas since 1947. It draws more than a thousand diners each year despite the claim of food writer Andrew Beahrs that raccoon being cooked in broth, onions, celery, and carrots is at best "a smell that's trying its damndest to smell good. . . . Raccoon fat is pretty awful."[52]

In spite of the natural abundance of the Delta region, many residents of the area have historically gone hungry and suffered from disease related to malnutrition. In *Deep South: A Study of Social Class and Color Caste in a Southern City,* a classic study of Natchez, Mississippi, and the surrounding rural countryside in the 1930s, Allison Davis and a small team of social scientists reported that African American sharecroppers lived in a state of "semistarvation" during portions of each year when their gardens were not producing and they were not able to secure credit to purchase food.[53] One informant told the researchers that between January and March most tenants had very little to eat and were forced to "strap it," or "tighten their belts," over their empty stomachs. Even when they were able to secure food with loans advanced to them before the cotton harvest, black sharecroppers were able to purchase "jes' unhough tuh let dem live, an' dat's all."[54] Many poor southerners during the time subsisted largely on what became known as the "three Ms" of meal, meat, and molasses, a diet that led to niacin deficiency and a disease known as pellagra.[55]

Pellagra causes a variety of adverse health effects ranging from skin lesions to lethargy, diarrhea, dementia, and even death. The illness was common in the South in the early twentieth century. There were 15,381 cases and 1,531 deaths from pellagra in Mississippi in 1915.[56] Initially, public health officials were mystified by the condition. They did not discover that the cause was a dietary deficiency and not a germ until a 1915 experiment conducted among inmates in a Mississippi penitentiary who were fed a restricted diet and then developed symptoms of the illness.[57] Although the disease can be cured or prevented with niacin-rich foods, such as yeast and greens, inadequate dissemination of public health information and widespread poverty meant that the condition was eradicated slowly. In Arkansas in 1927, 657 deaths were attributed to pellagra, a number that was reduced to 184 by 1938.[58]

Although poverty was the overwhelming cause of the illness, the dietary preference of poor southerners accustomed to a limited diet also played a role.[59] Planters and public health officials tended to downplay the socioeconomic dimension of the problem, blaming the victims for their illness, claiming that they made poor choices spending their meager monetary resources.[60] Writing in 1926, home economist Dorothy Dickens compared the shopping habits of African American sharecroppers in Mississippi to "an unsupervised six-year old . . . on his first visit to a cafeteria."[61]

In the years following the Great Depression, pellagra disappeared due to greater prosperity in the region as well as the development of vitamin-enriched foods. However, health problems related to poor nutrition persist still today. The South continues to be plagued by illnesses related to a high fat and carbohydrate diet associated both with poverty and with many traditional southern recipes, especially the regional taste for fried food. Harry Watson points out that the South is the center of the United States' "stroke belt," "diabetes belt," and "obesity and heart disease belts."[62] Mississippi currently has the highest percentage of overweight citizens in the United States. Seven out of ten adult residents are either overweight or obese.[63]

The history of deprivation and diet-induced illness in the Delta is often sublimated by a celebratory food culture. John Egerton has argued that "no other form of cultural expression, not even music, is as distinctively characteristic of [the South] as the spreading of a feast of native food and drink before a gathering of family and friends."[64] Although positive associations of the region's foodways are predominant in the minds of many, for those who historically assumed a secondary position at the southern table due to segregation, poverty, and associated health concerns, the food history of the region also contains many less savory memories. Elizabeth Engelhardt offers this reminder: "We are so quick today to romanticize southern food as nourishing, hearty and comforting that if nothing else it is worth remembering a major portion of the southern food story was one of decidedly unromantic, painful loss."[65]

Keeping distressing as well as positive memories of the food history of the Delta and the larger South alive has been difficult not only for those who are heirs of the region's abundance but also for those whose descendants battled against hunger and poverty. Anne Yentsch argues that many contemporary southern African American cookbooks engage in culinary mythmaking as well as culinary history. She notes, for example, the abundance of decadent desserts described in many cookbooks that claim to be rooted in a genuine tradition. If read as literal histories, cookbooks of this variety contain misleading testimony about a past of material abundance that did not exist for the majority of rural black people who rarely could have afforded to eat expensive sweets.[66] The same desire to invent a less ambiguously positive food history is also present in much contemporary food writing about the region, causing Anthony Stanonsis to claim that works such as the *Cornbread Nation* series "tend to overemphasize food's ability to unite southerners across racial lines."[67]

In the twenty-first century, residents of the Lower Mississippi Delta, like other Americans, eat at restaurants more than at any time in history, often at national chains that do not serve regionally specific foods. In 2006, residents of Arkansas reported eating half of their meals in restaurants.[68] Ironically, despite its rich agricultural history, many areas of the Delta are now classified as "food deserts" by the US Department of Agriculture, a designation that means that rural residents need to travel more than ten miles to get to a grocery store.[69] Fewer people have family gardens, and the vast majority of agricultural land is devoted to the production of commodity crops that are shipped out of the area rather than to food for local people. Delta residents living in areas designated as food deserts have limited access to places where they can buy fresh produce and greater proximity to processed foods sold at convenience stores and fast-food restaurants. Jennifer Hoskins, a resident of Lambert, Mississippi, reports, "It's really hard, because, you know, when I was coming up, we had greens and gardens and all that. But now you have to buy produce. So, it's real hard for the kids. I mean, and the majority of them, they eat like pizzas. And that's obesity."[70]

The changing reality of available foods may be having an impact on local food preferences. A study conducted by Beth A. Latshaw revealed that in 1995 only 16 percent of people residing in the South claimed they "often eat" foods traditionally identified as southern, a figure not much higher than the 13 percent of southerners who claim they "never eat it."[71] For many this trend is alarming, and various attempts have been made to revive southern heritage foods by projects such as the Southern Seed Legacy initiative.[72] The Southern Foodways Alliance, founded in 1999 at the University of Mississippi, has done remarkable work in documenting and preserving the foodways of the Delta and other regions of the South through its rich oral history project, among other initiatives. Preservation, however, can become a kind of performance as celebrants of southern foodways seek out obscure and artisanal products that are inaccessible or, in an age of greater culinary homogenization, perhaps even unappealing to many local residents. Marketing also plays a large role in perceptions of current Delta foodways. Visitors to the Memphis airport who purchase a MoonPie, a marshmallow and cookie sandwich manufactured in

Tennessee, from the large display in the gift shop instead of a more nationally recognizable treat are being convinced to use their tourist dollars to ingest what is sold as a regionally specific taste experience. Food studies scholar Elizabeth Engelhardt acknowledges, "Scholars (including me), media, advertisers, and artists not only excavate food practices, we actively shape them as well."[73]

Well-heeled foodies travel to Mississippi each summer to attend a conference sponsored by the Southern Foodways Alliance, where they consume and analyze southern food, paying a conference registration fee in excess of $500 for the privilege. The irony of the marketing and fundraising necessary to sustain the bighearted organization devoted to preserving food traditions created largely by poor people is not lost on the members who cling to the guiding principles that "honest regional food should be affordable to all."[74] However, socioeconomic inequalities in the Delta remain stark, and there is invariably a class divide between the scholars, journalists, and food enthusiasts analyzing southern foodways and the vast majority of Delta eaters.

Contemporary Delta food consists of the daily habits of people living in the region who have greater access than ever before to a wide variety of international food, national brands of processed food, and food prepared by chain restaurants. It also consists of a set of historical recipes and cherished local ingredients. The food of the Delta today merges concrete realities with the imagination. Many traditional recipes that resemble foods eaten in the antebellum Delta are still prepared in locally owned restaurants and in many homes, particularly on Sundays, on holidays, and at large family gatherings. Daily food practices, however, often collide with wishful thinking about what they should be and recollections—both historically founded and nostalgically reinvented—about what they once were.

Notes

1. Marcie Cohen Ferris, "The Edible South," *Southern Cultures* 15, no. 4 (2009): 6.
2. Rayna Green, "Mother Corn and the Dixie Pig: Native Food in the Native South," *Southern Cultures* 14, no. 4 (2008): 117.
3. Morris S. Arnold, "Arkansas Colonial Fare," *Arkansauce* 2 (2012): 7.
4. Michael B. Dougan, "Food and Foodways," *Encyclopedia of Arkansas*, accessed June 2, 2012, http://www.encyclopediaofarkansas.net/encyclopedia/entry-detail.aspx?search=1&entryID=4032.
5. Arnold, "Arkansas Colonial Fare," 6.
6. Antoine-Simon Le Page du Pratz, *History of Louisiana Or Of The Western Parts Of Virginia And Carolina: Containing A Description Of The Countries That Lie On Both Sides Of The River Mississippi* (1763; reprint, Whitefish, MT: Kessinger Publishing, 2004), 128. For more information about *sagamité*, see also Richard Campenella, *Bienville's Dilemma: The Historical Geography of New Orleans* (Layfayette: Center for Louisiana Studies, 2008).
7. Jean-Baptiste Le Moyne de Bienville, quoted in Daniel H. Usner, *Indians, Settler, and Slaves in a Frontier Exchange Economy* (Chapel Hill: University of North Carolina Press, 1992), 194.

8. To read more about the history and significance of corn, see Nicholas P. Hardeman, *Shucks, Shocks, and Hominy Blocks: Corn as a Way of Life in Pioneer America* (Baton Rouge: Louisiana State University Press, 1981); Paul Weatherwax, *Indian Corn in Old America* (New York: Macmillan Company, 1954); and Betty Fussell, *The Story of Corn* (New York: North Point Press, 1992).
9. Bethany Ewald Bultman, "A True and Delectable History of Creole Cooking," *American Heritage* (1986): 66–73; Marcelle Bienvenu, Carl A. Brasseaux, and Ryan A. Brasseaux, *Stir the Pot: The History of Cajun Cuisine* (New York: Hippocrene Books, 2005).
10. James McWilliams, *A Revolution in Eating: How the Quest for Food Shaped America* (New York: Columbia University Press, 2005), 8; Thomas Wessel, "Agriculture, Indians, and American History," *Agricultural History* 50, no. 1 (1976): 9–20.
11. John Gerard, *The Herball, or Generall Historie of Plantes* (London: John Norton, 1597). The full text is available here, http://caliban.mpiz-koeln.mpg.de/gerarde/index.html, accessed April 1, 2015.
12. Shannon Lee Dawdy, "'A Wild Taste': Food and Colonialism in Eighteenth Century Louisiana," *Ethnohistory* 57, no. 3 (2010): 393.
13. Ibid., 402.
14. Linda Murray Berzok, *American Indian Food* (Westport, CT: Greenwood Press, 2005), 21–25.
15. Green, "Mother Corn and the Dixie Pig," 115.
16. Charles Joyner, *Down by the Riverside: A South Carolina Slave Community* (Urbana: University of Illinois Press, 1984), 91; Gene Bourg, "New Orleans Foodways," in *The New Encyclopedia of Southern Culture: Foodways,* ed. Charles Reagan Wilson and John T. Edge (Chapel Hill: University of North Carolina Press, 2007), 83–88. For more information about southern African American foodways, see also Robert L. Hall, "Africa and the American South: Culinary Connections," *Southern Quarterly* 44, no. 2 (2007): 19–52; and Frederick Douglass Opie, *Hog and Hominy: Soul Food from Africa to America* (New York: Columbia University Press, 2008).
17. Jessica Harris, *High on the Hog: A Culinary Journey from Africa to America* (New York: Bloomsbury, 2011), 129.
18. Dougan, "Food and Foodways."
19. John Egerton, "Fried Chicken," in *The New Encyclopedia of Southern Culture,* ed. Wilson and Edge, 141–43; Judith A. Carney and Richard Nicholas Rosomoff, *In the Shadow of Slavery: Africa's Botanical Legacy in the Atlantic World* (Berkeley: University of California Press, 2009), 177–79.
20. For more information about the attitudes of European colonists toward corn, see Jennifer Jensen Wallach, *How American Eats: A Social History of US Food and Culture* (New York: Rowman & Littlefield, 2013).
21. Charles Reagan Wilson, "Cornbread," in *The New Encyclopedia of Southern Culture,* ed. Wilson and Edge, 152–54; Charles Reagan Wilson, "Biscuits," in *The New Encyclopedia of Southern Culture: Foodways,* ed. Wilson and Edge, 122–25; Joe Gray Taylor and John T. Edge, "Southern Foodways," in *The New Encyclopedia of Southern Culture,* ed. Wilson and Edge, 1–13.
22. Carney and Rosomoff, *In the Shadow of Slavery,* 56.
23. Wilson, "Cornbread," 152; Loyal Jones, "Corn," in *The New Encyclopedia of Southern Culture,* ed. Wilson and Edge, 151–52; Susan McLellan Plaisted, "Corn," in *The Oxford Companion to American Food and Drink,* ed. Andrew F. Smith (Oxford: Oxford University Press, 2007), 168–69; Bienvenu, Brasseaux, and Brasseaux, *Stir the Pot,* 81.

24. Charles Reagan Wilson, "Pork," in *The New Encyclopedia of Southern Culture,* ed. Wilson and Edge, 88–91; Bruce Kraig, "Pigs," in *The Oxford Companion to American Food and Drink,* ed. Smith, 457– 58; Sam Bowers Hilliard, *Hog Meat and Hoecake: Food Supply in the Old South, 1840–1860* (Carbondale: University of Illinois Press, 1972).
25. Dorothy Brackin, *Arkansas Soul Foods* (Jacksonville, AR: self-published, 1969).
26. For more about the significance of the term "soul food," see William C. Whit, "Soul Food as Cultural Creation," in *African American Foodways: Explorations of History and Culture,* ed. Anne L. Bower (Urbana: University of Illinois Press, 2007), 45–58.
27. Jessica Harris, quoted in Warren St. John, "Greens in Black and White: Staking Claims to the Origins of What's on the Southern Table," *New York Times,* October 6, 2004, D1.
28. John T. Edge, quoted in Miriam Wolf, "The Southern Activist," accessed June 3, 2012, http://www.johntedge.com/culinate/.
29. Willie Morris, *North toward Home* (New York: Dell Publishing, 1967), 386–87.
30. Charles Gayarré, "A Louisiana Sugar Plantation of the Old Régime," *Harpers* (March 1887): 606–21.
31. Rebecca Sharpless, *Cooking in Other Women's Kitchens* (Chapel Hill: University of North Carolina Press, 2010), 11–31.
32. Mrs. A. P. Hill, *Mrs. Hill's New Cook Book* (New York: Carleton Publisher, 1870), 12.
33. Craig Claiborne, *A Feast Made for Laughter: A Memoir with Recipes* (New York: Doubleday, 1982), 31.
34. Natalie V. Scott, *Mirations and Miracles of Mandy* (New Orleans: Robert H. True Company, 1929), 1.
35. For more information about African Americans and food stereotypes, see Psyche Williams-Forson, "Chickens and Chains: Using African American Foodways to Understand Black Identities," in *African American Foodways,* ed. Bower, 127–28; and Psyche Williams-Forson, *Building Houses Out of Chicken Legs: Black Women, Food, and Power* (Chapel Hill: University of North Carolina Press, 2006).
36. Dave Reed, "The Mississippi Barbecue" (New York: M. Witmark & Sons, 1904), Sam DeVint Collection of Illustrated American Sheet Music, Series 3, Box 79, Archives Center of the National Museum of American History, Washington, DC.
37. Richard Wright, *Black Boy* (1945; reprint, restored ed., New York: HarperPerennial, 2006).
38. Federal Writers' Project, America Eats, "Mullet Salad," U.S. WPA Records, Box A830, Library of Congress, Washington, DC.
39. Allison Davis, Burleigh B. Gardner, and Mary R. Gardner, *Deep South: A Social Anthropological Study of Caste and Class* (Chicago: University of Chicago Press, 1941), 82.
40. Marcie Cohen Ferris, "Feeding the Jewish Soul in the Delta Diaspora," *Southern Cultures* 10, no. 3 (Fall 2004), 53–54. See also Marcie Cohen Ferris, *Matzoh Ball Gumbo: Culinary Tales of the Jewish South* (Chapel Hill: University of North Carolina Press, 2005).
41. Sam Eifling, "What Do Arkansans Eat?," *Oxford American,* April 5, 2010, accessed June 3, 2012, http://www.oxfordamerican.org/articles/2010/apr/05/sam-eifling-asks-what-do-arkansans-eat/.
42. Southern Foodways Alliance, "Chinese Grocers in the Mississippi and Arkansas

Deltas," accessed June 2, 2012, http://southernfoodways.org/documentary/oh/chinese_grocers/index.shtml.

43. Dougan, "Food and Foodways"; The Delta Cultural Center, "A Land Promised: Immigrants in the Arkansas Delta," accessed June 3, 2012. http://www.deltaculturalcenter.com/education_programs/DCC_promised_lessonplans.pdf.
44. Amy Evans, "Hot Tamales," in *The New Encyclopedia of Southern Culture,* ed. Wilson and Edge, 184–85.
45. Taylor and Edge, "Southern Foodways," 4–5; Davis, Gardener, and Gardener, *Deep South,* 385–86.
46. Taylor and Edge, "Southern Foodways," 6; Hilliard, *Hog Meat and Hoecake,* 104–11, 203–4; Joe Gray Taylor, *Eating, Drinking, and Visiting in the South: An Informal History* (Baton Rouge: Louisiana State University Press, 1982), 25–26, 111–12.
47. For a more extensive discussion of twentieth-century changes in food habits in the South, see Scott Holzer, "The Modernization of Southern Foodways: Rural Immigration to the Urban South during World War II," *Food and Foodways* 6, no. 2 (1996): 97–107.
48. John T. Edge, "Clarence Sanders," in *The New Encyclopedia of Southern Culture,* ed. Wilson and Edge, 261–62; Sara Rath, "Piggly Wiggly," in *The Oxford Companion to American Food and Drink,* ed. Smith, 456–57.
49. J. Michael Luster, "Poke Sallet," in *The New Encyclopedia of Southern Culture,* ed. Wilson and Edge, 228–29; Marcia Camp, "Poke Springs Eternal," *Arkansauce* 2 (2012): 16–17.
50. Federal Writers' Project, America Eats, "Recipes from Arkansas," reprinted in Mark Kurlansky, *The Food of a Younger Land* (New York: Riverhead Books, 2009), 128.
51. Federal Writers Project, America Eats, "The Possum Club of Polk County Arkansas," reprinted in Kurlansky, *The Food of a Younger Land,* 159; Roy Vail, "Polk County Possum Club," in *Encyclopedia of Arkansas History & Culture,* ed. Guy Lancaster, accessed June 2, 2012, http://www.encyclopediaofarkansas.net/encyclopedia/entry-detail.aspx?entryID=5517.
52. Andrew Beahrs, *Twain's Feasts: Searching for America's Lost Foods in the Footsteps of Samuel Clemens* (New York: Penguin Press, 2010), 65; John Spurgeon, "Gillett Coon Supper," accessed June 3, 2012, http://www.encyclopediaofarkansas.net/encyclopedia/entry-detail.aspx?entryID=3827.
53. Davis, Gardner, and Gardner, *Deep South,* 379.
54. Ibid., 382.
55. Edward H. Beardsley, *A History of Neglect: Health Care for Blacks and Mill Workers in the Twentieth-Century South* (Knoxville: University of Tennessee Press, 1987), 54–58; Hardeman, *Shucks, Shocks, and Hominy Blocks,* 149–50.
56. Taylor, *Eating, Drinking, and Visiting in the South,* 145.
57. Kenneth Bridges, "Pellagra," accessed June 3, 2012, http://www.encyclopediaofarkansas.net/encyclopedia/entry-detail.aspx?entryID=2230; National Institutes of Health, "Dr. Jonas Goldberger and War on Pellagra," accessed June 3, 2012, http://history.nih.gov/exhibits/Goldberger/index.html.
58. Elliott West, ed., *The WPA Guide to 1930s Arkansas.* (Lawrence: University of Kansas, 1987), 294–95.
59. Beardsley, *A History of Neglect,* 57; Davis, Gardner, and Gardner, *Deep South,* 384.
60. Ted Ownby, *American Dreams in Mississippi: Consumers, Poverty, and Culture, 1830–1998* (Chapel Hill: University of North Carolina Press, 1999), 63.

61. Dorothy Dickins, "Negro Food Habits in the Yazoo Mississippi Delta," *Journal of Home Economics* 18, no. 9 (1926): 524.
62. Harry L. Watson, "Front Porch," *Southern Cultures* 18, no. 2 (Summer 2012): 3.
63. Debbie Elliott, "Mississippi Losing the War with Obesity," National Public Radio, accessed June 2, 2012, http://www.npr.org/2011/05/19/136018514/mississippi-losing-the-war-with-obesity.
64. John Egerton, *Southern Food: At Home, on the Road, in History* (New York: Alfred A. Knopf, 1987), 2.
65. Elizabeth Engelhardt, *A Mess of Greens: Southern Gender and Southern Food* (Athens: University of Georgia Press, 2011), 129.
66. Ann Yentsch, "Excavating the South's African American Food History," in *African American Foodways: Explorations of History and Culture*, ed. Ann L. Bower (Urbana: University of Illinois Press, 2007), 59–60.
67. Anthony J. Stanonsis, "Just like Mammy Used to Make: Foodways in the Jim Crow South," in *Dixie Emporium: Tourism, Foodways, and Consumer Culture in the American South*, ed. Anthony J. Stanonsis (Athens: University of Georgia, 2008), 209.
68. Dougan, "Food and Foodways."
69. US Department of Agriculture, "Access to Affordable and Nutritious Food: Measuring and Understanding Food Deserts and Their Consequences," 2009, accessed June 2, 2012, http://www.ers.usda.gov/Publications/AP/AP036/AP036fm.pdf.
70. Public Broadcasting Newshour, "Mississippi 'Food Deserts' Fuel Obesity Epidemic," October 21, 2010, accessed June 3, 2012, http://www.pbs.org/newshour/bb/health/jan-june10/food_06-03.html.
71. Beth A. Latshaw, "Food for Thought: Race, Region, Identity, and Foodways in the American South," *Southern Cultures* 15, no. 4 (2009): 113.
72. Southern Seed Legacy, "Preserving the Cultural and Genetic Diversity of Southern Agriculture," accessed June 3, 2012, http://pacs.unt.edu/southernseedlegacy.
73. Engelhardt, *A Mess of Greens*, 6.
74. Southern Food Alliance, "SFA Statement of Values 2011," accessed June 2, 2012, http://southernfoodways.org/about/mission.html.

CHAPTER 17

The Literature of the Delta

LISA HINRICHSEN

The South has long been a site of fantasy, conjuring up images of white-columned plantation homes, fertile fields, beautiful belles, and Arcadian visions of magnolias and moonlight. Yet its literature self-consciously navigates these bland clichés, presenting a far more nuanced, multifaceted, and multicultural portrait of the area that questions such oversimplified narratives and homogenizing myths, instead testifying to the complexities of intellectual, cultural, and imaginative life. In reading the literature of the Mississippi Delta, deemed "the most southern place on earth" by historian James Cobb, as well as that of the larger swath of the Mississippi River Delta region, which spans from southern Illinois to New Orleans, we see reflected a variety of "Souths," from the eastern sensibility of St. Louis, an early home to two of the best-known poets of the twentieth century, Marianne Moore and T. S. Eliot, to the steamy, diverse streets of New Orleans, where numerous writers of varied international origins reside. Defined by both cultural vibrancy and debilitating poverty, and marked by a long and complex history of trade, migration, cultural exchange, and slavery, the literature of the Delta is born of the intricacies of a complex, polymorphous history and culture. The region is full of a cacophony of different stories that mirror this demographic diversity and historical complexity, from early foreign-language texts and the earliest stirrings of a proud regionalism trying to assert itself against northern literary dominance to contemporary self-reflexive historical novels and multicultural postmodern literary production.

The region's diverse literary origins can be found in early travel narratives and its rich newspaper culture. Though Spanish explorer Álvar Núñez Cabeza de Vaca passed the mouth of the Mississippi in 1528 before being swept into the Gulf of Mexico to Galveston Island, later recording his harrowing journey in *Relacion* (1542), Hernando de Soto and his troops "discovered" the Mississippi on May 8, 1541, reaching the river again a year later in retreat, where, struck by fever, de Soto died and was buried. Other early descriptive texts include Le Page du Pratz's *Histoire de la Louisiane* (1758) and the travel writings of Meriwether Lewis and William Clark (1804–1806),

Jacques Marquette (1673–1674), Baron de Lahontan (1701), Samuel S. Forman (1789–1790), Jonathan Carver (1778), Louis Hennepin (1880), Zadok Cramer (1801), and others. Various other explorers and non-natives wrote of the awe that the Mississippi River inspired, including Charles Dickens (1841), Frances Trollope (1832), and William Makepeace Thackeray (1852). Though travel writing about the Mississippi was largely displaced by the advent of the Civil War and steamboat travel was diminished by the rise of the railroad, Samuel Clemens—better known as Mark Twain—revived interest in the river in *Life on the Mississippi* (1883), *The Adventures of Huckleberry Finn* (1884), and *The Adventures of Tom Sawyer* (1876). Late nineteenth-century travel writing about the Mississippi also included Nathaniel Bishop's two books about his small-craft voyages along the seaboard and down the Mississippi, *Voyage of the Paper Canoe* (1878) and *Four Months in a Sneak-Box* (1879), and Clifton Johnson's explorations of rural life in the Delta, *Highways and Byways of the Mississippi Valley* (1906). The Mississippi River continues to inspire a rich tradition of travel writing and literary nonfiction: Jonathan Raban's *Old Glory: A Voyage Down the Mississippi* (1998) and Eddy L. Harris's *Mississippi Solo: A River Quest* (1998) are notable contemporary additions to the genre.

Early literary works such as Julien Poydras de Lallande's poem "*La Prise du Morne du Baton Rouge*" (1779) and Paul Louis Le Blanc de Villeneufve's drama *La Fete du Petit-Ble, ou L'Heroism de Poucha-houmma* (1814) attest to the multilingual nature of early Delta literature, especially in multiethnic Louisiana. Early literary production along the Mississippi included work in Spanish, French, and other languages, including the Francophone *Les Cenelles* (1845), the first anthology of poetry by writers of color ever published; Creole folk songs and tales, which mingled elements of African folklore with French, German, Indian, Acadian, and Caribbean influences; slave songs, which were later transcribed into anthologies such as *Slave Songs of the United States* (1867); and periodicals such as *Le Moniteur de Louisiane* (1794–1815), *El Misisipí* (est. 1808), *El Mensagero Lusianes* (1809) and the bilingual paper *L'Abeille* (1827–1923). George Washington Cable's collection of short stories *Old Creole Days* (1879) and his novel *The Grandissimes* (1880) drew national attention to Louisiana Creole culture, exploring the complexities of race and identity, while writers such as Ruth McEnery Stuart, Grace King, and Alice Dunbar-Nelson portrayed mixed-blood life from a feminine point of view. Kate Chopin depicted Creole life in her collections *Bayou Folk* (1894) and *A Night in Acadie* (1897), while also suggesting, as she did in *The Awakening* (1899), the ways many white women felt the pressures of southern paternalism. Though the area was rich in Native American cultures—Mississippi has its roots in an Anishinaabe (Ojibwe or Alogonquin) word, and Quapaw, Tunica, Natchez, Houma, and Choctaw tribes, among others, have their roots along the Mississippi—these cultures, based on oral storytelling, left few early written traces. Many tribes were dislocated to the west of the Mississippi by the Indian Removal Act of 1830; the destruction of tribal solidarity was accelerated by the General Allotment Act of 1887 and practices of forced assimilation through education.

Early English-speaking literature of the Delta includes southwestern humor in the form of Henry Clay Lewis's *Odd Leaves from the Life a Louisiana Swamp Doctor* (1843; under the pseudonym "Madison Tensas") and James R. Masterson's *Tall Tales of Arkansas* (1943); slave narratives such as Solomon Northup's 1853 account, *Twelve Years a Slave*; dialect verse such as that by Mississippian Irwin Russell; and local color writing, which drew attention to peculiarities of speech and distinctive local customs, by Grace King (1852–1932), George Washington Cable (1844–1925), Sherwood Bonner (1849–1883), Lafcadio Hearn (1850–1904), and Ruth McEnery Stuart (1852–1917). Some of this early local writing found interested markets abroad: for example, the Arkansas stories of the transplanted New Englander Thomas Bangs Thorpe, in *The Mysteries of the Backwoods, Or, Sketches of the Southwest* (1846) and *Colonel Thorpe's Scenes in Arkansaw* (1858), were nationally popular and were translated into several European languages. While such works often drew attention to the area's unique wildlife and natural resources, they also worked to reinforce southern "distinctiveness"—here understood as both "backwoods" and backwards—internationally.

The Civil War inspired hundreds of poems, songs, and stories, and its compulsive remembrance has been an enduring trope of southern literature. Two of its most noted generals, Ulysses S. Grant and William Tecumseh Sherman, wrote most of their memoirs in St. Louis. Some notable literature of the Confederacy remained largely unknown for a generation or longer, hidden away in journals, letters, and diaries, such as those by Delta writers Sarah Morgan Dawson (1842–1899) and Kate Stone (1841–1907). The Civil War, as Drew Gilpin Faust argues in *Mothers of Invention*, "made thousands of white women of all classes into authors—writers of letters and composers of journals recording the momentous and historic events as well as creators of published songs, poetry, and novels" (161). Published posthumously in 1913 under the title *A Confederate Girl's Diary*, Dawson's personal account of the war years in Louisiana's capital city of Baton Rouge and in New Orleans is structured by her keen eye for detail. Likewise, Stone's memoir, *Brokenburn: The Journal of Kate Stone* (1955), which recalls life in a northern Louisiana plantation, is also marked by a facility for realistic description: Stone chronicles everyday details of life from 1861 to 1868, narrating not only details about plantation management, farming, and slave conduct, but also the intricacies of her domestic and imaginative life.

This penchant for realism would fade as the plantation romance rose in popularity. Though the most famous of these, John Pendleton Kennedy's *Swallow Barn* (1832) and William Gilmore Simms's *Woodcraft* (1852), take place outside of the Delta, books such as George Washington Cable's *The Cavalier* (1901) and *Kincaid's Battery* (1908) presented romanticized views of

Delta plantations, so as to insist on mythic fantasies of social and racial order. In envisioning a quixotic version of the past in order to legitimate present-day southern "difference," these writers presented what C. Vann Woodward, in *American Counterpoint,* later called "the compensatory dream of aristocracy, the airs of grace and decorum left behind, secretly yearned for but never realized" (6). This vision stood against a world in which blacks had begun to protest new forms of subjugation—the Jim Crow legislation that prohibited racial intermingling in public spaces, the recourse to lynching to terrorize African Americans—in attempts to dislocate white claims to political and social power. Though many of these plantation romances, with their sentimental prose and belabored plots, are now forgotten or dismissed as Lost Cause propaganda, they set the ground for southern literature's preoccupation with memory, history, and the presence of the past, which would be repeated, ironized, critiqued, and parodied as the twentieth century progressed in works by Delta writers: William Faulkner's *The Unvanquished* (1938), which interjects anxieties about the Great Depression into a Civil War tale, Barry Hannah's *Airships* (1978), which mingles the Vietnam War and the Civil War, and Shelby Foote's *Shiloh* (1952), which presents the war through a few selected participants in one battle, as well as his massive three-volume *The Civil War: A Narrative History* (1958, 1963, 1974).

The widespread economic, cultural, and material devastation created by the Civil War, the aftereffects of which lingered into the twentieth century, led H. L. Mencken (1880–1956) to publish a powerful indictment of southern culture, "The Sahara of the Bozart," which first appeared in 1917 in the New York *Evening Mail* and was later reprinted in his book *Prejudices, Second Series* (1920). Here Mencken, whose brash tone and polemical claims shocked and enraged southerners, maintained that the South was "almost as sterile, artistically, intellectually, culturally, as the Sahara Desert" (70). Attributing the decline of southern culture to the class of poor whites who, he charged, had seized control of the South after the Civil War, Mencken saw a South where "there is not a single picture gallery worth going into, or a single orchestra capable of playing the nine symphonies of Beethoven, or a single opera-house, or a single theater devoted to decent plays" (71). Though he especially singled out Georgia and Virginia, Mencken's attacks on the South hit hard in the Delta: in 1931, the Arkansas legislature, goaded by claims in this essay and others, passed a motion to pray for the soul of Mencken.

Both enraged and inspired by Mencken's incendiary claims, young southern writers sought to rework and override public perception of southern literary poverty. Mencken especially infuriated a group of writers deemed the Southern (Vanderbilt) Agrarians, who looked back to the past through a particularly conservative lens, extolling traditionalism and a romantic, land-based view of southern rootedness. Like his compatriots Allen Tate, Robert Penn Warren, John Crowe Ransom, John Gould Fletcher, and Andrew Lytle, the

Delta writer Stark Young, a native of Como, Mississippi, and author of *So Red the Rose* (1934), portrayed antebellum plantation society as the ideal of cultivated leisure, a lost alternative to the materialism and postwar renegotiation of labor and race relations afflicting modern life after World War I.

The burst of literary creativity that marked the 1930s was taken as something of a "Southern Renascence," with the region's cultural achievement vaulting over its continued economic floundering. In the years between the two world wars, the South began to surrender economic and social habits formed a century earlier, and many of the region's best writers, including many in the Delta, were drawn to narrating this upheaval. The freedom associated with the experimental methods of international modernist art coincided with a sense of gathering complaint, criticism, and skepticism in the modernizing South; most of this writing, though it sought to tell about the South from the southern standpoint, was not celebratory of now archaic notions of southern distinctiveness. Of all of the modern Delta writers of this period, William Faulkner is the best known. Though he lived most of his life in Oxford, Mississippi, he wrote his first novel, *Soldier's Pay* (1926), while living in Pirate's Alley in New Orleans; later, he would briefly spend time in Hollywood, though he was quickly drawn back to the Delta. The consummate novelist of the period, Faulkner set about excavating in exquisite modernist prose the rotten foundations of the plantation system that had defined his Mississippi. His leading novels explore the many facets of the past's survival in the present: *The Sound and the Fury* (1929) portrays the psychology of grief felt by those who lost their privileged status in the process of modernization; *As I Lay Dying* (1930) explores the hard lives of small-time farmers on an epic and catastrophic journey; *Light in August* (1932) exposes the contradictions of absolutist racial beliefs; *Absalom, Absalom!* (1936) and *Go Down, Moses* (1942), with its centerpiece "The Bear," trace the moral outrages secreted in the plantation past. In transcribing and translating the tensions of the South's belated and abrupt modernization into the fictional world of Yoknapatawpha, Faulkner astutely rendered complex histories of political, racial, and economic inequity, bringing both past and present into relief.

Despite, or perhaps because of, the hardships of the Great Depression and World War II, the 1930s and the 1940s were a fertile time for Delta writers, who created masterpieces from the ambivalence and anxiety that surrounded their confrontation with the blighted historical foundations of southern distinctiveness and who drew energy from the shifting status quo of the present-day South. A native of Columbus, Mississippi, the playwright Tennessee Williams explored mid-century anxiety, mental illness, and social change in *The Glass Menagerie* (1945; set in St. Louis), the demise of the Old South in *A Streetcar Named Desire* (1947; set in New Orleans), and themes of sexuality, desire, and self-destruction in *Summer and Smoke* (1948), *Cat on a Hot Tin Roof* (1955), and *Suddenly Last Summer* (1958). New Orleans, where Tennessee

Williams centered many of his plays, and Greenville, Mississippi, became hubs for the literate and creative. Greenville native William Alexander Percy (1885–1942), a lawyer, planter, and poet, is best remembered as author of the nostalgic memoir *Lanterns on the Levee* (1941). Other Greenville writers include James Robertshaw, Ernest D. Elliott, Charles Bell (poet and author of *Delta Return,* 1956), Shelby Foote, Walker Percy, Ellen Douglas, Hodding Carter Jr., David L. Cohn, Clarence Brannon, Bern Keating, Ben Wasson, Anne Metcalfe Clark, Mary Berkeley McNeilly Finke, William Alexander Attaway, Henry Tillinghast Ireys, Jay Milner, Jane Taylor Overton, Sinclair O. Lewis, and Emma Harrington.

As the twentieth century progressed, the Delta enriched its wealth of African American and female writing. A native of Jackson, Mississippi, Eudora Welty, in *A Curtain of Green* (1941), *The Wide Net* (1943), *Delta Wedding* (1946), *Losing Battles* (1970), and *The Optimist's Daughter* (1972), gives voices to marginalized characters and female experience and examines the limits of certain modes of being southern. Welty's short stories explore themes of memory and desire, examining the heavy undergrowth of repression that stunts and saddens women's lives in the early twentieth-century South; pieces like "Why I Live at the P.O.," "The Petrified Man," "June Recital," and "Clytie" compose a complex taxonomy of female suffering. Ellen Douglas, a native of Natchez, Mississippi, takes up Welty's critique of southern culture and her interest in femininity in works such as *Black Cloud, White Cloud* (1963), *Can't Quit You, Baby* (1988), and *Where the Dreams Cross* (1968). In *The Rock Cried Out* (1979), Douglas squarely confronts racial violence as she tells the story of Alan McLaurin, who returns to his native Mississippi after several years in the North and finds himself drawn into the civil rights struggle. Other notable female-authored texts include Elizabeth Spencer's *Fire in the Morning* (1948); Ellen Gilchrist's work, including *Victory over Japan* (1984) and *In the Land of Dreamy Dreams* (1981); Frances Parkinson Keyes's *The River Road* (1945), *Dinner at Antoine's* (1948), and *Blue Camellia* (1957); New Orleans native Shirley Ann Grau's haunting stories of abortion, miscegenation, destruction, and death in collections such as *The Black Prince, and Other Stories* (1955) and *The Keepers of the House* (1964); Lillian Hellman's numerous plays and screenplays, including *The Little Foxes* (1941), and her memoir, *Pentimento* (1973); and Greenwood, Mississippi, native Donna Tartt's *The Secret History* (1992) and *The Little Friend* (2002).

Though many African American writers left the Delta in the twentieth century to migrate north, notable African American writers such as Langston Hughes, born in Joplin, Missouri, drew on the rhythms of slave songs and black spirituals in collections such as *The Weary Blues* (1926) and *Fine Clothes to the Jew* (1927), exploring the sounds of jazz and blues, while responding to the history of the South's racial horrors. Hughes's career bespeaks of the confluence of the Southern Renascence and the Harlem Renaissance that blossomed in the 1920s and the ways that both movements emerged from within a context of demographic and economic change. His work, in poems such as "Wide River" and "The Negro Speaks of Rivers," reflects on the Mississippi as a site and a source of enslavement, for its bountiful waters and periodic flooding created the rich Delta soil that was the ground for African American labor. Exploring the emotional complexity of this historically charged relationship with the land has been a theme of much African American writing of the twentieth century, especially as recent black writers have returned to reclaim southern soil.

Other Delta writers, such as Arna Bontemps, born in Alexandria, Louisiana, drew from the South despite settling elsewhere later in their lives. Bontemps excavated African American history in three novels: *God Sends Sunday* (1931), *Black Thunder* (1936), and *Drums at Dusk* (1939). Richard Wright, a native of Natchez, Mississippi, rendered African American experience through the red-tinged lens of rage, fostered in the bigotry he experienced growing up in the Delta in his semi-autobiographical *Black Boy* (1945). In work such as *Uncle Tom's Children* (1938), *Native Son* (1940), and *The Long Dream* (1958), as well as autobiographical essays like *The Ethics of Living Jim Crow* (1937), Wright overturns myths of black subservience and constructs complex portraits of black masculinity. Modern black autobiographies such as Wright's *Black Boy* testify to the influence of the slave narrative on the first-person writing of African Americans. Beginning with Margaret Walker's *Jubilee* (1966) and extending through such contemporary novels as Ernest J. Gaines's *The Autobiography of Miss Jane Pittman* (1971) and Valerie Martin's *Property* (2003), the "neo-slave narrative" has become one of the most widely read and discussed forms of African American literature. These autobiographical and fictional descendants of the slave narrative confirm the continuing importance and vitality of its legacy: to probe the origins of psychological as well as social oppression and to critique the meaning of freedom for black and white Americans alike from the founding of the United States to the present day. Through this form, African American Delta writers reveal the ongoing cultural and psychological costs of the history of slavery and segregation, illuminating the insidious ways in which racial bias maintains a presence in American culture. Yet their work surmounts a mere recitation of injury and injustice to focus instead on the ways in which individual dignity and self-worth can be achieved and maintained.

The 1950s and 1960s brought massive changes to the Delta, and to the South as a whole: waves of immigration and emigration, greater civil rights for minorities, the waning of traditional patriarchal and aristocratic southern mores, and further loss of southern distinctiveness in the face of mass-market products and chains. The tempestuous violence of the civil rights era—including the assassination of Martin Luther King in Memphis and forced desegregation in Little Rock, Arkansas—led Delta writers to revisit in their fiction the racial brutality central to southern experience: Alex Haley's *Roots*

(1967), Ernest Gaines's *The Autobiography of Miss Jane Pittman* (1971), and, later, Lewis Nordan's *Wolf Whistle* (1993) turn back to the past to probe the origins and consequences of racial violence. Willie Morris, born in Jackson, Mississippi, but raised in Yazoo City, chronicles the years of desegregation in the 1940s, '50s, and '60s in *North toward Home* (2000); in other work, such as *The Ghosts of Medgar Evers* (1998), he directly reflects on racism in the area through one of the most incendiary murders in civil rights history. Welty's short story "Where Is the Voice Coming From?" (1963), written the night Medgar Evers was killed, eerily channels the personality of his fictive murderer through the first person. Anne Moody's *Coming of Age in Mississippi* (1968), set in her hometown of Centerville, recalls a life immersed in political activism and civil rights work in the Delta, while Will Campbell's *Brother to a Dragonfly* (1977) frames civil rights work as a spiritual practice. Alice Walker, who worked and lived in Jackson, Mississippi, testifies to the complexities of being black and female in books such as *In Love and Trouble* (1973), *The Color Purple* (1982), *In Search of Our Mother's Gardens* (1983), and *Meridian* (1976). And New Orleans writers such as Tom Dent and Kalamu ya Salaam, publications such as *BLKARTSOUTH, The Black River Journal, Bamboula, Callaloo,* and *Nkombo,* and organizations such as the Southern Black Cultural Alliance, the Congo Square Writers Union, and the Free Southern Theatre contributed to the flourishing of African American culture in the area.

The southern economic boom of the 1970s reignited old fears about cultural authenticity and "the Americanization of Dixie," as John Edgerton put it in the title of his popular 1974 book, which charted the myriad ways in which the South now mirrored the rest of the nation. Walker Percy's *The Moviegoer* (1961), *The Last Gentleman* (1966), and *Lancelot* (1977) reveal psychic struggles with melancholia, madness, and amnesia, brought on by contemporary rootlessness and alienation, and dramatize how material prosperity, which was concentrated in urban centers, threatened local cultures and created battles over space, place, and cultural power. In direct resistance to the homogenizing impact of mass culture, however, contemporary Delta writers such as Ellen Gilchrist, the poet Brenda Marie Osbey, and the playwright Beth Henley turn toward oral narratives and folk traditions as means of maintaining a distinctive regional identity.

The literature of the present-day Delta continues to engage in evolving debates over the idea of a distinctive southern identity while dismantling exclusionary mythologies and exhausted tropes of southern belonging. As early as 1956, in her famous essay "Place in Fiction," Eudora Welty argued that southern writing had reached a level of diversity that transcended regional classification: southern writers are able, Welty argued, to choose or not choose to write about slavery, the Civil War, Reconstruction, and limited tropes of "southern" writing (a sense of place, community, and so forth). Her premise has held true: while contemporary southern authors derive inspiration from traditional motifs of history, place, race, and community, they also place new emphasis on social class, sexuality, and gender, challenging conservative notions regarding the thematic range of southern literature. Recent fiction of the Delta questions and complicates what it means to be "southern," stressing the social construction of both communal and individual identity. Though they continue high modernism's concerns with concepts of individual alienation and the loss of communal values, contemporary writers in the Delta extend their experimentation with new narrative forms and methods and dwell upon but frequently stop short of the overt postmodern techniques found in other American literatures. Writing from a revitalized realist tradition (frequently termed "dirty realism" or "grit lit"), Frederick Barthelme (*Moon Deluxe*, 1983; *The Brothers*, 1993; *Natural Selection*, 1990) investigates the fragmentation of the American family within a media-driven world; much of his fiction evokes a New South of shopping malls, pop culture, brand names, and television screens. Peter Taylor's work, including his Pulitzer Prize–winning *A Summons to Memphis* (1986) and his collection *The Old Forest and Other Stories* (1985), investigates the complexity of family life in the New South; Richard Ford (*Women with Men*, 1997; *The Sportswriter*, 1986; *Independence Day*, 1995) also explores contemporary southern masculinity. The work of Larry Brown (*Dirty Work*, 1989) reveals the everyday struggles of the world of the perennially poor. Writers such as Charlaine Harris, Anne Rice, Tom Franklin, and Poppy Z. Brite draw on an ongoing appetite for the southern gothic in works that take on the supernatural, forming fiction that is at once identifiably "southern" and yet is also distinctly subversive in its treatment of patriarchy, gender roles, and the coherence of the nuclear family. The boisterous, bawdy hold of southern humor continues in John Kennedy Toole's sardonic *A Confederacy of Dunces* (1980), which explores the eccentric characters of the French Quarter; Lewis Nordan's *The All-Girl Football Team* (1986) and *Music of the Swamp* (1991); and Jack Butler's hilarious *Jujitsu for Christ* (1986) and *Living in Little Rock with Miss Little Rock* (1995). The urban landscape of Memphis is revealed in John Grisham's best-selling fiction, including *The Firm* (1991) and *The Rainmaker* (1995), James Conaway's *Memphis Afternoons: A Memoir* (1993), Beecher Smith's *The Guardian* (1999) and *Monsters from Memphis* (1997), and James Williamson's *The Architect* (2007). And poets such as Jo McDougall, Natasha Trethewey, Yusef Komunyakaa, Brenda Marie Osbey, Sybil Kein, and Mona Lisa Saloy keep southern verse alive with contemporary concerns. And, in deconstructing and decoding ways of being southern, contemporary writers in the Delta have frequently found themselves reckoning with how the past and the present engage each other in everyday life. Katherine Stockett's bestselling *The Help* (2010), for example, controversially turns back to the civil rights period to imagine a narrative about black domestic labor as a source for white female self-amplification, in the process diminishing black female agency.

Yet contemporary writing in the Delta does not fully set itself free of past themes and subject matters: it continues to examine issues of race, gender, class, family, community, and religion, but does so from changed perspectives and from previously marginalized voices. Twentieth-century Native American authors such as Louis D. Owens, Diane Glancy, LeAnne Howe, and Carter Revard underscore in their work a continued native presence in the Delta, seeking to correct endemic silences regarding the diversity that has always been there. In works like *Ponca War Dancers* (1980), *Shell Shaker* (2002), and *Pushing the Bear* (1996), these authors utilize themes of dispossession and subjugation, sovereignty and hybridity, drawing attention to issues of spirituality, the land, and historical memory: themes which resonate and overlap with southern literature as a whole but come into a unique focus when traced from a native perspective. Tim Gautreaux, author of *The Clearing* (1999) and *The Missing* (2009), details Cajun Louisiana, while Louis Maistros's *The Sound of Building Coffins* (2009) underscores the history of Italian immigration to New Orleans.

Recent years have also ushered in a burst of literature dedicated to commemorating the destruction wrought by Hurricane Katrina (2005). While the environmentally vulnerable Lower Mississippi River Valley has historically lent itself to a wealth of writing about its cyclic floods in texts such as George Washington Cable's "Belle Demoiselles Plantation" (1874), Robert Penn Warren's "Blackberry Winter" (1947), the "Old Man" sections of Faulkner's *The Wild Palms* (1939), Hodding Carter's *Flood Crest* (1947), Lyle Saxon's *Father Mississippi* (1927), William Alexander Percy's *Lanterns on the Levee* (1941), and Eudora Welty's "At the Landing" (1943), the plethora of literature that emerged after Katrina attests anew to the power of the Mississippi. The growing collection of post-Katrina literature includes John Biguenet's acclaimed plays, *Rising Water* (2006) and *Shotgun* (2009); Tom Piazza's novel *City of Refuge* (2008) and his post-Katrina polemic *Why New Orleans Matters* (2005); James Lee Burke's novel *The Tin Roof Blowdown* (2007), as well as his short-story collection *Jesus Out to Sea* (2007); Jesmyn Ward's novel *Salvage the Bones* (2011); Chris Rose's essay collection *1 Dead in Attic* (2006); inventive memoirs such as Jerry Ward's *The Katrina Papers: A Journal of Trauma and Recovery* (2009) and Joshua Clark's *Heart like Water* (2007); poetry by Raymond McDaniel, Patricia Smith, and Katie Ford; as well as numerous collections, such as Jarret Lofstead and Joe Longo's *Life in the Wake: Fiction from Post-Katrina New Orleans* (2007). A wave of graphic novels has also emerged, forming innovative visual and narrative responses to the problem of writing trauma: Brad Benischek's *Revacuation* (2007), Mat Johnson's *Dark Rain* (2011), and Josh Neufeld's *A.D.: New Orleans after the Deluge* (2010) are notable. Many of these plays, poems, essays, memoirs, and graphic novels make political as well as aesthetic statements, highlighting the failed crisis management, the endemic racial and economic inequity, the lasting environmental damage, and the mix of carelessness and compassion that defined the human response to the disaster.

The changes that global capitalism has created in the Delta can be seen in the way its newest literature reflects a recent surge of immigrants to the South from Southeast Asia, the Caribbean, and Mexico. In revealing southern issues from an outsider's perspective, an outpouring of new novels since the 1990s helps to realign racial conflicts previously thought of in terms of black-white binaries and challenge assumptions that one must be born in the South to understand it. Texts such as Robert Olen Butler's 1994 collection of short stories, *A Good Scent from a Strange Mountain*, which focuses on Vietnamese lives, nod to how the Mississippi Delta has attracted immigrants from around the world and work to reconceptualize exclusionary and exceptionalist notions of nation and region by placing the South in transnational perspective, reflecting the cross-cultural connections that define the "Global South." Dave Eggers's post-Katrina novel, *Zeitoun* (2009), by centering on a Syrian immigrant and his American-born family, explores, among other things, multicultural New Orleans and the meaning of Katrina as a world event. Erna Brodber's diasporic novel *Louisiana* (1994) underscores connections between the US South and the Caribbean while critiquing Western constructions of history and knowledge. Cynthia Shearer's *The Celestial Jukebox* (2004), set in the invented Mississippi town of Madagascar, underscores the global diversity of the area through intertwining the stories of a Chinese American grocer and a young African immigrant and setting them against a soundtrack of American melodies that form a global pop culture. As Delta writers continue to examine issues of region and nation from a global viewpoint, the US South will increasingly be understood in solidarity with other Global Souths: developing countries with similar economic histories of belated capitalist modernization, slavery, and exploitation of labor and raw materials. The future of Delta literature rests in this growing demographic diversity and in continued meditations on the social and historical complexity that continues to shape the region.

References

Bone, Martyn. *The Postsouthern Sense of Place in Contemporary Fiction.* Baton Rouge: Louisiana State University Press, 2005.

Brinkmeyer Jr., Robert H. *The Fourth Ghost: White Southern Writers and European Fascism, 1930–1950.* Baton Rouge: Louisiana State University Press, 2009.

Duck, Leigh Anne. *The Nation's Region: Southern Modernism, Segregation, and U.S. Nationalism.* Athens: University of Georgia Press, 2006.

Faust, Drew Gilpin. *Mothers of Invention: Women of the Slaveholding South in the American Civil War.* Chapel Hill: University of North Carolina Press, 1996.

Gray, Richard J. *The Literature of Memory: Modern Writers of the American South.* Baltimore: Johns Hopkins University Press, 1997.

Guinn, Matthew. *After Southern Modernism: Fiction of the Contemporary South.* Jackson: University Press of Mississippi, 2000.

Hobson, Fred C. *The Southern Writer in the Postmodern World.* Athens: University of Georgia Press, 1991.

Hobson, Geary, Janet McAdams, and Kathryn Walkiewicz, eds. *The People Who Stayed: Southeastern Indian Writing after Removal.* Norman: University of Oklahoma Press, 2010.

Jones, Anne Goodwyn, and Susan Van D'Eldon Donaldson, eds. *Haunted Bodies: Gender and Southern Texts.* Charlottesville: University of Virginia Press, 1997.

Jones, Suzanne W., and Sharon Monteith, eds. *South to a New Place: Region, Literature, Culture.* Baton Rouge: Louisiana State University Press, 2002.

King, Richard H. *A Southern Renaissance: The Cultural Awakening of the American South, 1930–1955.* New York: Oxford University Press, 1980.

Kreyling, Michael. *Inventing Southern Literature.* Jackson: University Press of Mississippi, 1988.

Mencken, H. L. *Prejudices: A Selection.* Edited by James T. Farrell. Baltimore: Johns Hopkins University Press, 2006.

Romine, Scott. *The Real South: Southern Narrative in the Age of Cultural Reproduction.* Baton Rouge: Louisiana State University Press, 2008.

Smith, Jon, and Deborah Cohn. *Look Away! The U.S. South in New World Studies.* Durham: Duke University Press, 2004.

Smith, Thomas Ruys. "The Mississippi River as Site and Symbol." In *The Cambridge Companion to American Travel Writing,* edited by Alfred Bendixen and Judith Hamera, 62–78. Cambridge: Cambridge University Press, 2009.

Woodward, C. Vann. *American Counterpoint: Slavery and Racism in the North-South Dialogue.* Oxford: Oxford University Press, 1983.

Yaeger, Patricia. *Dirt and Desire: Reconstructing Southern Women's Writing, 1930–1990.* Chicago: University of Chicago Press, 2000.

Contributors

L. Dyann Arthur resides in the Pacific Northwest. Arthur began playing piano at five and guitar at eleven, and her extensive musical training includes classical voice, Afro-Cuban rhythms for percussion and congas, and jazz piano, alongside decades of performing. She attended Seattle Central Community College and Cornish School for the Allied Arts and graduated magna cum laude from Berklee College of Music in Boston with concentrations in musicology, music theory, and history. She now runs MusicBox Project, a 501(c)(3) arts organization documenting and preserving music history while advancing avenues of education, creation, and performance. She is currently producing and editing a documentary film that will be developed into web-based educational content.

William Clements is a former editor of *Arkansas Review* and of another Delta-oriented periodical, the now-defunct *Mid-South Folklore*. A fellow of the American Folklore Society, he also coedited *An Arkansas Folklore Sourcebook* (University of Arkansas Press, 1992) with the late W. K. McNeil.

Janelle Collins is former general editor and current managing editor of *Arkansas Review: A Journal of Delta Studies*. She is an associate professor of English and chairs the Department of English, Philosophy, and World Languages at Arkansas State University.

Randel Tom Cox is an associate professor of geology in the Earth Sciences Department at the University of Memphis. His primary research interests involve tectonic and climatic geomorphology and active tectonics. Cox's ongoing research involves landscape evolution of the Mississippi Valley and surrounding region, documentation of prehistoric earthquakes in the central United States and Central America, and the effects of volcanic hotspots on moving crustal plates.

Fiona Davidson is a geography professor and director of the Fulbright's European Studies Program at the University of Arkansas, Fayetteville. She comes from a diverse background in political and economic geography, cartography, and European, American, and Mediterranean regional geography. She has written numerous articles and chapters on regional politics both in Europe and the United States. Her expertise has been sought by various agencies and political parties across the United States, the European Union, and the United Kingdom. She has taught and lectured at universities in Rome (Italy), the United Kingdom (Glasgow, Cambridge, Newcastle), and the United States.

David H. Dye is a professor of archaeology in the Department of Earth Sciences at the University of Memphis. He specializes in Mississippian iconography, religion, and warfare. The author of *War Paths, Peace Paths: An Archaeology of Cooperation and Conflict in Native Eastern North America* (2009), he has also edited several books, including *Towns and Temples along the Mississippi* (1990) and *Cave Archaeology of the Eastern Woodlands: Papers in Honor of Patty Jo Watson* (2008). In addition, he was a principal photographer for the Art Institute of Chicago's exhibition catalog, *Hero, Hawk, and Open Hand: American Indian Art of the Ancient Midwest and South* (2004).

William L. Ellis is an assistant professor of music at Saint Michael's College, Colchester, Vermont, where he specializes in American roots music. An award-winning writer and musician, he is the guitar-playing son of banjo composer Tony Ellis and the godson of bluegrass patriarch Bill Monroe. Ellis records and performs traditional blues and gospel styles and was a music journalist before earning his PhD in ethnomusicology from the University of Memphis. He also holds an MM in classical guitar performance from the University of Cincinnati–College Conservatory of Music. Ellis is the author of an upcoming book on the music of Reverend Gary Davis, to be published by the University Press of Mississippi.

William Ferris, a widely recognized leader in southern studies, African American music, and folklore, is the Joel R. Williamson Eminent Professor of History at the University of North Carolina at Chapel Hill and the senior associate director of UNC's Center for the Study of the American South. His recent publications include the acclaimed *Give My Poor Heart Ease: Voices of the Mississippi Blues* (2009) and *The Storied South: Voices of Writers and Artists* (2013).

Gary D. Glass Jr. is a doctoral candidate in sustainable development, holds a master's degree in analytical processes and a graduate certificate in community processes from the Department of Rural Sociology at the University of Missouri. He is a graduate assistant at the University of Missouri's Office of Social and Economic Data Analysis and serves as a field manager for several state and regional health and policy analysis projects.

John J. Green is director of the University of Mississippi Center for Population Studies. He received his PhD in rural sociology from the University of Missouri. Green has published widely in the fields of rural social sciences and public health and is editor of the journal *Community Development* (2012–2015).

Tracy Greever-Rice is interim director of the University of Missouri Office of Social and Economic Data Analysis. She holds a PhD in rural sociology and a master's degree in community development from the University of Missouri. She is the principal investigator for the Missouri Census Data Center, Kids Count in Missouri, and numerous state-level policy analysis projects.

Ruth A. Hawkins is director of Arkansas Heritage Sites at Arkansas State University and chairman of the Culture and Heritage Committee of the Mississippi River Parkway Commission. She holds a bachelor of journalism from the University of Missouri–Columbia, a master of arts in political science from Arkansas State University, and a PhD in higher education administration from the University of Mississippi. Hawkins is the author of *Unbelievable Happiness and Final Sorrow: The Hemingway-Pfeiffer Marriage.*

Lisa Hinrichsen is an associate professor of English and director of Graduate Studies at the University of Arkansas. She teaches widely in the history of twentieth-century American literature, but her research and more specialized courses focus on the literature of the US South, psychoanalytic theory, autobiographical literature, and modernist poetics. She has published articles on William Faulkner, Robert Frost, Bobbie Ann Mason, and Elizabeth Madox Roberts. She is the author of *Possessing the Past: Trauma, Imagination, and Memory in Post-Plantation Southern Literature* (LSU Press, 2015).

Stephen A. King earned his PhD in speech communication at Indiana University in 1997. He is currently a professor and the chairperson of the Department of Communication Studies at Eastern Illinois University. King's research intersects rhetoric, intercultural communication, and popular music. He is the author of two books, *Reggae, Rastafari, and the Rhetoric of Social Control* (2002) and *I'm Feeling the Blues Right Now: Blues Tourism and the Mississippi Delta* (2011). King's work has also been published in a variety of journals, including the *Southern Communication Journal, Popular Music and Society, Journal of Popular Culture,* and *Caribbean Studies,* as well as in edited books such as *The Resisting Muse: Popular Music and Social Protest* (2006) and *Popular Music and Human Rights, volume 1* (2011).

Seth C. McKee received his PhD in government from the University of Texas in 2005 and a master's degree in economics (1998) and a bachelor's degree in political science (1996) from Oklahoma State University. His primary area of research focuses on American electoral politics and, especially, party system change in the American South. He has published research on such topics as political participation, vote choice, redistricting, party switching, and strategic voting behavior. McKee is the author of *Republican Ascendancy in Southern U.S. House Elections* (Westview Press, 2010) and the editor of *Jigsaw Puzzle Politics in the Sunshine State* (University Press of Florida, 2015).

Tom Paradise is a geography and geosciences professor and former director of the King Fahd Center for Middle East Studies at the University of Arkansas, Fayetteville. He comes from a diverse background in the environmental sciences, cartography/GIS, architectural and stone conservation, cultural heritage management, and Middle East and North Africa geography. His notable research in Petra, Jordan, spans twenty-five years, and his expertise on stone architectural deterioration has been requested by countries across the Mediterranean and Middle East, including Vatican City, Italy, Tunisia, Jordan, Morocco, and Egypt. He is a consultant to a number of documentary production companies, including Nova (PBS), the Discovery, ARTE, NatGEO, and Smithsonian channels. Paradise has taught abroad at universities in Rome, Venice, and Amman, as well as in the United States, in Georgia, Hawaii, Arizona, and California. He has lived in Rome, Amman, San Francisco, and Hawaii and currently lives in Fayetteville, Arkansas.

Mary Sue Passe-Smith has been a lecturer in geography at the University of Central Arkansas for ten years. She holds a master of arts in geography with a hazards/GIS (geographic information system) emphasis from the University of Arkansas and teaches GIS and other geography courses to undergraduates. Her master's thesis and subsequent research interests involve the use of GIS to model how topography and land-use change might affect large- and small-scale atmospheric phenomena such as tornadoes and the study of and modeling of human vulnerability to hazardous meteorological/climatological events.

Mikko Saikku is the McDonnell Douglas Professor of American Studies at the University of Helsinki. He is also a docent of environmental history at the University of Tampere, Finland. His research has concentrated on North American environmental history and the history and culture of the US South. His publications include *This Delta, This Land: An Environmental History of the Yazoo-Mississippi Floodplain* (University of Georgia Press, 2005) and *Encountering the Past in Nature: Essays in Environmental History* (with Timo Myllyntaus, Ohio University Press, 2001).

Jennifer Jensen Wallach holds a PhD in Afro-American studies from the University of Massachusetts. She is an associate professor of history at the University of North Texas, where she teaches African American history and US food history. She is the author of *Closer to the Truth than Any Fact: Memoir,*

Memory, and Jim Crow (2008), *Richard Wright: From Black Boy to World Citizen* (2010), and *How America Eats: A Social History of U.S. Food and Culture* (2013), and the coeditor of *Arsnick: The Student Nonviolent Coordinating Committee in Arkansas* (2011), *American Appetites: A Documentary Reader* (2014), and the forthcoming *Routledge History of American Foodways*. She is also the series editor of the University of Arkansas Press series Food and Foodways.

Jeannie M. Whayne is a professor of history at the University of Arkansas and curator of American History at the Crystal Bridges Museum of American Art. She received her BA, MA, and PhD at the University of California, San Diego. She publishes in the area of agriculture, Arkansas, and southern history. In addition to books she coauthored or edited, she has two other books, including the newly published *Delta Empire: Lee Wilson and the Transformation of Agriculture in the New South* (Louisiana State University Press, 2011). *Delta Empire* won the J. G. Ragsdale Award for the best nonfiction book on an Arkansas history topic. An earlier book, *A New Plantation South: Land, Labor, and Federal Favor in Twentieth-Century Arkansas,* published with the University of Virginia Press in 1996, won the Arkansiana Award. She has published many articles, given dozens of conference papers, and has had fellowships at the Smithsonian Institution and the Carter Woodson Institute. In 2010 the Agricultural History Society named her a "Fellow of the Society" and in 2011 elected her vice president and president-elect. She is a distinguished lecturer with the Organization of American History. She has won numerous awards for both her teaching and publications.

Index

Note: Page numbers followed by an f, m, p, or t indicate figures, maps, photographs, or tables respectively.

A

B

C

D

E

G

H

I

J

N

Y

Z

www.ingramcontent.com/pod-product-compliance
Lightning Source LLC
Chambersburg PA
CBHW020459270326
41926CB00008B/666

* 9 7 8 1 5 5 7 2 8 6 8 7 1 *